INTERNATIONAL UNION OF CRYSTALLOGRAPHY
MONOGRAPHS ON CRYSTALLOGRAPHY

IUCr BOOK SERIES COMMITTEE

A. Authier, *Paris*
A.M. Glazer, *Oxford*
J.P. Glusker, *Philadelphia*
A. Hordvik, *Tromsö*
K. Kuchitsu, *Nagaoka*
J.H. Robertson (*Chairman*), *Leeds*
V.I. Simonov, *Moscow*

INTERNATIONAL UNION OF CRYSTALLOGRAPHY
BOOK SERIES

This volume forms part of a series of books sponsored by the International Union of Crystallography (IUCr) and published by Oxford University Press. There are three IUCr series: IUCr Monographs on Crystallography, which are in-depth expositions of specialized topics in crystallography; IUCr Texts on Crystallography, which are more general works intended to make crystallographic insights available to a wider audience than the community of crystallographers themselves; and IUCr Crystallographic Symposia, which are essentially the edited proceedings of workshops or similar meetings supported by the IUCr.

IUCr Monographs on Crystallography
1. *Accurate molecular structures: Their determination and importance*
   A. Domenicano and I. Hargittai, *editors*
2. *P.P Ewald and his dynamical theory of X-ray diffraction*
   D.W.J. Cruickshank, H.J. Juretschke, and N. Kato, *editors*
3. *Electron diffraction techniques, Volume 1*
   J.M. Cowley, *editor*
4. *Electron diffraction techniques, Volume 2*
   J.M. Cowley, *editor*
5. *The Rietveld method*
   R.A. Young, *editor*

IUCr Texts on Crystallography
1. *The solid state: From superconductors to superalloys*
   A. Guinier and R. Jullien, *translated by* W.J. Duffin
2. *Fundamentals of crystallography*
   C. Giacovazzo, *editor*

IUCr Crystallographic Symposia
1. *Patterson and Pattersons: Fifty years of the Patterson function*
   J.P. Glusker, B.K. Patterson, and M. Rossi, *editors*
2. *Molecular structure: Chemical reactivity and biological activity*
   J.J. Stezowski, J. Huang, and M. Shao, *editors*
3. *Crystallographic computing 4: Techniques and new technologies*
   N.W. Isaacs and M.R. Taylor, *editors*
4. *Organic crystal chemistry*
   J. Garbarczyk and D.W. Jones, *editors*
5. *Crystallographic computing 5: From chemistry to biology*
   D. Moras, A.D. Podjarny, and J.C. Thierry, *editors*

P. P. Ewald and his Dynamical Theory of X-ray Diffraction

A Memorial Volume for Paul P. Ewald
23 January 1888 – 22 August 1985

Edited by

D.W.J. CRUICKSHANK,
H.J. JURETSCHKE,
and
N. KATO

INTERNATIONAL UNION OF CRYSTALLOGRAPHY
OXFORD UNIVERSITY PRESS
1992

Oxford University Press, Walton Street, Oxford OX2 6DP
Oxford   New York   Toronto
Delhi   Bombay   Calcutta   Madras   Karachi
Kuala Lumpur   Singapore   Hong Kong   Tokyo
Nairobi   Dar es Salaam   Cape Town
Melbourne   Auckland   Madrid
and associated companies in
Berlin   Ibadan

Oxford is a trade mark of Oxford University Press

Published in the United States
by Oxford University Press Inc., New York

© The contributors listed on pp. ix–x, 1992
© IUCr pp. 125–7, 1968; pp. 129–34, 1969; pp. 136–44, 1979; pp. 146–8, 1986.

All rights reserved. No part of this publication may be reproduced, stored in a retrieval system, or transmitted, in any form or by any means, without the prior permission in writing of Oxford University Press. Within the UK, exceptions are allowed in respect of any fair dealing for the purpose of research or private study, or criticism or review, as permitted under the Copyright, Designs and Patents Act, 1988, or in the case of reprographic reproduction in accordance with the terms of licences issued by the Copyright Licensing Agency. Enquiries concerning reproduction outside those terms and in other countries should be sent to the Rights Department, Oxford University Press, at the address above.

A catalogue record for this book is available from the British Library

Library of Congress Cataloging in Publication Data
(Data available as request)

ISBN 0–19–855379–X

Typeset by
Graphicraft Ltd., Hong Kong
Printed in Great Britain by
Bookcraft Ltd., Midsomer Norton, Avon

Preface

Paul P. Ewald died in 1985 at the age of 97. He was the last of the pioneers involved in the events of 1912 which gave birth to X-ray crystallography. Remarkably, his thesis on crystal optics, which provided an exact formulation for electromagnetic waves of any wavelength, preceded the experimental discovery of X-ray diffraction. In 1913 he introduced the sphere of reflection in the reciprocal lattice, and by 1917 his dynamical theory of crystal optics had been extended in detail to X-ray diffraction. He was the prime mover in the founding of the International Union of Crystallography in 1947, the first editor of *Acta Crystallographica* 1948–1960, and President of the Union 1960–1963. His dynamical theory came to be fully appreciated only many years after its formulation when experimental techniques had greatly advanced and perfect crystals, especially silicon, could be grown. In consequence, conference organizers sought lectures by him right into his old age.

After his death the Union founded the Ewald Prize, and the first award was made in Perth in August 1987 to J.M. Cowley and A.F. Moodie for their achievements in electron diffraction and microscopy. At the Perth Congress a commemorative microsymposium was held on Ewald's scientific achievements. It was organized by Professor N. Kato, and six short talks were presented by N. Kato, G. Hildebandt, D.W.J. Cruickshank, R. Colella, D.H. Templeton, and H.J. Juretschke. It was then proposed, with the encouragement of Professor Th. Hahn and Professor M. Nardelli, successive Presidents of the Union, that these talks should form the core of a Memorial Volume. Clearly, the talks themselves needed to be expanded and formalized, and much other material had to be selected and added. The speakers agreed that D.W.J. Cruickshank, H.J. Juretschke, and N. Kato should become editors for the proposed volume. The volume offered here is the outcome of the joint efforts of both authors and editors.

Part A is an introduction consisting of a chapter by Kato which introduces and surveys the dynamical theory and its development in modern science and technology. The presentation treats the crystal as a continuous dielectric medium, as is now usual, rather than as the periodic array of discrete dipoles considered by Ewald.

Part B, 'Ewald as seen by others', is essentially biographical and covers Ewald's working life in Germany, the UK, and the USA and his involvement in the building of the crystallographic community (Hildebrandt, Bethe, Perutz, Hodgkin, Kamminga, and Juretschke).

Part C, 'Aspects of Ewald's work and their legacies', discusses topics in his scientific work: the reciprocal lattice (Cruickshank), lattice sums (Templeton), X-ray topography (Authier and Capelle), multiple scattering and the phase problem (Colella), and electron scattering (Cowley and Moodie). A commentary on Ewald's four classic papers on the 'foundations of crystal optics' (Juretschke) is followed by a contribution on the optical extinction theorem (Bullough and Hynne).

Part D, 'Ewald speaks for himself', starts with an English translation by Juretschke of Ewald's 1913 paper on X-ray interferences in which the reciprocal lattice and the sphere of reflection were introduced from the dynamical theory. There follows his 'personal reminiscences' of 1968, in which he wrote of his childhood and early interest in science. We then reprint from *Acta Crystallographica* his 1969 'Introduction to the dynamical theory' and his 1979 Review of his papers on crystal optics. The Introduction and Review are complementary, and were published when he was 81 and 91 years old. Also reprinted is his last, posthumous, paper on the so-called correction of Bragg's law. This was published, 98 years after his birth, in the P.P. Ewald Memorial Issue of *Acta Crystallographica* in November 1986. The present book closes with a Bibliography of Ewald's publications, compiled by Juretschke.

Clearly, the book is not a systematic biography or assessment of Ewald's entire body of work. Rather, it is a sampler, to be dipped into, or read in chapters (each with its own set of references), allowing the modern reader to gain an overview of Ewald's accomplishments and personality as seen by diverse observers. There is a little repetition and overlap between the different chapters, but the editors have preferred to leave this so as not to destroy the internal balance of the individual contributions. At the same time, not all of Ewald's work or its modern consequences have been included and there are major omissions, such as the flourishing extension of his methods in today's infra-red crystal optics. The circumstances of the origin of this volume, as well as constraints of time and of potential contributors' ability to work under these constraints, all helped to shape the final version. Perhaps a more systematic account of Ewald's life and work—which the subject well deserves—will in the future redress any of this volume's imbalances.

Diversity also manifests itself occasionally in the scientific notation used in the various chapters. Uniformity was not possible, as the reprinted Ewald papers of 1969 and later use lower case k_o and upper case K for the wave vector magnitudes $1/\lambda$ in vacuum and crystal, respectively, whereas his 1913 paper used K_o and K for $2\pi/\lambda$. Cruickshank, in summarizing the 1913 paper in Chapter 6, uses K_o and K for $1/\lambda$ to achieve an exactly reciprocal lattice. Kato, in Chapter 1, uses K and k for $1/\lambda$, with bold-face symbols **K** and **k** for the wave vectors; whereas Bullough and Hynne, in Chapter 12, use k_o and k for $2\pi/\lambda$ in vacuum and crystal, respectively.

Ewald's unique contribution to the theory of X-ray scattering is to be found in his four papers on the Foundations of Crystal Optics, published in German in 1916 (Parts I and II), 1917 (Part III), and 1937 (Part IV). Parts I and II were translated into English by L.M. Hollingsworth in 1970, but were published only as a US Air Force document. Parts III and IV have been translated since the Perth meeting by H.J. Juretschke. We are pleased to report that these translations were published in 1991 by the American Crystallographic Association in its Monograph Series. We also draw attention to the fine memoir on Ewald by H.A. Bethe and G. Hildebrandt published in *Biographical Memoirs of Fellows of the Royal Society* (1988) **34**, 135–76. A part of this is reproduced, with permission from the Royal Society, in the chapter on Ewald's life in Cambridge and Belfast.

We are grateful to the International Union of Crystallography for permission to reproduce Ewald's articles from *Acta Crystallographica*. We acknowledge permissions from the IUCr and other publishers for the reproduction of various figures, whose sources are indicated in the figure captions.

Special appreciation is due Mrs Ella Ewald, whose presence and keen encouragement provided a powerful incentive for moving the work forward, and other members of the Ewald family, particularly Rose Bethe of Ithaca, New York, and Arnold Ewald of Sydney, Australia, who helped unstintingly in searching through family papers for pertinent references or photographs. We were delighted that it was possible to present an advance copy of the typescript to Ella for her 100th birthday on 27 May 1991.

N. Kato wishes to thank the Yamada Foundation of Japan for generous support which enabled him to travel to critical editorial meetings abroad.

Finally, we thank the staff of the Oxford University Press for much help in the production of this volume, which is published as one of the Monographs on Crystallography in the IUCr/OUP Book Series.

Manchester, UK
New York, USA
Nagoya, Japan
August 1991

D.W.J.C.
H.J.J.
N.K.

Contents

List of Contributors		ix
	A Introduction	1
1.	The significance of Ewald's dynamical theory of diffraction *Norio Kato*	3
	B Ewald as seen by others	25
2.	Paul P. Ewald—The German period *G. Hildebrandt*	27
3.	Paul Ewald in Cambridge and Belfast *H.A. Bethe and G. Hildebrandt* *Max F. Perutz* *Dorothy C. Hodgkin*	35
4.	Paul P. Ewald and the building of the crystallographic community *Harmke Kamminga*	39
5.	Paul P. Ewald—a personal appreciation *Hellmut J. Juretschke*	44
	C Aspects of Ewald's work and their legacies	51
6.	The reciprocal lattice imbedded in Fourier space *D.W.J. Cruickshank*	53
7.	Lattice sums, the Madelung constant, and Paul Ewald *David H. Templeton*	61
8.	X-ray topography *A. Authier and B. Capelle*	64
9.	Multiple diffraction of X-rays and the phase problem *R. Colella*	71
10.	Paul Ewald and the dynamical theory of electron scattering *J.M. Cowley and A.F. Moodie*	79
11.	Commentary on Ewald's fundamental papers of the dynamical theory of X-ray diffraction *Hellmut J. Juretschke*	90
12.	Ewald's optical extinction theorem *R.K. Bullough and F. Hynne*	98

D Ewald speaks for himself — 111

1913: Contributions to the theory of the interferences of X-rays in crystals. — 113
Phys. Z. (1913), **14**, 465–72
(Translated by Hellmut J. Juretschke)

1968: Personal reminiscences. — 124
Acta Cryst. (1968), A**24**, 1–3

1969: Introduction to the dynamical theory of X-ray diffraction. — 128
Acta Cryst. (1969), A**25**, 103–8

1979: A review of my papers on crystal optics 1912 to 1968. — 135
Acta Cryst. (1979), A**35**, 1–9

1986: The so-called correction of Bragg's law. — 145
Acta Cryst. (1986), A**42**, 411–13

P.P. Ewald: Bibliography — 149
Compiled by Hellmut J. Juretschke

Index of names — 157

Index of subjects — 160

Contributors

Authier, A.
Laboratoire de Minéralogie–Cristallographie, Université P. et M. Curie et associé au CNRS, 4 Place de Jussieu, F-75252 Paris Cedex 05, France

Bethe, H.A.
Newman Laboratory of Nuclear Studies, Cornell University, Ithaca, New York 14853, USA

Bullough, R.K.
Department of Mathematics, University of Manchester Institute of Science and Technology, Manchester M60 1QD, England

Capelle, B.
Laboratoire de Minéralogie–Cristallographie, Université P. et M. Curie, 4 Place de Jussieu, F-75252 Paris Cedex 05, France

Colella, R.
Department of Physics, Purdue University, West Lafayette, Indiana 47907, USA

Cowley, J.M.
Facility for High Resolution Electron Microscopy, Center for Solid State Science, Arizona State University, Tempe, Arizona 85287, USA

Cruickshank, D.W.J.
Department of Chemistry, University of Manchester Institute of Science and Technology, Manchester M60 1QD, England

Hildebrandt, G.
c/o Fritz-Haber-Institut der Max-Planck-Gesellschaft, D-1000 Berlin 33, Germany

Hodgkin, Dorothy C.
Chemical Crystallography Laboratory, 9 Parks Rd, Oxford OX1 3PD, England

Hynne, F.
Chemistry Laboratory III, H.C. Ørsted Institute, Universitetsparken 5, DK 2100 Copenhagen Ø, Denmark

Juretschke, Hellmut J.
Department of Physics, Polytechnic University, Brooklyn, New York 11201, USA

Kamminga, Harmke
Wellcome Unit for the History of Medicine, Department of History and Philosophy of Science, University of Cambridge, Free School Lane, Cambridge CB2 3RH, England

Kato, Norio
Department of Physics, Meijo University, Nagoya 468, Japan

Moodie, A.F.
Department of Applied Physics, Royal Melbourne Institute of Technology, Melbourne, Victoria 3001, Australia

Perutz, Max F.
  MRC Laboratory of Molecular Biology, University Medical School, Hills Rd, Cambridge CB2 2QH, England
Templeton, David H.
  Department of Chemistry, University of California, Berkeley, California 94720, USA

Paul Ewald: a portrait taken in 1960.
(Photograph by Lotte Meitner-Graf, London.)

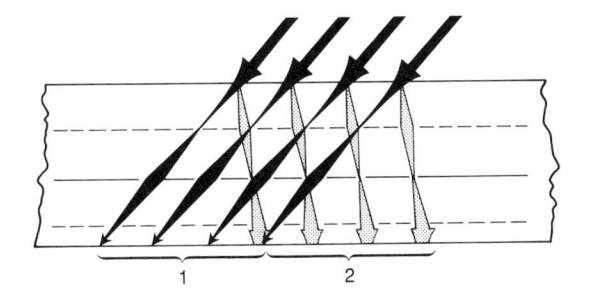

Pendellösung — Ewald's prediction of reversible energy exchange between diffracted beams with depth in a perfect crystal, giving rise to thickness dependent fringe patterns. The name draws attention to the analogy with the temporal energy interchange between coupled sympathetic pendulums.

A
Introduction

1

The significance of Ewald's dynamical theory of diffraction

Norio Kato

Fig. 1 Photograph of P.P. Ewald (right) and A.J.C. Wilson (left) taken in Munich, 1962.

1. Prologue

Professor Ewald passed away at the age of 97, on 22 August 1985 at Ithaca, the university town for Cornell where he spent the last period of his life. When I received the news, I fell into deep grief. After a year or so, however, considering his age in particular, I began to think more about his scientific achievements themselves and what we could do to develop them. It was very natural, therefore, that a 'microsymposium' with this aim had been planned in the 1987 Congress of the IUCr (International Union of Crystallography) at Perth, Australia.

Of course, his greatness is not confined to his science. At one time he was Secretary General and Vice-President of IUPAP (International Union of Pure and Applied Physics), and he largely inaugurated the IUCr in 1948, foreseeing the importance of crystallography in modern science. He was an international figure and was loved and respected by many scientists all over the world. When he died, obituaries appeared in scientific journals of many countries and even in newspapers. Also, many condolences came to Mrs Ewald and his students and collaborators from his Brooklyn period. The tributes of Professors M. Perutz and D. Hodgkin are recorded at the end of the obituary Memoir of the Royal Society of London (H.A. Bethe and G. Hildebrandt (1988); see the Obituaries at the end of this chapter; see also Chapter 3 of this volume). I shall not repeat their words dedicated to his remarkable life. In one sentence, he was the father and later the grandfather of our community of crystallography.

Let me start with the photograph of Professor Ewald (Fig. 1). This was taken in 1962 at Munich, where an international meeting was held to commemorate the fiftieth anniversary of M. von Laue's discovery of X-ray diffraction. By that time, the editorship of *Acta Crystallographica* (in brief, *Acta*), which also had been inaugurated by Professor Ewald, had been transferred from him to Professor A.J.C. Wilson, who is on the left side of this photograph. Kamminga's chapter in this volume describes Ewald's early roles in our community.

Regarding his scientific life work, namely, the foundation of crystal optics or, more specifically, the dynamical theory of X-ray diffraction, it would be interesting to read his own personal history. The following is an extract from his 40-minute talk entitled 'the origin of the dynamical theory of X-ray diffraction', which was delivered in 1961, at Kyoto.[†] At the beginning of the microsymposium mentioned above we actually listened to his tape-recording of this talk, and it recalled his humour and charm. In this transcript, brackets [ ] indicate the present author's comments.

Ladies and gentlemen:

I must first apologize for being late. But I had to sleep.

Let me go back to the year 1913. As I said this morning [in the Opening Ceremony] Laue's discovery was made 1912, and by 1913 Laue had been made Professor of Physics in Zurich University. Friedrich in the spring of 1913 took up a position in the gynaecological clinic in Freiburg with Professor König for measuring the intensities of doses in therapeutic X-ray work, and I became Sommerfeld's experimental assistant. It was fortunate that this period did not last very long. In this way the apparatus of Friedrich is still preserved and in the Deutsches Museum in Munich.

[The lecture continues with a humorous description of his service in German army. He volunteered for medical work with X-ray equipment. During this period at Königsberg he wrote his paper on optics as a development of his thesis, and then he was moved to a little town in Lithuania. There was no battle in this area. The season was spring. He describes the environment very poetically.]

So that gave me time to think over the dynamical theory.... It was not quite simple to get the idea of the dynamical theory carried out. There were a lot of things which were novel, and for which my previous work was a very good starting point. My previous work consisted of my dissertation—my thesis—and this paper which I had finished in Königsberg.

The subject of the dissertation was given to me by Sommerfeld, my teacher. I went to him, after having heard his courses for two years, and told him that I would like to enter into the work of Doctor's Thesis under his direction. He pulled out the drawer of his desk and took a sheet out. There were about ten or fifteen different topics for theses, which ranged from propagation of radio waves, the setting-in of turbulence, and the self-induction of coils of certain cross-sections and all kinds of applied problems, mainly for solving differential equations with boundary conditions.

Quite at the end of the list, there stood a topic to investigate whether an anisotropic arrangement of resonators would lead to crystal optical properties. When I saw this last topic, I said 'This it is I want to have'; although Sommerfeld told me that he could not give me much help with this topic. But I later on remembered that even as a boy it had been my idea that one should use light to investigate the structure of solids and delve deeper into the mystery of the solid state. So, unknowingly at the time, I picked this subject which was preformed in my mind and I have never been sorry to have done it.

[The lecture continues with a lengthy account of the theory and its problems, especially the exact cancellation of the incident wave in the interior of a crystal—a most remarkable result. For X-rays, the entity in the interior is a set of strong waves, each with a different refractive index.]

When I wrote up the paper for my Habilitationsschrift [1917]—for becoming a Lecturer in Munich—and I sent it to Sommerfeld, I know that he remarked to my mother that it might be very nice but it was utterly un-understandable, and besides it would never have any practical value. Well, it seemed so for a long time but then the practical value appeared by and by.

[The lecture then describes the early experiments which showed the practical value.]

Well... I would like to mention perhaps that the theory of electron diffraction—the Bethe theory—was formed directly in connection with my dynamical theory for X-rays. It was Bethe's dissertation under Sommerfeld, and Sommerfeld and I had been skiing together in the mountains and we had talked it over. Later on, Bethe came as my assistant in Stuttgart and so we were close in touch. So, in a way, I feel that this is an excuse for talking at this Conference on Electron Diffraction about the dynamical theory for X-rays.

Thank you.

[†] The formal text can be seen in Ewald's paper (1962a). The present text was transcribed directly from the tape recording to preserve his style of talking.

This is the start of his famous research work on the dynamical theory of diffraction. More details of his life in this period are described in the chapters by Hildebrandt and Juretschke.

2. Ewald's dissertation and X-ray dynamical theory

The title of his dissertation under A. Sommerfeld was 'Dispersion und Doppelbrechung von Elektronengittern (Kristallen)'. In homogeneous and isotropic media, visible rays with a fixed frequency ν can propagate with the wavelength λ_m, which differs from λ in vacuum. The refractive index $n = \lambda/\lambda_m$ depends on ν, the phenomenon being called dispersion.

If the medium is anisotropic, or in crystals lower than cubic in symmetry, double refraction and various associated phenomena are observable. The refractive index depends also on the direction of propagation. The phenomenon was discovered by Erasmus Bartholinus [1669, Copenhagen] in Iceland spar. The theory was given by Augustin J. Fresnel [1821] by introducing a tensor ε for the dielectric constant. It is one of the masterpieces of classical physics and constitutes the core of crystal optics. The refractive index n along a line having the direction cosines (α,β,γ) is given by **Fresnel's equation of wave normals** (Born and Wolf, 1980),

$$\frac{\alpha^2}{(1/n)^2 - 1/\varepsilon_1} + \frac{\beta^2}{(1/n)^2 - 1/\varepsilon_2} + \frac{\gamma^2}{(1/n)^2 - 1/\varepsilon_3} = 0, \quad (1)$$

where ε_1, ε_2, and ε_3 are the principal values of the tensor ε, and the co-ordinate axes are fixed along its principal axes. In isotropic media ($\varepsilon_i = \varepsilon$), however, the well-known relation $n = \sqrt{\varepsilon}$ must be used instead of eqn (1). In any case, if we assign complex quantities for ε_i, the refractive index n and the wave number k ($= n/\lambda$) also become complex. The imaginary component of k describes the attenuation of the wave. The same is true in the dynamical theory described below.

The theory is 'phenomenological' in the sense that it does not answer why a set of ε_i values is to be attached to the individual crystal and why it is ν dependent, etc.

Advanced scientists around the turn of the century believed that matter was composed of molecules or at least of electrons embedded in a positively charged medium. With this recognition, the microscopic (molecular) theory of crystal optics seemingly attracted Ewald and his teacher Sommerfeld. The dissertation appeared in 1912, and two papers under the title *Zur Begründung der Kristalloptik* were published (Ewald, 1916a,b).

Soon after Ewald finished his dissertation, the famous experiment of X-ray diffraction by crystals was conducted by Knipping and Friedrich, under the guidance of M. Laue. The work led W.L. Bragg to establish the famous Bragg's law [1913].

The success of Laue stimulated Ewald to extend his optical theory to the X-ray region. He published the work (1913) in a preliminary version which was short but full of fundamental new concepts in diffraction physics. The work was extended and published in a more complete version (Ewald 1917) as the third part of *Zur Begründung der Kristalloptik*, which is the Bible of the dynamical diffraction theory. The English translation by Juretschke of the 1913 paper is included in this volume.

In closing this section, the author must mention that Ewald wrote a chapter on his own thesis in *Fifty years of X-ray Diffraction* (Ewald 1962b). The following are some extracts which describes his fundamental approach to crystal optics.

Having this model [a system of resonators] in mind, two features must be explained: First, the existence of an index of refraction, and its dependence on frequency. This is the same as the question: how is it that the presence of the scattered wavelets changes the wave velocity from its free space value c to a value q? And second: How does refraction and reflection arise at the surface of body?

It should not be assumed that division of the problem into that of dispersion and that of refraction was understood at the beginning of Ewald's investigation—it developed clearly only in the course of the work.

In this sense, the following introduction to the dynamical theory is somewhat different from his approach. The form of presentation is for the sake of conciseness.

3. The dynamical theory of X-ray diffraction

In this section, the author intends to explain the outline of the dynamical theory as an introduction to other chapters. Many important concepts originated by Ewald are left for discussion in the chapter by Cruickshank and the commentary chapter by Juretschke on Ewald's early papers. The contribution of Bullough and Hynne describes ordinary optics, which also may help in understanding the Ewald approach. The chapter by Templeton about the so-called lattice sum is also mathematically relevant, although the physical subject (Madelung constant) is entirely different from crystal optics.

Nowadays, we have several versions of the dynamical theory which are suitable for specific problems (for example, Darwin 1914, 1922; Bethe 1928; von Laue 1931; Cowley and Moodie 1957, 1959a,b; Hamilton 1957; Takagi 1962, 1969; Taupin 1964; Werner and Arrott 1965; Kuriyama 1973). Here, however, we shall follow a standard track in X-ray diffraction, which is close to classical optics as briefly explained in the previous section.

3.1 The kinematical theory

Before discussing the dynamical theory, it seems appropriate to begin with the more tangible kinematical theory. It is a specific version of the theory of elastic scattering within the Born approximation. Needless to say, the approximation is valid only when the total of scattered waves is sufficiently weak. Otherwise, multiple scattering has to be taken into account. That is the dynamical theory.

In the kinematical theory, the diffracted wave is assumed approximately as the total sum of wavelets scattered by individual electrons in the crystal. The elementary process is the dipole (Thomson) scattering of the electromagnetic wave.

Because of the spatial periodicity of crystals, a strongly directional (diffracted) wave is created when the incident wave satisfies the Bragg condition

$$2d \sin \theta_B = \lambda, \quad (2)$$

where d is the spacing of the net plane concerned and θ_B is the Bragg angle, which is a particular glancing angle θ to the net plane.

The geometrical aspect given by eqn (2) is well illustrated by the celebrated Ewald construction

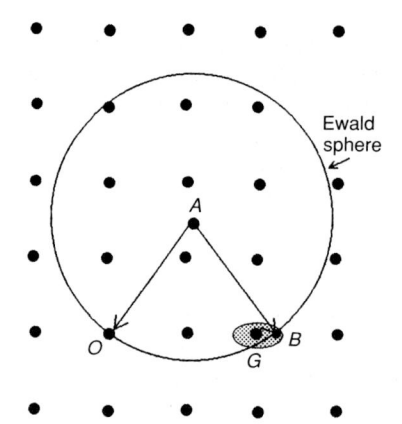

Fig. 2 The Ewald construction: $\mathbf{K}_e = \overrightarrow{AO}$ and $\mathbf{K}_g = \overrightarrow{AB}$ are the wave vectors of the incident and diffracted waves in vacuum. An appreciable wave is expected for a small $\mathbf{q} = \overrightarrow{GB}$.

together with the concept of the reciprocal lattice. In Fig. 2, $\overrightarrow{AO}$ and $\overrightarrow{AB}$ represent the wave vector $\mathbf{K}_e$ of the incident wave and $\mathbf{K}_g$ of the diffracted wave, respectively. Their magnitude must be $1/\lambda$, as we are concerned with elastically scattered waves in vacuum. Therefore, the end-point B always lies on the Ewald sphere. When B is identical to the reciprocal lattice point G ($\overrightarrow{OG} = 1/d$), the condition (2) follows immediately. Then, the diffracted wave is strongest.

The kinematical theory, of course, involves more than geometrical aspects. I shall not discuss the details except for a few comments. The amplitude per unit incident amplitude can be given in the form[†]

$$A(\mathbf{q}) = r_e \cdot C \cdot F_g \cdot S(\mathbf{q})/V, \quad (3)$$

where r_e is the Thomson scattering amplitude (due to a single electron), C the polarization factor inherent to the scattering of electromagnetic vector waves, V the unit cell volume and F_g is the crystal structure factor, which is defined by the Fourier transform

$$F_g = \int \rho(\mathbf{r}) \exp(2\pi i \mathbf{g} \cdot \mathbf{r}) \, dV \quad (4)$$

of the electron density $\rho(\mathbf{r})$, the integration being taken over the unit cell, and $\mathbf{g}$ is a reciprocal lattice vector. Finally, $S(\mathbf{q})$ is called 'shape function', where $\mathbf{q} = \overrightarrow{GB}$ in Fig. 2. It is also the Fourier transform

---

[†] We are concerned with the asymptotic form of the diffracted wave, $A(\mathbf{q}) [\exp(2\pi i KR)]/R$, at a large distance R from the crystal. Its vectorial character is suppressed for simplicity, except for formally introducing the polarization factor C.

of the crystal shape in the case of perfect crystals. As the mathematics of Fourier transforms is well developed, the kinematical theory is very powerful for determination of the structure of matter.

Through the shape function $S(\mathbf{q})$, we see that, strictly speaking, the diffracted wave is not a plane wave but a bundle of plane waves. Not only that, but the intensity for an incident plane wave depends upon the direction of $\mathbf{K}_e$. Rotating the crystal, we have an intensity profile $I(\theta)$, as the Ewald sphere cuts different parts of $S(\mathbf{q})$. The calculation can show easily that angular width of the profile for a perfect crystal is of the order of

$$\Delta\theta_{kin} = \lambda/L\cos\theta_B, \quad (5a)$$

where L is the thickness perpendicular to the net plane. In distorted crystals, the width will be much wider. The Ewald construction is very useful also in such general cases. It enables us to grasp the nature of diffracted waves created by the crystal.

By integrating $I(\theta)$ with respect to θ, we can obtain a kind of integrated intensity, which is also an important quantity, for example, for determination of the structure factor. The expression for a slab of thickness L perpendicular to the net plane concerned is given by

$$R_{kin} = r_e^2(1/\sin 2\theta_B)\, \lambda^3(C|F_g|/V)^2\, (L/\sin\theta_B). \quad (5b)$$

3.2 The proper wave in the crystal

The dynamical theory can be easily understood by elementary concepts of optics such as reflection, refraction, absorption, and so forth. One needs, however, to assume an optical model of the crystal, as in the case of Fresnel's theory of crystal optics mentioned in Section 2. In X-ray optics, it is proper to assume von Laue's model (1931); namely, the polarizability $\chi(\mathbf{r})$, and consequently the dielectric constant $\varepsilon(\mathbf{r})$ ($= 1 + 4\pi\chi$), are spatially periodic but scalar.

How can we justify von Laue's model and what expression should we take for χ in the context of elementary dipole scattering due to electrons? The Ewald theory indeed amounted implicitly to a justification covering the range from visible rays to X-rays. The theory was, of course, classical, and Ewald's microscopic model was primitive in his original theory; namely, it was an orthorhombic arrangement of electron resonators. Now we know that, in general cases, the expression for χ is given by

$$\chi(\mathbf{r}) = -r_e(\lambda/2\pi)^2\, \rho(\mathbf{r}). \quad (6a)$$

In perfect crystals, where $\rho(\mathbf{r})$ has a three-dimensional spatial periodicity, the Fourier coefficient of $\chi(\mathbf{r})$ is given by

$$\chi_g = -r_e(\lambda/2\pi)^2\, (F_g/V), \quad (6b)$$

which will be used often in the following.

Incidentally, quantum electrodynamics also leads to expressions identical to (6a,b) (Molière 1939a,b,c), provided that one neglects a small correction due to anomalous dispersion. What is important is that $4\pi\chi$ is negative and extremely small; 10^{-5} in order of magnitude.

Now, we shall proceed to consider crystal waves in more detail. To distinguish them from the vacuum waves $\mathbf{K}_e$, $\mathbf{K}_g$ discussed in the kinematical theory, lower case symbols $\mathbf{k}_o$, $\mathbf{k}_g$, etc., are used to denote the crystal waves.

Once we admit the above model, any plane wave of the form

$$\mathbf{d}(\mathbf{r}) = \mathbf{d}\, \exp 2\pi\, i(\mathbf{k}\cdot\mathbf{r}) \quad (7)$$

can no longer be a proper wave in the crystal. It is allowed only in a special case.

In summary, the expression (7) must be replaced by a wave of the form

$$\mathbf{d}(\mathbf{r}) = \sum_g \mathbf{d}_g \exp 2\pi\, i(\mathbf{k}_g\cdot\mathbf{r}), \quad (8a)$$

in which the so-called momentum condition

$$\mathbf{k}_g = \mathbf{k}_o + \mathbf{g} \quad (9)$$

is satisfied. We use an electric displacement vector as the crystal wave. This is partly traditional. Then, the transverse condition for each component wave in the summation of (8a),

$$(\mathbf{d}_g \cdot \mathbf{k}_g) = 0, \quad (10)$$

follows from the general relation div $\mathbf{d}(\mathbf{r}) = 0$. We shall not derive these results but an attempt will be made to explain their implications.

First of all, (8a) can be rewritten in the form

$$\mathbf{d}(\mathbf{r}) = \mathbf{A}(\mathbf{r}) \exp[2\pi\, i(\mathbf{k}_o\cdot\mathbf{r})], \quad (8b)$$

where

$$\mathbf{A}(\mathbf{r}) = \sum_g \mathbf{d}_g \exp[2\pi\, i(\mathbf{g}\cdot\mathbf{r})]. \quad (8c)$$

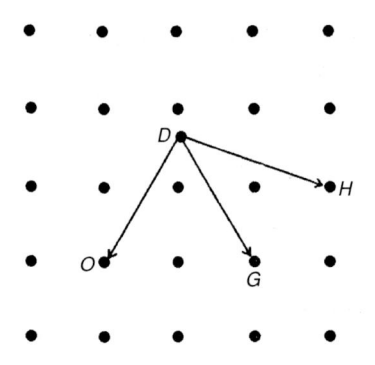

Fig. 3 A representation of the Ewald crystal wave as a physical entity.

Thus, the wave (8a) can be interpreted as a vector wave whose amplitude has the same lattice periodicity as the crystal.

Incidentally, the similar but scalar wave is familiar in the electron case and is called a Bloch wave in solid-state physics. There, in fact, the scalar potential $V(\mathbf{r})$ plays the same role as the polarizability $\chi(\mathbf{r})$. Ewald knew much earlier that (8a) is the proper wave in the crystal.

In Fig. 3, $\overrightarrow{DO}$, $\overrightarrow{DG}$ ... denote $\mathbf{k}_o$, $\mathbf{k}_g$ ... respectively, where the $\{G\}$ are reciprocal lattice points, including the origin O. Each component wave to be summed in (8a) is called a $\{G\}$-wave. The figure is very similar to Fig. 2. However, Fig. 2 shows the relation of vacuum waves so that the condition of energy conservation $|\mathbf{K}_e| = |\mathbf{K}_g| = K$ is strictly satisfied, whereas Fig. 3 shows the relations among the $\{G\}$-waves in the infinite crystal. Needless to say, their magnitudes are not necessarily K.

One can rewrite the relation (9) as $\mathbf{k}_o = \mathbf{k}_g - \mathbf{g}$. In this sense, all $\{G\}$-waves are equivalently treated in the crystal. This is one of the essences of the dynamical theory. The G-wave is a reflected wave of the O-wave in the sense of the Bragg reflection on the net plane specified by $\mathbf{g}$, and at the same time the O-wave is a reflected wave of the G-wave on the reverse side of the net plane($-\mathbf{g}$). Thus, behind the simple relation (9), there is hidden an endless process of multiple reflections. Moreover, the Bragg condition is a matter of degree. Any G-wave creates other $\{H\}$-waves to some extent and vice versa.

In conclusion, the wave having the form (8a) is a physical entity in the crystal. Each component wave does not exist individually. The total set must also be under a self-consistent dynamical balance. Self-consistency is an important concept in modern physics. Probably, Ewald was one of a few pioneers who recognized its importance and knew how to handle the problem.

3.3 The dynamical condition

Now, we shall go further in discussing the wave vectors $\{\mathbf{k}_g\}$ and the amplitudes $\{d_g\}$ in eqs (8a–c). To avoid notational complexity, we shall again ignore the vectorial character of the wave field.

Any wave motion can be described by a wave equation. For X-rays, this is the Maxwell equation, of course. Suppressing vectorial characters, the wave equation may be written in the form

$$[\Delta - (1/q)^2 (\partial^2/\partial t^2)]d(\mathbf{r},t) = 0,$$

where Δ is the Laplacian $[(\partial^2/\partial x^2) + (\partial^2/\partial y^2) + (\partial^2/\partial z^2)]$ in a rectangular co-ordinate system], and q is the phase velocity in the medium. As in the case of ordinary optics, it is assumed that $q = c/n$, c being the velocity of light. Moreover, if we are concerned only with a time-harmonic wave with frequency ν, the wave equation must have the form

$$[\Delta + (2\pi K)^2 (1 + 4\pi \chi)]d(\mathbf{r}) = 0, \quad (11)$$

where K is defined by ν/c. For the later argument, we also define k by ν/q.

It can be shown that any single plane wave having the form of (7) is a solution of the wave equation (11), provided that

$$(\mathbf{K}^2 - K^2) D = 0 \quad \text{(vacuum)}, \quad (12a)$$

$$(\mathbf{k}^2 - k^2) d = 0 \quad \text{(isotropic media)}. \quad (12b)$$

It follows immediately that

$$\mathbf{K}^2 = K^2, \quad (13a)$$

$$\mathbf{k}^2 = k^2. \quad (13b)$$

These are dispersion relations in the respective media. There is no directional dependence of $|\mathbf{K}|$ and $|\mathbf{k}|$. The relations (13a,b) are drawn in Fig. 4, where the refractive index $n = k/K$ is assumed less than unity, which is the case for X-rays.

One can rewrite (12b) in the form

$$(\mathbf{k}^2 - K^2)d = K^2(n^2 - 1)d = K^2(4\pi \chi)d. \quad (12c)$$

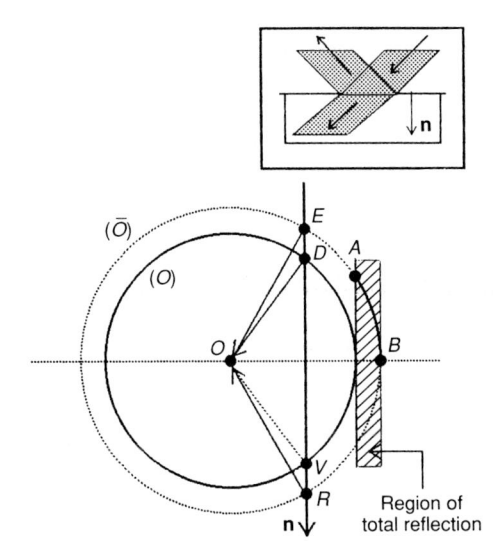

Fig. 4 Reflection and refraction at the boundary of an isotropic medium. The inset figure illustrates the experimental arrangement in real space. $\mathbf{K}_e = \overrightarrow{EO}$ is the wave vector of the incident wave. $\mathbf{k}_o = \overrightarrow{DO}$ and $\mathbf{K}_R = \overrightarrow{RO}$ are the wave vectors of the refracted and reflected waves. $\overrightarrow{VO}$ is a virtual wave vector (Section 3.4). E, D, V, and R must line up parallel to $\mathbf{n}$, the normal of the boundary surface. When E lies on the arc AB, no real wave can exist in the medium. Then total reflection occurs. Spheres $(\bar{O})$ and (O) are the dispersion surfaces for the waves in vacuum and the isotropic medium, respectively. The same figure can be used also for the one-wave case in crystalline media.

The condition for wave propagation in the vacuum (12a) is modified by the presence of polarizability in the medium.

It is also instructive to assume deliberately a plane wave in the case of perfect crystals, where χ is spatially periodic so that it has the form of a Fourier series; namely, $\sum_g \chi_g \exp[2\pi i(\mathbf{g} \cdot \mathbf{r})]$. Inserting the expression (7) with the suffix o on $\mathbf{k}$ and d into (11), we shall see that many terms with the spatial variation $\{\exp[2\pi i(\mathbf{k}_g \cdot \mathbf{r})]\}$ appear in addition to the original wave with the form $\exp[2\pi i(\mathbf{k}_o \cdot \mathbf{r})]$.

To deal with this awkward situation, it seems promising to take the wave that has the form (8a) from the beginning. Then, any G-wave created from other waves might be balanced with (or cancelled by) the originally existing G-wave in (8a). As mentioned above, the creation of a G-wave can be interpreted physically as the Bragg reflection of other $\{H\}$-waves. Their amplitudes will be self-consistently balanced. Moreover, we may know what wave vectors $\{\mathbf{k}_g\}$ are allowed under the condition of balance.

To see this more closely, we shall consider the case in which only two waves (O and G) exist in the crystal, for reasons which will be explained soon. Then, by separating out the terms proportional to $\exp[2\pi i(\mathbf{k}_o \cdot \mathbf{r})]$, we shall have

$$(\mathbf{k}_o^2 - K^2)d_o = 4\pi K^2(\chi_o d_o + \chi_{-g} d_g).$$

The propagation of the O-wave is modified by d_o itself through χ_o as in the case of (12c). Similarly, if the G-wave exists, it modifies the condition of the propagation of the O-wave through χ_{-g}. Usually, the above equation is rewritten in a slightly neater form as

$$(\mathbf{k}_o^2 - k^2)d_o = 4\pi K^2 \chi_{-g} d_g, \qquad (14a)$$

where χ is replaced by χ_o in the definition of k.

Similarly, the propagation of the G-wave can be described by

$$(\mathbf{k}_g^2 - k^2)d_g = 4\pi K^2 \chi_g d_o \qquad (14b)$$

If the point G is far from the modified Ewald sphere of radius k, so that $(\mathbf{k}_g^2 - k^2)$ is large, the amplitude d_g would be negligibly small. For this reason, our treatment assuming two waves is justified when only one G point is close to the k-sphere.

Eliminating d_o and d_g from (14a,b), we shall have the secular equation

$$(\mathbf{k}_o^2 - k^2)(\mathbf{k}_g^2 - k^2) = (4\pi)^2 K^4 \chi_g \chi_{-g} \qquad (15)$$

When $\chi_g \chi_{-g}$ is zero, as in a non-crystalline homogeneous medium, either $\mathbf{k}_o^2$ or $\mathbf{k}_g^2$ is equal to k^2. Therefore (15) amounts to a generalization of the relation (13b). The suffix o or g is a matter of notation.

It is important to note that $\mathbf{k}_o$ and $\mathbf{k}_g$ are already connected by the momentum condition (9). Therefore in a crystal where χ_g is non-zero, (15) can be regarded as the condition to fix $\mathbf{k}_o$ or $\mathbf{k}_g$. No longer can the point D in Fig. 3 (dispersion point) be an arbitrary point in reciprocal space. It must lie on the double surface, (1) and (2), as illustrated in Fig. 5. This surface is called the dispersion surface (in brief, D-surface) or the equi-energy surface for possible waves with a fixed frequency ν. In fact, near the interaction region of the O- and G-waves, the D-surface is a hyperboloid with diameter

$$d_{BB} = 4\pi K |\chi_g|/\cos\theta_B \qquad (16)$$

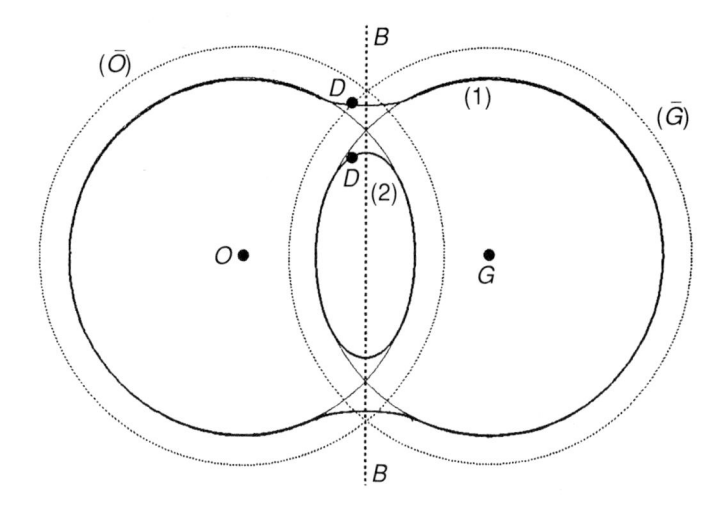

Fig. 5 The dispersion surface, (1) and (2), of the crystal wave in the two-wave case. $(\bar{O})$ and $(\bar{G})$ are the dispersion surfaces for the vacuum O- and G-waves, and have radius K. D is the dispersion point. BB (the broken line) is the Brillouin zone boundary.

along the bisector plane BB of $\overline{OG}$. (In solid-state physics, this plane is called the Brillouin zone boundary. In (16), the approximations $k = K$ and $|\chi_g \chi_{-g}| = |\chi_g|^2$ are employed.) The important point is that the magnitude of $|\mathbf{k}_o|$, or the wavelength, has a directional dependence and takes two-fold values for a fixed direction. Consequently, the wave vector $\mathbf{k}_g$ also has the same characters.

The relations (14a,b) also fix the amplitude ratio of d_o and d_g. One of them, however, is arbitrary. This is natural because an Ewald solution, or Bloch wave (8a), of any field strength can exist in the crystal. In fact, the amplitude d in (12b) was also arbitrary for the same reason.

In conclusion, the Ewald–Bloch wave (8a) is a proper wave if it satisfies the dynamical balance (14a,b) in the crystal. All properties of the wave are fixed by the D-point. Ewald often called it the 'tiepoint' to emphasize this character. It remains to see how the D-point is determined by the incident wave and what waves come out from the crystal into the vacuum (air). These are the subjects of the next subsection.

3.4 Basic results of the dynamical diffraction

Meanwhile, it is assumed that the incident wave is a plane wave having a wave vector $\mathbf{K}_e$. (The subscript e derives from the German word *einfallend* for incident.) The theory is called **plane wave theory**. For simplicity, we shall mainly consider a plane-bounded half-crystal, unless otherwise stated.

We begin with the one-wave case as a preparation, returning to Fig. 4 where the wave vectors $\mathbf{K}_e$ and $\mathbf{k}_o$ are denoted by $\overrightarrow{EO}$ and $\overrightarrow{DO}$. The point D is the dispersion point of the wave in the medium. Incidentally, E is that of the incident wave. The point D can be determined from E by a simple relation called 'tangential continuity', which implies that the tangential components of $\mathbf{K}_e$ and $\mathbf{k}_o$ are identical. This is required because otherwise the matching condition $[D(\mathbf{r}_e) = d_o(\mathbf{r}_e)]$ is not satisfied at every entrance position $\mathbf{r}_e$ on the boundary. The construction shown in Fig. 4 leads to Snell's law [1621] of refraction.

We see immediately that $\mathbf{K}_R = \overrightarrow{RO}$ and $\overrightarrow{VO}$ also are possible candidates as wave vectors. The former is identified with the mirror reflection in the vacuum. The latter is virtual because we have no reason for a crystal wave to arrive at the boundary from the crystal side. Such a wave becomes real only when another surface reflects the crystal wave.

Here again, the author must apologize for the short-cut treatment using an 'artificial' plane boundary, which Ewald did not like because we do not know where to locate the boundary in the atomic model. In his mind, there were regularly arranged electrons (resonators) in a half-space and the incident wave plus the scattered waves in the whole space, and nothing more. He arrived at essentially the same results as mentioned above, but only after considerable mathematical manipulation. The incident wave going through the crystal is exactly cancelled by a wave generated in the crystal and only a refracted wave penetrates through the crystal. This result is known as the Ewald–Oseen

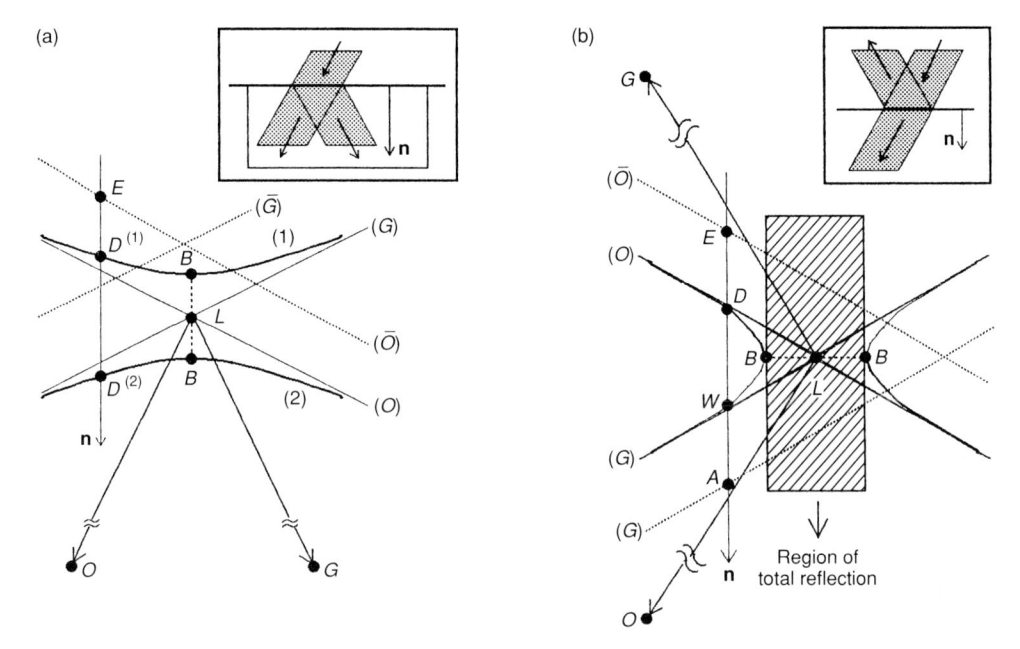

Fig. 6 The construction of the wave vectors for two-wave cases. The experimental arrangement in real space is shown in the inset. The notations are the same as in Fig. 5. $(\overline{O})$ and $(\overline{G})$ are parts of the spheres of radius K with centres at O and G, respectively, and (O) and (G) are parts of the spheres of radius k. (Figures 6 (a,b) are practically the same as Figs. 16 and 17 of Ewald's 1917 paper.) (a) The symmetric Laue case: $D^{(1)}$ and $D^{(2)}$ are two dispersion points for the crystal wave. $\mathbf{k}_o^{(1)} = \overrightarrow{D^{(1)}O}$, $\mathbf{k}_g^{(1)} = \overrightarrow{D^{(1)}G}$, $\mathbf{k}_o^{(2)} = \overrightarrow{D^{(2)}O}$, and $\mathbf{k}_g^{(2)} = \overrightarrow{D^{(2)}G}$ are the wave vectors relevant to the crystal wave. (b) The symmetric Bragg case: $\mathbf{K}_e = \overrightarrow{EO}$ and $\mathbf{K}_g = \overrightarrow{AG}$ are the wave vectors relevant to the vacuum wave. $\mathbf{k}_o = \overrightarrow{DO}$ and $\mathbf{k}_g = \overrightarrow{DG}$ are the wave vectors of the crystal wave. It should be noted that only one D-point is excited from one E. The point W is virtual, similarly to the point V in Fig. 4, because the associated ray propagates towards the surface from the inner side of the crystal. The hatched region corresponds to total reflection similarly to Fig. 4.

extinction theorem in optics (Born and Wolf, 1980; see Bullough and Hynne's chapter in this volume).

An interesting phenomenon may occur when the point E lies on the arc AB. Then, no intersection point exists for the normal $\mathbf{n}$ through E and the D-surface. This is the case of **total reflection**. Inside the crystal the wave attenuates rapidly along the normal because then the wave vector has an imaginary component along $\mathbf{n}$.

All of above arguments are rather elementary. Similar principles can be applied also to the case of dynamical diffraction. Next, we shall consider the two-wave case with the use of Figs 6(a) and 6(b), where the incident direction ($\mathbf{K}_e$) changes as the point E moves on the dispersion surface $(\overline{O})$ of vacuum. The drawings are an enlargement of the upper part of the D-surface near the Brillouin zone boundary in Fig. 5, which illustrates the global features. The neglect of the lower part implies neglect of the mirror reflection and virtual waves corresponding to $\overrightarrow{RO}$ and $\overrightarrow{VO}$ in Fig. 4. This approximation is called **linearization of dispersion surface** (see part 3 of Section 4.3). In fact, the scale of Fig. 5 is wrong. The deviation of the D-surface from the (O) and (G) spheres is of the order of the diameter d_{BB} [eqn(16)]. Therefore, if one draws it with a length of 1 cm, K must be of the order of 1 km!

The Laue case (see Fig. 6(a)) First, we shall treat the case when both O- and G-waves propagate through the crystal, downwards in the figure. In other words, the normal $\mathbf{n}$ is assumed to have a direction between $\overrightarrow{LO}$ and $\overrightarrow{LG}$. Then, two real dispersion points $D^{(1)}$ and $D^{(2)}$ always result from one point E by tangential continuity. This implies that a kind of double refraction occurs in the crystal. An important difference from the ordinary

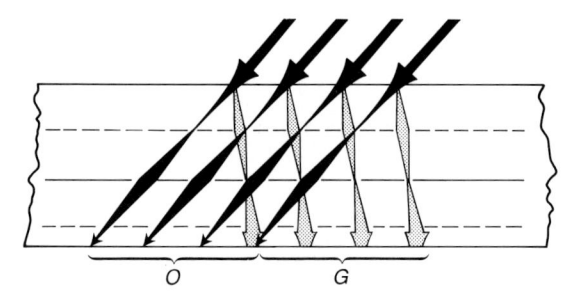

Fig. 7 Pendellösung fringes in the plane wave theory. Adapted from Fig. 81 of Ewald's 1927 paper.

crystal optics associated with eqn (1) is that the two refracted waves have the same polarization mode. Therefore, one can expect a set of thickness fringes of sinusoidal form both in the O- and G-wave fields as the result of interference of the waves associated with $D^{(1)}$ and $D^{(2)}$. This remarkable result is illustrated in Fig. 7. The fringes were named *Pendellösung fringes* by Ewald, because the mathematical structure is similar to that in a coupled pendulum.

In the special case where the normal $\mathbf{n}$ lies on BB in Fig. 6(a) (the exact Bragg condition), it turns out that the fringe spacing is

$$\Lambda = d_{BB}^{-1} = r_e^{-1}(\pi V/\lambda) \cos \theta_B / C |F_g|. \quad (17)$$

Here, d_{BB} is the diameter given by (16) and the polarization factor C is inserted for the sake of generality with electromagnetic waves. The distance Λ amounts to 24.7 μm in the case of the Si(220) reflection, for $\lambda = 1$ Å X-rays and $C = 1$.

For the two-wave case, it is convenient to consider the σ and π modes of polarization with respect to the plane of incidence. Then, one can assign $C = 1$ and $C = |\cos 2\theta_B|$ to the respective modes. In fact, for X-ray problems, we have to consider two similar D-surfaces, each corresponding to one polarization mode. To avoid complexity, however, only the D-surface of one mode is drawn in Fig. 6.

In considering actual experiments we need to add at least one exit surface from which the crystal waves come out into the vacuum. They are easily handled by the principle of tangential continuity if the exit surface is sufficiently large. The total of the vacuum G-waves constitutes the diffracted wave, which is essentially two plane waves. In that case, we may observe the fringes for wedge-shaped crystals. In general, however, the total of the vacuum G-waves is a bundle of plane waves, for example when the crystal size is finite and polyhedral. Once the diffracted wave is obtained, we can interpret it by the Ewald construction explained in the section on the kinematical theory. The difference between the two theories is reduced essentially to the functional form of $S(\mathbf{q})$.

The fringes were observed first for electron waves by Heidenreich (1942) and independently by Kinder (1943) in the electron-microscopic image of MgO smoke crystal (a polyhedral crystal). Also, double-refracted spots were observed in electron diffraction (Cowley and Rees 1946). For X-rays, the interesting experiment is the measurement of the **rocking curve**, which is the intensity profile $I(\theta)$ explained in Section 3.1 for a large crystal. To be successful in this experiment, one needs an extremely parallel and monochromatic incident wave. Therefore, precise experiments were much delayed. We shall return to this subject in Section 4.2.

The Bragg case (see Fig. 6(b)) When the net plane (the Brillouin zone boundary) is nearly parallel to the crystal surface, it may happen that the normal $\mathbf{n}$ lies in the external angle of OLG. Then, the G-wave comes out into the vacuum from the entrance surface. The wave vector $\mathbf{K}_g$ (not that of the mirror reflection) is given by $\overrightarrow{AG}$ and those of relevant waves are explained in the caption. Through the same consideration as discussed in Fig. 4, total reflection occurs for a certain range of the glancing angle θ. The famous top-hat profile is expected as the rocking curve (Fig. 8). Here, absorption is neglected. The theory including absorption was worked out by Prins (1930). The theoretical details are described in the textbook of Zachariasen (1945). Again, the crucial experimental test came later because of difficulties in producing the sophisticated incident wave. Renninger (1955) was one of the pioneers in this type of work.

In the geometry of Fig. 6(b), the angular width of the total reflection is given by

$$\Delta\theta_{dyn} = (d_{BB}/\sin \theta_B)/K = \lambda/\Lambda \sin \theta_B, \quad (18a)$$

where Λ is given by eqn (17). The numerical value is 3.20 seconds of arc under the same conditions as mentioned above for Λ. In the Laue case, we did not show the angular width because the profile $I(\theta)$ involves the Pendellösung oscillation, which

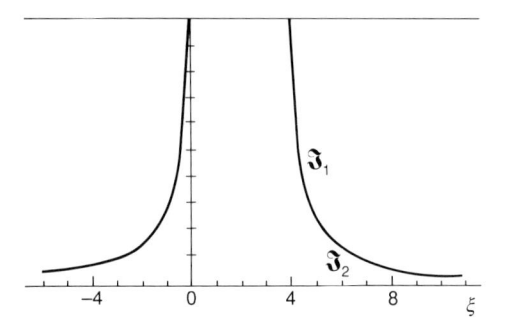

Fig. 8 The top-hat rocking curve in the plane wave theory. (Reproduction of Fig. 19 of Ewald's 1917 paper.) ξ is essentially proportional to $\theta - \theta_B$ in the present notation. The centre of the total reflection is shifted from the kinematical peak at θ_B as a result of the refractive index $n = k/K < 1$. The exact coincidence of the left flank with $\xi = 0$ is due to the Ewald model (point scatterers), for which $\chi_o = \chi_g$.

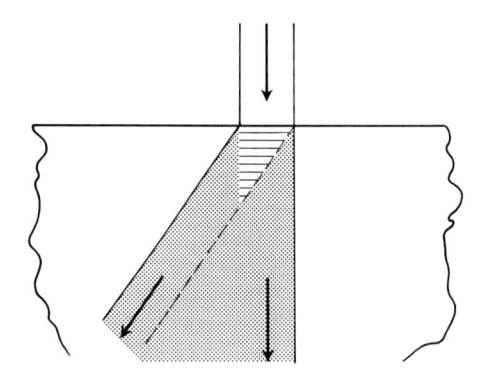

Fig 9. The wave field for a finite wave front. (Reproduction of Fig. 21 of Ewald's 1917 paper.)

depends upon the crystal thickness. However, if the oscillation is averaged out, the result is very similar to (18a) for sufficiently thick crystals.

It is easily anticipated that the integrated intensity R_{dyn} is of the order of magnitude of $\Delta\theta_{dyn}$. The exact integration of the top-hat reflectivity (Fig. 8) gives the expression

$$R_{dyn} = (4/3)\Delta\theta_{dyn} = 8/3 \cdot r_e(\lambda^2/\pi V)C|F_g|/\sin 2\theta_B \quad (18b)$$

for sufficiently thick crystals. It is worth noting that R_{dyn} is independent of the actual crystal thickness.

It is interesting to compare the expressions $\Delta\theta_{kin}$ (5a) and $\Delta\theta_{dyn}$ (18a) for the angular width of diffraction. According to the kinematical theory, the width tends to zero as the actual thickness L increases. This situation, however, cannot occur in practice, because X-rays attenuate rapidly, at least in the range of total reflection. If we use an effective thickness $T = \Lambda \tan \theta_B$ for the diffraction instead of the thickness L, we can see that $\Delta\theta_{kin}$ (5a) gives the same answer (18a) as is obtained by the dynamical theory. (Incidentally, the dynamical theory predicts that the attenuation distance is $(1/\pi) \Lambda \tan \theta_B$ under the exact Bragg condition at which the maximum attenuation occurs.)

The same principle is applicable also to understanding the difference between R_{kin} (5b) and R_{dyn} (18b). For a sufficiently thin crystal, the kinematical expression R_{kin} must be true. If the thickness increases, however, only the top thin layer is effective in diffraction. As in the case of the angular width, R_{dyn} can be derived by using $T = (1/\pi) \Lambda \tan \theta_B$ for L in the expression (5b), except for a numerical factor close to unity.

In this context, it is worth pointing out that the integrated intensity is much smaller than the value expected from the kinematical theory when the crystal is thick and perfect. This situation was experimentally recognized as **extinction** in the very early stage of X-ray crystallography. Because of its importance, and the complexity of its nature, study of this phenomenon still continues. Nevertheless, its essence is understood in terms of the concept of effective crystal volume discussed here.

So far, the plane wave theory has been discussed. Ewald knew its limitation at the very beginning. Figure 9 is the reproduction of Fig. 21 in his 1917 paper. In this context, he quoted the essence as follows:

Only within the strongly [horizontally] striped region of Fig. 21 do incident and interference beams coexist, as required for the applicability of the [plane wave] theory.... the incident wave must strictly be treated as a spherical wave. Even if it can be thought of as resolvable into a packet of plane waves each of which is reflected according to the amplitude expressions given above, nevertheless, the fixed phase relationships between them do not allow the simple superposition of energy... (English translation by Juretschke (Ewald 1991))

Ewald's underlying idea was extensively used in the development of **spherical wave theory** (Kato

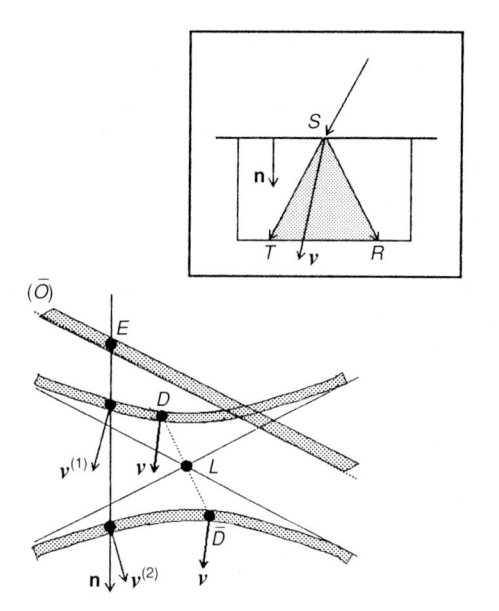

Fig. 10 The ray picture in spherical wave theory. The inset figure illustrates real space. The main figure shows the ray directions (v) in reciprocal space, which are normal to the dispersion surface at each dispersion point.

1960, 1961a,b). In fact, the latter theory deals with essentially the case that the incident wave front is infinit-esimal on the crystal surface, whereas the plane wave theory assumes implicitly an infinitely wide wave front. In Fourier or **k** space, the situation is exactly opposite. In this sense, the two theories are complementary.

The author does not intend to discuss the detail of the theory. Nevertheless, it would be worth interpreting the result in terms of **ray optics**. Here, the ray implies a wave packet, which is a bundle of plane waves in ordinary optics. It moves with a group velocity which, in general, is given by grad [$v(\mathbf{k})$]. Consequently, its direction is normal to the dispersion (or equi-energy) surface. This is the golden rule of rays, and is true also for the wave packet of the Ewald–Bloch wave which consists of O- and G-waves, except that one needs to recall that the dispersion surface is two-fold.[†]

Now, we shall return to our specific problem, namely the case in which the incident wave has a sufficiently narrow wave front, and consequently is a superposition of plane waves as Ewald mentioned. This implies that a wide range of the $(\bar{O})$-surface is excited, as illustrated in Fig. 10. For simplicity, only the Laue case will be discussed. As already seen in Fig. 6(a), an incident (plane) wave specified by E will excite the dispersion points $D^{(1)}$ and $D^{(2)}$. The associated rays propagate in the different directions $v^{(1)}$ and $v^{(2)}$, in general. They would not make interference fringes, as they are separated in real space. In any case, when a wide region of the D-surface is excited, we can expect the wave field to have the triangular form STR in the crystal (see the inset figure and also Fig. 11(a)). Sometimes, this is called the Borrmann fan. We may also expect interference fringes because the wave field consists of two fields that have a definite phase relation. If one takes a direction v in real space, the field can be interpreted as the result of interference of two rays propagating along this direction. These rays correspond to the conjugate points D and $\bar{D}$ on the different branches of the D-surface.

The pattern of fringes for a wedge-shaped crystal is schematically drawn in Fig. 11(a); these are the Pendellösung fringes observed under ordinary conditions of the X-ray source (Kato and Lang 1959). Figure 11(b) is a photograph of this phenomenon. Later, Authier (1960a,b, 1961) observed the two separate beams along $v^{(1)}$ and $v^{(2)}$ with the use of a pseudo-plane wave confined moderately both in real and **k** space. This experiment directly demonstrated double refraction in X-ray cases.

The concept of 'rays' is nothing new. It has been well investigated in the classical crystal optics of visible light. It is especially useful when the exact wave field is not mathematically available. Even when the exact field is obtained, as in the case of the spherical wave theory, the interpretation discussed above is very instructive.

Before closing this section, it is worth mentioning that the dynamical theory discussed in this section is a typical example of two-state physics. The naming of 'Pendellösung fringes' shows clearly that Ewald recognized this at the very beginning of his investigation. There are many examples in modern physics; up and down spins, an electron and hole pair, etc. (see Feynman et al. 1965).

[†] For the electromagnetic wave, the energy flow vector can be calculated by the time- and space-averaged Poynting vector. The expression for the Ewald wave is simply the sum of the Poynting vectors of the {G}-waves in many-wave cases. Its direction proves also to be normal to the dispersion surface. The details were discussed by Ewald (1958) and Kato (1958).

THE SIGNIFICANCE OF EWALD'S DYNAMICAL THEORY OF DIFFRACTION

It was inevitable, however, that the theory did not attract many experimentalists because ideally perfect crystals as assumed in the theory were not practically available. Also, the crucial test was not very successful because experimental techniques were still primitive.

In 1925 on the Ammersee, a picturesque spot near Munich, a symposium was organized by Ewald himself on the intensity of X-ray reflections (Bragg et al. 1926). In the list of participants one can find many names of famous physicists of the time. Naturally, dynamical phenomena, particularly extinction effects, were a main theme.[†] The attendees seemed to reach a conclusion that real crystals are not perfect, and that a plausible theory for the correction to the kinematical theory would be one which deals with the intensity balance neglecting wave-optical coherence. In fact, before the symposium, such a theory had been worked out by Darwin (1922). Also the phrase 'mosaic crystals' became popular among crystallographers.

After the symposium, the general interest in dynamical phenomena gradually died out and the subject was left in the hands of specialist: M. Renninger under Ewald, E. Kossel, the Japanese school under S. Nishikawa, and the Dutch school under J.A. Prins. The textbooks of Compton and Allison (1934) and James (1948) describe this period. The dynamical people had played the role of Greek tutors for the sons of Roman noblemen. Indeed, structure analysis in the following years was like a Roman empire. In the USA and Sweden, some precise measurements of the rocking curves and the reflectivity from good crystals were attempted but the main interest was in X-ray spectroscopy.

Fortunately, electron diffraction was discovered (C.J. Davisson and L.H. Germer, and G.P. Thomson [1927]; S. Kikuchi under S. Nishikawa [1928]; see Goodman 1981) and attention then focused on electron diffraction, in which the dynamical theory was indispensable. Since that time, the dynamical interaction between electron, later neutron, and X-ray diffraction has continued up to the present. The chapter by Cowley and Moodie gives a flavour of important coupling phenomena.

Bethe (1928) published his dynamical theory of electron diffraction, which became the classic in

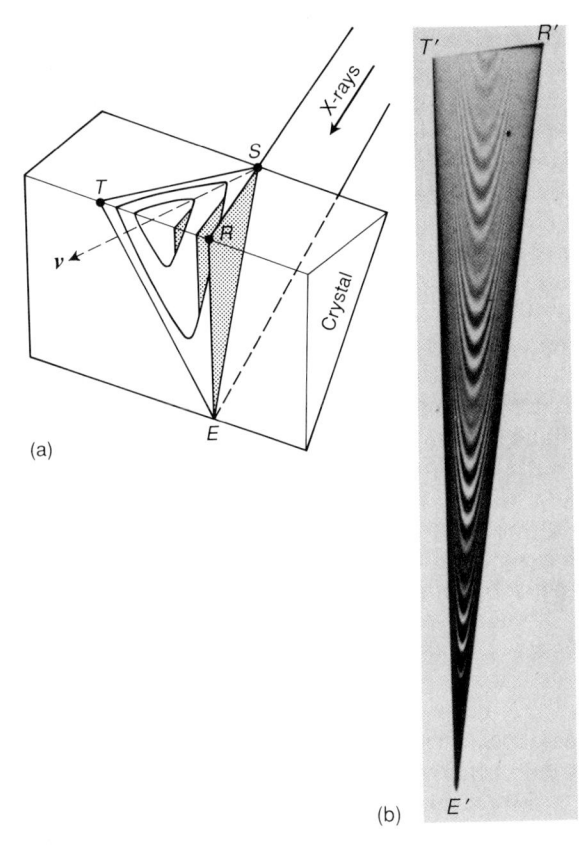

Fig. 11 (a) Schematic diagram of Pendellösung fringes in spherical wave theory. The triangle *TRE* is the intensity field on the exit surface. (b) Pendellösung fringes for (440) reflection of Si. The experimental set-up is shown in Fig. 11(a). *T'R'E'* is the projection of the intensity field, *TRE* of Fig. 11(a), along the direction of the Bragg reflection.

Probably, the present subject is the best example that has been investigated both theoretically and experimentally, and also both in real and reciprocal (**k**) space.

4. The historical development and the future

Ewald's theory was very elegant and complete, so that it is easily imagined that the theory itself was highly appreciated by theoretical physicists in those days. Born discussed the theory for more than 20 pages in his classical textbook on solid state (1923).

[†] Another topic was the quantum-mechanical calculation of the charge density $\rho(r)$ of atoms to obtain the structure factor $|F_g|$.

this field. The core of this theory strongly resembles the Bethe–Sommerfeld band theory of electrons in metals. Von Laue (1931) reformulated the X-ray theory in a similar style. Ewald himself generalized his theory, and published the fourth paper of *Zur Begründung der Kristalloptik* (1937). In this context, we must mention Molière (1939a,b,c), who justified the practical adequacy of von Laue's model as mentioned earlier. He started from a microscopic model described in a modern language which had not been available in Ewald's early period, and he developed the dynamical theory based on semi-classical quantum electrodynamics. An excellent review has been given by Dederichs (1972).

The new dawn came when a new phenomenon was discovered. Borrmann (1941, 1950) found that X-rays penetrate through good crystals (quartz, calcite) when the Bragg condition is satisfied. The phenomenon is now called the Borrmann effect or anomalous transmission. It was derived theoretically by Zachariasen in his book (1945) without knowing about the experiment of Borrmann. Von Laue explained it independently (1949), with more physical insights. It is a little ironic that the phenomenon was rediscovered by Campbell (1951a,b) in the USA without knowing about the work of the German group.

The real renaissance came rapidly in a decade after Borrmann. Solid-state technology, particularly for semiconductor devices, required crystals of an increasing degree of perfection, and urgently needed some experimental methods to characterize crystal perfection. Knowledge of crystal defects such as point defects, dislocations, and stacking faults had been accumulated over the years. Consequently, 'mosaic crystals' were no longer accepted as the physical model of real crystals.

Dash (1956, Schenectady) grew Si single crystals by the necking method (a variation of the Czockralski method). Other industrial laboratories also could grow larger crystals by similar methods. Some of them seemed nearly free from dislocations, at least as judged by the use of etch-pits and decoration techniques. This achievement eliminated the crucial obstacle in the study of dynamical phenomena.

A stimulus came also from electron microscopy.

W. Bollmann (Switzerland) and the Cambridge school led by P.B. Hirsch succeeded in the direct observation of stacking faults and dislocations in metals. The result was reported at the 1957 Montreal Congress of the IUCr. Soon after, three attendees, Borrmann *et al.* (1958), Lang (1958), and Newkirk (1958) observed dislocations independently by their own X-ray topography. It was the good fortune of the present author to find Pendellösung fringes (Fig. 11b) in Si with Lang (Kato and Lang 1959).

In the following three decades, the developments in research related to dynamical phenomena were fast and so vast that it is impossible to review them in a short chapter. For this reason, I would like to pick up a few *ad hoc* topics, with emphasis on future developments. In fact, our future is very promising because of the availability of synchrotron radiation sources.

Readers who are interested in further details of various aspects of dynamical phenomena will find numerous literature references in reviews and books which the author could not mention specifically in this section; for example, von Laue (1960), James (1963), Batterman and Cole (1964), Kato (1974), Brümmer and Stephanik (1976), Authier (1977), and Pinsker (1978).

Recent developments in science and technology are supported by computer technology, not only in theoretical calculations but also in sophisticated instrumentation. Here, together with crystal growers, we may be proud of the fact that the core of computers is built on Si perfect crystals; the structure of the Si crystal was determined by X-ray crystal analysis[†] and the perfection is characterized ultimately by dynamical phenomena.

4.1 The characterization of crystal perfection

Diffraction crystallography is a science that determines the structure of matter at the atomic level. It can be divided into two categories;

(1) structure inside the unit cell;
(2) structure outside the unit cell.

Dynamical phenomena are very useful for investigating structures of category (2) because many

---

[†] According to *Strukturbericht* [1913–1926; edited by P.P. Ewald and C. Hermann], P. Debye and P. Scherrer found the similarity between the Debye–Scherrer rings of Si and of diamond [1917]. The structure of diamond was determined by W.H. and W.L. Bragg [1914]. For semi-conductor scientists knowledge of the diamond structure is something as basic as air and water in daily life.

of the phenomena are very sensitive to lattice distortion (i.e. the arrangement of unit cells). Pendellösung fringes and the Borrmann absorption are good examples.

The last three decades, in fact, are characterized by a deeper theoretical understanding of diffraction in distorted crystals than in the previous period. Equations of the Takagi–Taupin type are useful for lattice distortions in general. For slight distortions, the Eikonal theory (Penning and Polder 1961; Kato 1963, 1964a,b; Kambe 1965, 1968) can be used. This is nothing other than a ray theory in distorted crystals. Extinction phenomena are now understood on a sound theoretical basis, that of optical coherence. (Kato 1980a,b; Al Haddad and Becker 1988; Guigay 1989). Numerical approaches have been advanced by the French school (Y. Epelboin), where the main interest is in topography, and have become practical in many laboratories. A different approach was also presented by Wilkins (1981), who was mainly concerned with extinction problems. All of these approaches are applied to understanding real crystals.

Another characteristic feature is that the topographic method is extensively used after the success achieved using this method in the direct observation of dislocations mentioned above, although the goniometric method represented by the measurement of rocking curves is also highly developed (see Section 4.2). Ewald realized the importance of topography at the very beginning. Figure 9 is the first drawing of the topographic behaviour of the crystal wave (Ewald 1917). The chapter by Authier and Capelle in this book describes the development of topographic research and the contributions of the French school.

It is interesting to observe that Ewald himself worked out little theory on distorted crystals, although he was interested in the subject and encouraged us. This is probably due to his taste for perfection in physical theory. Partly for this reason, I shall stop describing this subject. Readers may find many references in the review book of Tanner (1976) and the proceedings of the Durham conference (Tanner and Bowen 1980).

4.2 X-ray optical devices

One of the most remarkable achievements is the construction of **X-ray interferometers** of the Mach–Zehnder type (Bonse and Hart 1965a, 1966). We may discern three significant points in this work. First, it was the first experiment to demonstrate X-ray interference between O- and G-waves extracted into air. Pendellösung fringes also are a result of interference but between two waves belonging to the dispersion surfaces (1) and (2) inside the crystal so that their applicability is limited.

Second, it has potential usefulness in the measurement of physically important quantities. Deslattes and Henins (1973) used an X-ray interferometer to measure the lattice constant of Si and Avogadro's number with an accuracy of a few parts per million. Hart (1981) demonstrated how the lattice spacing of Si changes from position to position within a single crystal. Moreover, neutron interferometers working on the same principles are employed to detect physically fundamental phenomena far from crystallography, such as a gravitationally induced quantum phase (Colella *et al.* 1975). The Aharonov–Casher effect has also been investigated (Cimmino *et al.* 1989); this is the phase shift of neutron (magnetic dipole) rays surrounding a charged body, which is analogous to the famous Aharonov–Bohm effect.

Third, the work inaugurated a new concept in diffraction-optical devices, namely, the monolithic multi-crystal. The idea of using multiple reflections for X-ray optics can be traced back to DuMond (1937), but the proposed designs could not be put in practice until a monolithic crystal became available. For example, the channel-cut monochromator (Bonse and Hart 1965b), which amounts to a device using several reflections, is now used routinely at many ports of synchrotron sources. Kohra's group also developed several techniques both for topography and goniometry (Kohra 1972).

In the fundamental problem of verifying the plane-wave theory, the work of Renninger (1955) and Authier (1961) showed, for example, how this might be done, and the experiments became much easier than we had earlier anticipated. Hart and Milne (1986) observed the transition of fringes from the spherical wave case to the plane wave case with a monolithic double-crystal. Figure 12 is a recent example of measurements of rocking curves in the Laue case (Deutsch and Hart 1985). Here, the angular resolution is better than $\frac{1}{100}$ seconds of arc, a precision which no one could have imagined in Sommerfeld–Ewald's time.

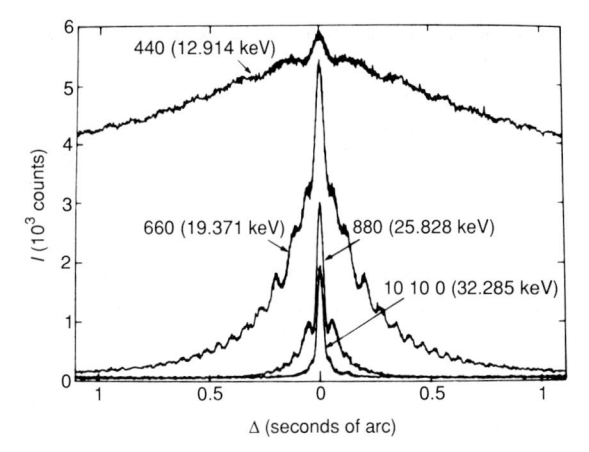

Fig. 12 The rocking curve for Si in the Laue case: Δ is essentially equivalent to $(\theta - \theta_B)$ in the present chapter (Deutsch and Hart 1985).

The function of all of these devices is ultimately characterized by the dynamical theory. The paper by Bonse and Graeff (1977) is worth consulting.

4.3 Studies related to crystal analysis

The topics in this section are related to crystal structures of category (1) mentioned above. Although these studies are still in the hand of specialists or in their very early stages, they are interesting from the viewpoint of crystallography as well as of solid-state physics.

1. **Absolute values of structure factors** $|F_g|$, eqn (4), can be determined accurately by means of the Pendellösung method. The attempt had been made at the beginning of the observation of the fringes (Kato and Lang 1959). Recently, the oscillatory nature of the rocking curve as shown in Fig. 12 has also been used. The basic principle is that the fringe spacing Λ, eqn (17), is inversely proportional to $|F_g|$, and the proportionality factor depends only physical constants such as e, m, c, and the wavelength λ. Therefore, $|F_g|$ values can be determined on an absolute scale.

The accuracy attained for Si is at present of an order of better than 5×10^{-4} for lower-order reflections. This makes it possible to know precisely the distribution of the bonding charge of Si in the crystal. Accurate theoretical calculations are very desirable for comparison. A similar method was applied also to neutron diffraction, and the nuclear scattering amplitude b has been determined with an accuracy of 2.5×10^{-5} (Shull 1968). This work is very important in neutron physics because this value is used as the standard of calibration for the b values of other nuclei. The current state of the art was reviewed recently by the present author (Kato 1988).

2. Surface structures can be studied by means of the **standing wave method**. In the geometry of the Bragg case, a standing wave parallel to the net plane is formed as the result of interference between the O- and G-waves, both inside and outside the crystal. Its spacing must be nearly the same as the spacing d of the net plane employed. Then, if any atom is located on the loop (anti-node) of the standing wave, it scatters strongly either secondary electrons or characteristic X-rays; the scattering is very weak if the atom is located on the node. Thus, we can obtain valuable information on the location of the atom concerned on the scale of d.

At an earlier stage, impurity atoms inside the crystal were studied (Batterman 1964), and it was possible to detect whether they were substitutional or interstitial. Even this case yields information near the surface because the crystal wave attenuates rapidly in the region of total reflection, as mentioned above. Recently, interest has focused on standing waves outside the crystal (Andersen et al. 1976; Cowan et al. 1980). Here, in favourable cases, we can determine the location of atoms in an adsorbed layer on the top of the crystal.

In these experiments, obviously it is important to know where the standing wave is formed relative to the crystal lattice. Such knowledge is entirely contained within the dynamical theory.

3. **Many-wave diffraction theory** is relevant to the phase problem. The two-wave theory discussed in Section 3 is by no means exactly self-consistent. The formal extension to the many-wave case is, in principle, not very difficult. To demonstrate this, we shall consider the three-wave case involving O, G-, and H-waves. In the same manner as in (14a,b), we have

$$(\mathbf{k}_o^2 - k^2)d_o = 4\pi\, K^2(\chi_{-g}d_g + \chi_{-h}d_h), \quad (19a)$$

$$(\mathbf{k}_g^2 - k^2)d_g = 4\pi\, K^2(\chi_g d_o + \chi_{g-h}d_h), \quad (19b)$$

$$(\mathbf{k}_h^2 - k^2)d_h = 4\pi\, K^2(\chi_h d_o + \chi_{h-g}d_g). \quad (19c)$$

The last additional terms in (19a,b) imply the creation of O- and G-waves from the H-wave respectively, and the new equation (19c) describes the

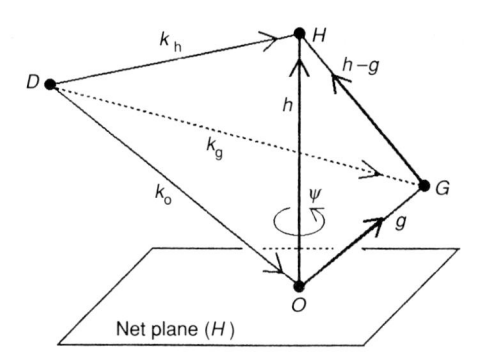

Fig. 13 The geometry of the wave vectors and reciprocal lattice vectors in a three-wave case. Anomalous reflection occurs when the azimuthal angle ψ satisfies the condition of simultaneous reflection; namely, O, G, and H lie nearly on a single Ewald sphere.

propagation of the H-wave. The set of three equations describes the dynamical balance among these three waves. Procedures to determine the dispersion surface, the dispersion points (actually three real and three virtual), and the amplitude ratios are analogous to the two-wave case.

Historically, the three-wave theory was first used to explain the Umweganregung (indirect excitation) phenomenon discovered by Renninger (1937a,b). He noticed that the (222) reflection of diamond became appreciable at certain azimuthal directions of the incident wave when the crystal was rotated around the $H(222)$ direction. This technique is called Renninger plotting. The phenomenon was thought strange at first, because F_{222} should have a null magnitude, from the symmetry of the standard diamond structure composed of spherical atoms. Renninger, probably involving Ewald as his teacher, explained the essence as illustrated in Fig. 13. If the reciprocal lattice points G and H lie nearly on an Ewald sphere, the $H(222)$-wave can be excited through an indirect excitation, $\mathbf{k}_o \to \mathbf{k}_g \to \mathbf{k}_g + (\mathbf{h} - \mathbf{g}) = \mathbf{k}_h$. The necessary requirement must be that both F_g and F_{h-g} are not forbidden. Actually, one can list such pairs of reflection indices, namely, $(3\bar{1}1)$ and $(\bar{1}33)$; $(3\bar{1}3)$ and $(\bar{1},3,\bar{1})$; and their cyclic interchanges. Renninger could assign all of the observed anomalous reflections to these pairs.

The Umweganregung phenomenon is an extreme example in which one of the structure factors involved in (19) is zero. In general cases, where F_h is not forbidden, the intensity will be disturbed by other simultaneous reflections. This is a serious problem in collecting precise $|F_g|$ data by means of diffractometry.

Another interesting phenomenon is an anomalous Borrmann effect. As mentioned above, even in the two-wave case, the X-ray absorption is anomalously small when the Bragg condition is satisfied (ordinary Borrmann effect). Under three-wave conditions, this anomaly is either anomalously enhanced (Borrmann and Hartwig 1965) or depressed (Feldman and Post 1972). The theoretical background was discussed extensively by Post et al. (1977).

Ewald was continuously interested in the many-wave case and finally published his thoughts in two papers with Héno (Ewald and Héno 1968; Héno and Ewald 1968).

It should be mentioned also that many-wave cases are unavoidable in electron diffraction and microscopy. Therefore, numerous papers have been published on electron cases. Many of the results are valuable also for X-ray cases. Lamla's work (1938a,b) is a good example. A serious complication, common to electron and X-ray cases, lies in situations where the Bragg and Laue geometries cannot be separated, contrary to the two-wave case. In some geometries, the approximation called 'linearization of dispersion surface' (see Section 3.4) may not be allowed. This causes another complexity. Colella (1974) worked out an extensive theory, similar to Lamla's theory, with the proper boundary condition for X-ray cases.

The X-ray case is made more complicated than the electron case by the vectorial nature of the wave field. However, this does not cause any conceptual difficulty. As shown in eqn (10), $\mathbf{d}_g$ is transverse to $\mathbf{k}_g$ so that the problem can be handled by taking the double field components associated with each wave vector. The ray concept also can be established (Ewald 1958; Kato 1958).

Some many-wave phenomena were studied primarily from an academic viewpoint. Recently, however, much attention has been focused on application to the phase problem. As is well known, the phase problem is very important in crystal analysis, because once the phases of large structure factors are known we can determine the structure in a straightforward manner.

In the three- or four-wave case, the intensity profile of the Bragg reflection in Renninger plots is often asymmetric on both sides of the peak. Such asymmetry is theoretically predictable by assigning a proper (relative) phase to each of the structure factors involved in the simultaneous reflection. This experience suggests the possibility of determining the phase by a careful analysis of the intensity profile. Post (1977), Chapman *et al.* (1981), Chang (1982), and Juretschke (1982) published pioneer works in this field. Because of the importance of his work in crystallography, Post was awarded the Warren prize at the 1981 ACA (American Crystallographic Association) meeting. The present book includes a chapter by Colella, which describes mainly recent developments by his own group.

This field of research is actively developing. Interested readers will find many references in two extensive reviews by Chang (1984, 1987). Naturally, some aspects of this field are still under debate as can be seen even in a recent issue of *Acta*. It is very desirable to settle unsolved problems in handling the phase problem.

In closing this section, the author would like to mention that investigations apparently of an academic nature when first published can become valuable if they are fundamental. The history of Ewald's dynamical theory is a good scientific example on a large scale.

5. Epilogue

When I started research [1944] under R. Uyeda, my teacher at Nagoya University, P.P. Ewald was already a legendary person in my mind, probably because his famous papers had been published before my birth. On Uyeda's bookshelves there were old volumes of his personal *Phys. Rev.* and other journals plus soon a grand new journal, *Acta Cryst.* He used to tell us with a little pride that this *Acta* was a gift from Ewald. In fact, it was one of three sets sent regularly to Japan. At that time, if we wished to read a new volume of an American journal, we had to go to the American Cultural Center near the centre of the city. As it took nearly an hour by tram, you may easily imagine how grateful we were for this gift.

Several years later, I moved to a small institute near Tokyo, where I was guided by S. Nishikawa. There, I finished writing my Doctor's Thesis on a dynamical theory of electrons for finite polyhedral crystals, and I changed my research subject from electrons to X-rays. Meanwhile, Nishikawa died so that I had to work nearly alone.

Fortunately, I had the chance to work with A.R. Lang at Harvard from September 1957. Later, Lang told me that he consulted with Uyeda and Ewald when he was searching for a young scientist and I was one of a few suggested candidates.

Anyway, I went to the USA by ship. All passengers were Fulbright grantees. A week or so before the embarkation, I received a letter from Lang about the IUCr Congress at Montreal. After describing the electron microscopic work of P.B. Hirsch's group, he proposed with enthusiasm that similar work should be done also in X-ray diffraction. When I first met Lang at Harvard, however, I had nearly forgotten the topic after the two-week journey because everything was new for me.

Nevertheless, we started work following Lang's idea, and we succeeded in the direct observation of dislocations within a month or so. Also, by good fortune, we found the Pendellösung fringes in a wedge part of Si. As my Doctor's Thesis was on essentially the same subject in the electron cases, I could solve the problem fairly quickly. For this reason, henceforth we worked separately. Lang studied dislocations and I worked on fringes.

We needed many samples of Si, quartz, diamond, etc. The last two were supplied from Frondel's laboratory (Department of Mineralogy). As to Si, I remember that Lang drove with me on the Massachusetts Turnpike to visit W. Dash (General Electric, Schenectady), and we were given a few pieces of Si single crystal. At that time, J. Patel (now in Bell Laboratories) also was growing Si at Raytheon Manufacturing Company near Boston, and we received some also from him.

In size and form they resembled a small finger. One day, I lost a sliced sample in a sink and asked Lang the price. His answer was 'Priceless, but certainly more precious than gold.' Probably, no one could have predicted the present cost. It seems literally priceless.

In November 1957, the ACA Annual Meeting was held at Pittsburgh. Lang suggested to me that I should give the talk on the energy-flow problem in dynamical diffraction, which I had submitted to the Montreal Congress from Japan, but had been unable to deliver. When I finished my first English-speaking talk, a tall gentleman stood up and said 'You might not understand what he said', and went on to explain my theory, spending nearly the same time that I had used. He was Paul Ewald. After lunch, I shook his hand for first time, while he was walking with Elizabeth Wood (Bell Laboratories) and a few others on a sunny street.

In the middle of December, I visited Ewald at Brooklyn, and we discussed the ray and energy-flow problem, which I had had in mind since my doctoral thesis work. In January, next year, he sent me a letter and we continued to exchange letters on this subject. Finally, these discussions led to our coupled papers in *Acta Cryst.* (Ewald 1958; Kato 1958).

In rereading his letters recently, I recalled an overlap of ours with the Berlin group including M. von Laue and H. Wagner, his coworker. Ewald sent me a typed letter (most of his letters were handwritten) with a copy of his reply to von Laue (17 December 1958). In the letter, Ewald mentioned his own experience of an overlap as follows:

In 1923 Hermann and myself have clearly seen the Patterson series, but never published since necessary deconvolution methods were lacking. We cannot complain about that.

(Translation)

Two years later, I had a chance to visit Berlin and could talk with Wagner, Borrmann, and Hildebrandt in a very friendly way. At that time, Hildebrandt had nearly finished his experimental work on the bending of rays in distorted crystals (1959a,b). Probably, this work subconsciously influenced my Eikonal theory. Unfortunately, von Laue had died a few months earlier as the result of a car accident.

In January 1959, Ewald stopped at our flat off Oxford Street near Harvard University. By that time, I had submitted a paper on the experimental aspect of Pendellösung fringes with Lang, and I was working on the theory. Lang had decided to move to Bristol, England, and he was often absent from the USA. I explained my spherical wave theory with enthusiasm to Ewald. We scattered many sheets of paper on the desk in the kitchen, which he preferred to a small table in the living room. After much patience with my explanations in poor English, he finally said 'Better to write the paper if you believe it correct.' However, he suggested to me that Laue's paper on Kossel lines should be looked at. Curiously, he did not mention at all his own 1917 paper, in which my idea was already clearly stated, as mentioned in Section 3.4. In a later letter, he also suggested that I should read Debye's paper on the wave-optical theory of focusing by a lens. Much later, I could explain the fringe positions, which differ from those expected from a simple ray theory, along the lines of Debye's theory.

These are my reminiscences on the ray problem and Pendellösung fringes. I now realize how greatly Ewald encouraged people and how well he guided young scientists, but I came to feel it too late.

In January 1960, I also went to Bristol and returned to Japan around November.

In 1966, I stayed for a few days at Ewald's Connecticut house. I remember that he was making a model of the dispersion surface in the three-wave case with sheets of white paper and thin sticks. Undoubtedly, this study led to his joint papers with Héno (Ewald and Héno, 1968; Héno and Ewald 1968). He explained his ideas to me with enthusiasm.

I visited also his house at Ithaca, around 1975. He led me to his large basement library and said, 'Take any books except in this part.' I hesitated to do so but took a book irrelevant to the dynamical theory. Soon after, he sent me a copy of his own 1933 volume in the *Handbuch der Physik*. Also, on another occasion, he sent me the three volumes of von Laue's collected works. In a letter, he asked me 'Are you tired with dynamical theory?.' Probably, he worried about me doing other tasks, although in my mind I always stayed on his dispersion surface even if spreading out on it to some extent.

After the 1981 Ottawa Congress, I visited him again at Ithaca with my wife, staying at H. Bethe's house. Mrs Bethe and her daughter, Monica *San*, who speaks Japanese quite fluently, and studies O-No (Japanese classical plays) in a scholarly way, took care of us. On leaving, at the exit staircase, Ewald said: 'Next time, visit the widow of Mr Ewald.' His sense of humour had not left him.

When I received the final news, I fell deeply into grief so that I even lost the spirit to write an obituary to anyone. This is my excuse for now writing a rather lengthy epilogue.

Acknowledgements

The author would like to express his sincere thanks to Professors D.W.J. Cruickshank, H.J. Juretschke, A.R. Lang, and M. Hart for their encouragement and plentiful advice on this chapter.

Obituaries

Scientific journals

F. Balibar (Paris); *Bull. Minéral.* (1986), **109**, 329.
J.J. Dropkin and B. Post (New York); *Acta Cryst.* (1986), **A42**, 1–5.
G. Hildebrandt (Berlin); *Phys. Bl.* (1985), **41**, 412–13.
H.J. Juretschke (New York), A.F. Moodie (Melbourne), H.K. Wagenfeld (Melbourne), and H.A. Bethe (Ithaca); *Physics Today* (1986), May, 101–4.
S. Miyake (Tokyo); *J. Jpn. Cryst. Soc.* (1986), **28**, 240–2. (in Japanese).
Chronicles; *Kristallografiya* (July–August 1986); **31**, 830. Translation, *Soc. Phys. Crystallogr.* (1986), **31**, 494.
H.A. Bethe and G. Hildebrandt; *Biogr. Mem. Fellows Roy. Soc. (London)* (1988), **34**, 133–76.

Newspapers

The Times (London) 7 September 1985.
 Professor Paul P. Ewald. Pioneer work in X-ray crystallography.
New York Times; 7 September 1985.
 Paul P. Ewald, a pioneer in crystal analysis.

References

Al Haddad, M. and Becker, P.J. (1988). *Acta Cryst.*, **A44**, 262–70.
Andersen, S.K. Golovchenko, J.A., and Mair, G. (1976). *Phys. Rev. Lett.*, **37**, 1141–5.
Authier, A. (1960a). *Compt. Rend. Sci. Paris*, **251**, 2003–5.
Authier, A. (1960b). *Compt. Rend. Sci. Paris*, **251**, 2502–4.
Authier, A. (1961). *Bull. Soc. Fr. Minéral. Crystallogr.*, **84**, 51–89.
Authier, A. (1977). *Section topography*. In *X-ray optics* (*Topics in applied physics*, Vol. 22) (ed. H.J. Queisser). Springer, Berlin.
Batterman, B.W. (1964). *Phys. Rev.*, **A133**, 759–64.
Batterman, B.W. and Cole, H. (1964). *Rev. Mod. Phys.*, **36**, 681–717.
Bethe, H.A. (1928). *Ann. Phys. (Leipzig)*, **87**, 55–129.
Bonse, U. and Graeff, W. (1977). *X-ray and neutron interferometry*. In *X-ray optics* (*Topics in applied physics*, Vol. 22) (ed. H.J. Queisser). Springer, Berlin.
Bonse, U. and Hart, M. (1965a). *Appl. Phys. Lett.*, **6**, 155.
Bonse, U. and Hart, M. (1965b). *Appl. Phys. Lett.*, **7**, 238–40.
Bonse, U. and Hart, M. (1966). *Z. Phys.*, **188**, 154–64.
Born, M. (1923). *Atomtheorie des festen Zustandes*. Springer, Berlin.
Born, M. and Wolf, E. (1980). *Principles of optics*, 6th edn. Pergamon, Oxford.
Borrmann, G. (1941). *Phys. Z.*, **42**, 157–62.
Borrmann, G. (1950). *Z. Phys.*, **127**, 297–323.
Borrmann, G. and Hartwig, W. (1965). *Z. Kristallogr.*, **121**, 401–9.
Borrmann, G., Hartwig, W. and Irmler, M. (1958). *Z. Naturforsch.*, **13**a, 423–5.
Bragg, W.L., Darwin, C.G. and James, R.W. (1926). *Philos. Mag.*, **1**, 897–922.
Brümmer, O. and Stephanik, H. (1976). *Dynamische Interferenztheorie*. Akademische Verlag, Leipzig.
Campbell, H.N. (1951a). *Acta Cryst.*, **4**, 180–1.
Campbell, H.N. (1951b). *J. Appl. Phys.*, **22**, 1139–42.
Chang, S.L. (1982). *Phys. Rev. Lett.*, **48**, 163–6.
Chang, S.L. (1984). *Multiple diffraction of X-rays in crystals* (Springer series in solid-state sciences, Vol. 50) (ed. H.J. Queisser). Springer, Berlin.
Chang, S.L. (1987). *Crystallogr. Rev.*, **1**, 87–189.
Chapman, L.D., Yoder, D.R. and Colella, R. (1981). *Phys. Rev. Lett.*, **46**, 1578–81.
Cimmino, A., Opat, G.I., Klein, A.G. Kaiser, H., Werner, S.A., Arif, M. and Clothier, R. (1989). *Phys. Rev. Lett.*, **63**, 380–3.
Colella, R., (1974). *Acta Cryst.*, **A30**, 413–23.
Colella, R., Overhauser, A.W. and Werner, S.A. (1975). *Phys. Rev. Lett.*, **34**, 1472–4.
Compton, A.H. and Allison, S.K. (1934). *X-rays in theory and experiment*. Von Nostrand, New York.
Cowan, P.L., Golovchenko, J.A. and Robbins, M.F. (1980). *Phys. Rev. Lett.*, **44**, 1680–3.
Cowley, J.M. and Moodie, A.F. (1957). *Acta Cryst.*, **10**, 609–19.
Cowley, J.M. and Moodie, A.F. (1959a). *Acta Cryst.*, **12**, 353–9.
Cowley, J.M. and Moodie, A.F. (1959b). *Acta Cryst.*, **12**, 360–7.

Cowley, J.M. and Rees, A. (1946). *Nature (London)*, **158**, 550–2.
Darwin, C.G. (1914). *Philos. Mag.*, **27**, 675–90.
Darwin, C.G. (1922). *Philos. Mag.*, **43**, 800–29.
Dash, W.C. (1956). *J. Appl. Phys.*, **27**, 1193–5.
Dederichs, P.H. (1972). *Solid State Phys.*, **27**, 135–236.
Deslattes, R.D. and Henins, A. (1973). *Phys. Rev. Lett.*, **31**, 972–5.
Deutsch, M. and Hart, M. (1985). *Phys. Rev.*, **B31**, 3846–58.
DuMond, J.W.M. (1937). *Phys. Rev.*, **52**, 872–83.
Ewald, P.P. (1913). *Phys. Z.*, **14**, 465–72.
Ewald, P.P. (1916a). *Ann. Phys. (Leipzig)*, **49**, 1–38.
Ewald, P.P. (1916b). *Ann. Phys. (Leipzig)*, **49**, 117–43.
Ewald, P.P. (1917). *Ann. Phys. (Leipzig)*, **54**, 519–97.
Ewald, P.P. (1927). *Handbuch der Physik.*, **24**, 191–369.
Ewald, P.P. (1937). *Z. Kristallogr.*, (A)**97**, 1–27.
Ewald, P.P .(1958). *Acta Cryst.*, **11**, 888–91.
Ewald, P.P. (1962a). *J. Phys. Soc. Jpn.* **17**; Supplement B-II, 48–52.
Ewald, P.P. (1962b). *Fifty years of X-ray diffraction* (ed. P.P. Ewald), p. 39. Oosthoek, Utrecht.
Ewald, P.P. (1991). *On the foundations of crystal optics.* (ed. H.J. Juretschke) Monograph **10**, American Crystallographic Association, Buffalo, NY.
Ewald, P.P. and Héno, Y. (1968). *Acta Cryst.*, A**24**, 5–15.
Feldman, R. and Post, B. (1972). *Phys. Stat. Solidi*, (a)**12**, 273–6.
Feynman, R.P., Leighton, R.B. and Sands, M. (1965). *Lectures on physics*, Vol. III, Chapters 8–11. Addison–Wesley, Reading, MA.
Goodman, P. (ed.) (1981). *Fifty years of electron diffraction.* D. Reidel, Dordrecht.
Guigay, J. (1989). *Acta Cryst.*, A**45**, 241–4.
Hamilton, W.C. (1957). *Acta Cryst.*, **10**, 629–34.
Hart, M. (1981). *J. Cryst. Growth*, **55**, 409–27.
Hart, M. and Milne, A.D. (1968). *Phys. Stat. Solidi*, **26**, 185–9.
Heidenreich, R.D. (1942). *Phys. Rev.* (L), **62**, 291–2.
Héno, Y. and Ewald, P.P. (1968). *Acta Cryst.*, A**24**, 16–42.
Hildebrandt, G. (1959a). *Z. Kristallogr.*, **112**, 312–39.
Hildebrandt, G. (1959b). *Z. Kristallogr.*, **112**, 340–61.
James, R.W. (1948). *The optical principles of the diffraction of X-rays*, lst edn. G. Bell, London. (3rd edn 1962.)
James, R.W. (1963). *Solid State Phys.*, **15**, 53–220.
Juretschke, H.J. (1982). *Phys. Rev. Lett.*, **48**, 1487–9.
Kambe, K. (1965). *Z. Naturforsch.*, **20**a, 770–86.
Kambe, K. (1968). *Z. Naturforsch.*, **23**a, 25–43.
Kato, N. (1958). *Acta Cryst.*, **11**, 885–7.
Kato, N. (1960). *Z. Naturforsch.*, **15**a, 369–70.
Kato, N. (1961a). *Acta Cryst.*, **14**, 526–32.
Kato, N. (1961b). *Acta Cryst.*, **14**, 627–36.
Kato, N. (1963). *J. Phys. Soc. Jpn.*, **18**, 1785–91.
Kato, N. (1964a). *J. Phys. Soc. Jpn.*, **19**, 67–77.
Kato, N. (1964b). *J. Phys. Soc. Jpn.*, **19**, 971–85.
Kato, N. (1974). *X-ray diffraction* Chapters 3, 4 and 5 (ed. L.V. Azaroff). McGraw–Hill, New York.
Kato, N. (1980a). *Acta Cryst.*, A**36**, 763–9.
Kato, N. (1980b). *Acta Cryst.*, A**36**, 770–8.
Kato, N. (1988). *Aust. J. Phys.*, **41**, 337–49.
Kato, N. and Lang, A.R. (1959). *Acta Cryst.*, **12**, 787–94.
Kinder, E. (1943). *Naturwissenschaften*, **31**, 149.
Kohra, K. (1972). *Proceedings of the 6th int. conf. on X-ray optics and microanalysis* pp. 35–45. University of Tokyo Press, Tokyo.
Kuriyama, M. (1973). *Z. Naturforsch.*, **28**a, 622–6.
Lamla, E. (1938a). *Ann. Phys. (Leipzig)*, **32**, 178–89.
Lamla, E. (1938b). *Ann. Phys. (Leipzig)*, **32**, 225–41.
Lang, A.R. (1958). *J. Appl. Phys.* (L), **29**, 597.
Laue, M. von (1931). *Ergenbn. Exakt Naturwiss.*, **10**, 133–58.
Laue, M. von (1949). *Acta Cryst.*, **2**, 106–13.
Laue, M. von (1960). *Röntgenstrahl Interferenzen.* Akademische Verlag, Frankfurt.
Molière, G. (1939a). *Ann. Phys.*, **35**, 272–96.
Molière, G. (1939b). *Ann. Phys.*, **35**, 297–313.
Molière, G. (1939c). *Ann. Phys.*, **36**, 265–74.
Newkirk, J. (1958). *Phys. Rev.* (L), **110**, 1465.
Penning, P. and Polder, D. (1961). *Philips Res. Rep.*, **16**, 419–40.
Pinsker, Z. (1978). *Dynamical scattering of X-rays in crystals.* Springer, Berlin.
Post, B. (1977). *Phys. Rev. Lett.*, **39**, 760–3.
Post, B., Chang, S.L. and Huang, T.C. (1977). *Acta Cryst.*, A**33**, 90–7.
Prins, J.A. (1930). *Z. Phys.*, **63**, 477–93.
Renninger, M. (1937a). *Naturwissenschaften*, **25**, 43.
Renninger, M. (1937b). *Z. Phys.*, **106**, 141–76.
Renninger, M. (1955). *Acta Cryst.*, **8**, 597–606.
Shull, C.G. (1968). *Phys. Rev. Lett.*, **21**, 1585–9.
Takagi, S. (1962). *Acta Cryst.*, **15**, 1311–12.
Takagi, S. (1969). *J. Phys. Soc. Jpn.*, **26**, 1239–53.
Tanner, B.K. (1976). *X-ray diffraction topography.* Pergamon, Oxford.
Tanner, B.K. and Bowen, D.K. (eds.) (1980). *Characterization of crystal growth defects by X-ray methods.* Plenum, New York.
Taupin, D. (1964). *Bull. Soc. Fr. Minéral. Cristallogr.*, **87**, 469–511.
Werner, S.A. and Arrott, A. (1965). *Phys. Rev.*, A**140**, 675–86.
Wilkins, S.W. (1981). *Phil. Trans. Roy. Soc. (London)*, A**299**, 275–317.
Zachariasen, W.H. (1945). *Theory of X-ray diffraction in crystals.* Wiley, New York. (Available in Dover Publications.)

B
Ewald as seen by others

2

Paul P. Ewald—the German period

G. Hildebrandt

This chapter treats mainly Ewald's working life in the years between 1910 and 1937, beginning with the year when Ewald began his thesis and ending with the year when he left Germany. A few reminiscences will be added to touch on some important events for which Ewald returned to his native country after the Second World War.

Paul Peter Ewald was born in 1888 in Berlin. His father, a historian, died three months before his birth. His mother, a painter, occasionally liked to spend time in other countries such as France and England, so that young Paul had some knowledge of French and was nearly fluent in English even before his formal education began.

In 1900 Paul and his mother moved to Potsdam, near Berlin. There he took the Abitur, his final high-school examination, in 1905, with good success in most disciplines and disastrous results in (Prussian) history. Friends of the family had encouraged Ewald's interest in chemistry, so, to study this science, he went first to Caius College, Cambridge. After one year there, he spent two semesters in Göttingen, where the lectures of the famous David Hilbert turned his interests towards mathematics. Another year later we find Ewald as Pringsheim's pupil in Munich. However, it was not Pringsheim, but rather Sommerfeld who was eventually decisive for Ewald's future in mathematical physics. A friend of Ewald's had taken him to a Sommerfeld lecture on hydrodynamics, and sixty years later Ewald wrote: 'Since then I knew that not pure mathematics but instead theoretical physics was worth my love, this marvellous harmony between clear mathematical thinking and physical phenomena.'

Shortly after, Ewald asked Peter Debye, then Sommerfeld's assistant, to organize seminars for the younger students. Debye agreed, Sommerfeld contributed a box of cigars, and the seminars started. The first visiting speaker was Max Laue from Berlin. In this way, Ewald was responsible for initiating the Munich physics colloquium (Ewald 1968).

In 1910 Ewald decided to begin work on a thesis. From among a dozen topics offered to him by Sommerfeld, he chose the most difficult one: to find out whether an anisotropic arrangement of isotropic resonators (e.g. an orthorhombic lattice of dipoles) would be capable of producing optical double refraction in the medium. Ewald started his calculation with a very unconventional approach to the problem. Instead of following the fate of waves entering matter from the vacuum (as in all then existing dispersion theories), he considered an infinite lattice of point scatterers and searched for proper modes, that is, for waves which were consistent with the lattice and with each other. In other words, he discovered a self-consistent-field method. The result of the very difficult calculation was indeed double refraction as an intrinsic property of the anisotropic lattice. Only then did Ewald restrict his infinite array of oscillators to a parallel-sided crystal plate, and he was successful in joining his internal solutions to incident and emerging waves.

In spite of this success, which supported Sommerfeld's belief in the existence of space lattices, Ewald was a little doubtful about his unusual calculation because it left no place for the incident wave inside the lattice. (Ewald was quoted in the obituary by Dropkin and Post (1986) as saying: 'For the unbounded crystal no such wave is required, or even possible'; and in the half-crystal 'the surface of the body shields its interior from the incident wave' [extinction theorem]). A little earlier, Max Laue, well known for his expertise in theoretical optics, had left Max Planck in Berlin to join Sommerfeld's group. Hence Ewald was able to seek an interview with him to obtain advice. Laue agreed, and early in 1912 they took a walk together through the famous English Garden in Munich.

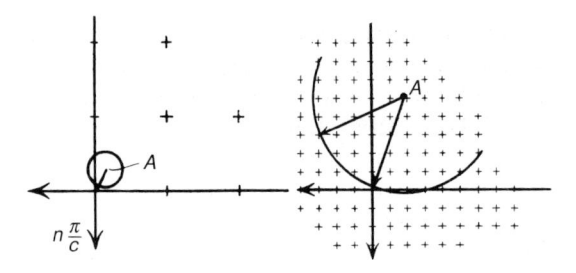

Fig. 1 The Ewald sphere in reciprocal space in the case of light optics (left) and X-ray optics (right). (Ewald 1917.)

Ewald explained his problem, but, to his surprise, realized that Laue seemed to be unaware of the concept of crystals as being built up by space lattices. Ewald provided the necessary information, and Laue then asked about the likely separations between the scatterers in such a hypothetical lattice. Ewald's answer, 'about ten times the likewise hypothetical X-ray wavelength', obviously inspired Laue to ask the most important question: 'But tell me: what happens to short waves in such a lattice?' (Dedication, in Ewald 1937).

This was not Ewald's problem, and he came away somewhat disappointed. Laue, however, obtained from this conversation the idea for his famous experiment, successfully performed in April 1912 with the help of Walter Friedrich and Paul Knipping. It definitively proved the existence of crystalline space lattices and the wave character of X-rays diffracted in such lattices.

At that time Ewald had already submitted his thesis and then moved back to Göttingen. It was there that he heard about Laue's discovery, during a lecture presented by Sommerfeld in June 1912. Ewald wrote: 'The evening after Sommerfeld's lecture I finally sat down to discuss the case of short wavelengths. It was hereby that the reciprocal lattice became an essential tool and that the "sphere of reflection" or "Ausbreitungskugel" was an obvious construction' (Ewald 1979a).

Already this first contribution of Ewald, published in 1913 in *Physikalische Zeitschrift* (Ewald 1913) [a translation appears in Part D], bore great significance for the future: the famous 'Ewald sphere' in reciprocal space (Fig. 1) not only provided a method for constructing diffracted vectors and for proving in an obvious way the identity of Bragg's law and Laue's formulation (see a further discussion by Cruickshank in Chapter 6 of this volume), but also, later on, was to provide a lucid scheme for understanding details of the dynamical theory such as the 'excitation error' (Anregungsfehler), the 'surface of dispersion' (Dispersionsfläche), and the like. Hence it was only as part of the much later developments that the full power of these ideas became evident.

The second Solvay conference in 1913 was entirely devoted to the topic 'The structure of matter'. Sommerfeld intended to present a lecture about the Munich results, which Ewald, by then back in Munich, had to work out in detail. During this preparation, as he said, 'the concept and name of structure factor appeared for the first time'; it was needed in connection with the discussion of missing spots in Laue's diagrams.

A year later Ewald and W. Friedrich published a paper on pyrite which criticized the interpretation of measurements by W.L. Bragg and presented the first precise 'structure refinement' (Ewald and Friedrich 1914).

The First World War interrupted Ewald's contributions to the new field of 'X-ray diffraction in crystals', but only for a short time. Ewald learned about the medical applications of X-rays in a Munich hospital, and practised these methods for several months. But then, as he said, 'I grew nervous lest the glorious war might come to an end without my having seen the front. So I applied for a position as "Field X-ray mechanic of the Army" and soon... got a gorgeous uniform that was often mistaken for that of an officer of the General Staff and treated with much respect' (Ewald 1962a). He was given a mobile X-ray station, with wagon and horses (Figs 2 and 3; cf. Ewald 1974), and was finally transferred to a village miles behind the Russian front. There he had plenty of time, as he puts it, 'to consider the case of short wavelengths, not in the purely geometrical sense, but as a dynamical problem of how an X-ray optical field could travel in the crystal'.

The result of these efforts was the set of three famous papers, published in *Annalen der Physik* (Ewald 1916a,b, 1917), that created the foundation of crystal optics for visible light and X-rays. The theoretical content of these papers is discussed in Chapter 1 of this volume by Kato. Perhaps one interesting detail might be added here. In 1918

Fig. 2 P.P. Ewald in his 'gorgeous' uniform in the First World War. (Ewald 1974; courtesy Arnold Ewald.)

Ewald combined these papers in a '*Habilitationsschrift*', a document that was required in Germany to obtain a lectureship. In the accompanying oral examination the candidate had to defend one of a number of theses proposed by him. It was only many years later that Ewald, when looking through old papers, rediscovered his theses. One of them read as follows: 'Provided the absorption of X-rays, like that of visible light, can be accounted for by dissipation of energy in the vibrating dipoles, then under special circumstances the diffracted X-rays will suffer no reductions whatsoever in an absorbing crystal' (Ewald's own translation, 1965).

Unfortunately, the faculty selected another thesis for discussion, and Ewald immediately forgot his idea—otherwise he would have been the discoverer of the Borrmann effect, as early as 1918!

Ewald now became a lecturer at Munich University, but only for 3 years. In 1921 he was offered an assistant professorship at the Technische Hochschule Stuttgart, and a year later he was promoted to a chair of Theoretical Physics at the same institution (Fig. 4).

However, while still in Munich, he had published four interesting papers. The first two, in 1920, presented a critical evaluation and discussion of measurements by Stenström and Hjalmar which clearly exhibited small deviations from Bragg's law in higher-order reflections. These deviations were in reasonable agreement with dynamical theory, and thus provided a first experimental indication of its validity, at least in this particular case (Ewald 1920*a,b*). The third paper (in 1921) discussed the question of how to describe in the reciprocal lattice crystal structures with compound bases (i.e. several atoms in the asymmetric unit). Ewald showed that this can be done simply by adding 'weights' to the existing reciprocal lattice points. It turned out that these weights were identical to the structure factors (Ewald 1921*a*). In a further paper, Ewald used his method of 1916 to calculate optical and electrostatic potentials (Ewald 1921*b*) (discussed by Templeton in Chapter 7 of this volume).

In 1923 Ewald wrote a book *Kristalle und Röntgenstrahlen*. It summarized the state of the art of

Fig. 3 Ewald's mobile X-ray station. (Ewald 1974; courtesy *Physics Today*.)

Fig. 4 Ewald in Stuttgart (Courtesy Prof. M. Renninger, Marburg.)

X-ray diffraction in crystals, but without any discussion of dynamical theory. Only a few lines deal with the results of the above-mentioned measurements of higher-order Bragg angles. The calculation of intensities was declared as being a still open problem. Ewald wrote: 'Because these questions are still too unclear I have omitted them entirely from the contents of this book' (Ewald 1923).

This book had no second edition in which this omission of theory could be rectified. Instead, in 1927, Ewald wrote an important chapter for the *Handbuch der Physik* with the (translated) title 'The investigation of the structure of matter with X-rays', which finally presented a detailed discussion of the entire theory of X-ray diffraction, including Ewald's dynamical theory (Ewald 1927). It was followed by an updated second edition in 1933 (Ewald 1933).

In 1924 Ewald became a co-editor of the *Zeitschrift für Kristallographie*. He began with volume 60, and continued to volume 102 in 1940. Volume 97 in 1937 was dedicated to the twenty-fifth anniversary of von Laue's 1912 discovery, and Ewald contributed the first paper. This was the fourth part of his series of papers on crystal optics, with the subtitle 'Establishing a generally valid dispersion relation especially for X-ray fields' (Ewald 1937).

Ewald had succeeded in solving the complicated case of an interference field with n rays in a lattice with an arbitrarily complicated basis. The result was an equation for the dispersion surface in the form of a determinant, and its solution $\Delta \cdot F = 0$ contained the cases without and with absorption. It was therefore valid also in what Ewald calls the 'Infraröntgengebiet', that is, in the wavelength region between the ultraviolet and X-rays where absorption dominates. Nowadays synchrotron radiation has made access to this region very easy. It might be challenging to try an experimental check on Ewald's results.

However, we are getting ahead of the chronology. Let us return to 1924. In that year Ewald discussed once more some recently determined deviations from Bragg's law. The new experiments by Bergen Davis used an extremely asymmetrical Bragg case on 'sound' pyrite crystals, much enlarging the effects. Davis achieved very good agreement with theoretical values from a simple classical Lorentz–Planck dispersion relation. Applying the two-beam formulation of his dynamical theory, Ewald provided a justification of this simplification within the correct theory, which yielded the same good agreement (Ewald 1924).

A year later, a relatively short but important paper appeared in *Physikalische Zeitschrift* (Ewald 1925). Ewald had solved the two-beam case for lattices with compound bases, and for this centrally important case he gave the result that for a perfect crystal the integrated diffracted intensity is proportional to the magnitude of the structure amplitude $|F_H|$, and not to $|F_H|^2$ as for the mosaic crystal. Ewald mentioned that the measurements of W.H. Bragg on calcite and on small diamond crystals, that is, on relatively perfect crystals, were in agreement with this conclusion.

This and other problems of crystal structure determination with X-rays were discussed at a small conference organized by Ewald in 1925. Meeting on the Ammersee, at his mother's home, it had the aim of bringing together the various approaches to theory, and the various leading 'diffractionists' from different countries. As can be seen in Fig. 5, the participants enjoyed this week of intense discussions (Bragg 1968).

By 1925 Ewald's department in Stuttgart had become a centre for X-ray diffraction and solid-state physics. Visitors came from all over the world.

Fig. 5 After the 1925 Ammersee conference. From left to right: P.P. Ewald, C.G. Darwin, H. Ott, W.L. Bragg, R.W. James. (Bragg 1968.)

Ewald understood well how to co-operate closely with his collaborators within a friendly, nearly familial, atmosphere. (The list included, for shorter or longer times, H.A. Bethe, U. Dehlinger, W. Ehrenberg, E. Fues, C. Hermann, H. Hönl, A. Kochendörfer, F. London, H. Mark, G. Meyer, M. Renninger, and M. Ruhemann.) Some of his papers from that period are the outcome of these effective collaborations. For example, in 1927 Hermann and Ewald discussed Friedel's law in absorbing crystals (Ewald and Hermann 1927). The most important outcome of the Ewald–Hermann collaboration, however, was the famous *Strukturbericht 1913–1928* (Ewald and Hermann 1931), which appeared in 1931 and was succeeded after the war by the *Structure Reports*. Another paper with Ehrenberg and Mark (Ehrenberg *et al.* 1928) was devoted to the intensity problem. It discussed the influence on measured intensities of temperature, absorption, crystal perfection, etc. Measurements on diamond in a double-crystal arrangement found not only the width of the diffraction spot but also its integrated intensity to be in good agreement with theory. In the same year, Ewald discussed the problems that arise in the transition from light optics to X-ray optics (Ewald 1928).

Together with M. Renninger (who at the time of the 1987 Perth conference was still, at the age of 83, working in Marburg; he died on 23 December 1987), Ewald wrote a paper in 1934 on the mosaic structure of rocksalt, which considered the influence of crystal imperfections (or, more correctly, of the size of coherent domains) on the integrated intensity (Ewald and Renninger 1934). The paper 'X-ray diffraction in diamond as a wave-mechanical problem' (1936, with M. Hönl) had as its main point the experimental observation of the 222 reflection which should actually be forbidden in the diamond lattice (for spherical atoms). The calculation resulted in a charge density of about one electron between the atomic cores, large enough to produce some 222 intensity (Ewald and Hönl 1936).

With E. Schmid in 1936, Ewald combined and compared optical total reflection of X-rays near grazing incidence with interference total reflection (in an asymmetric Bragg case). One interesting result was that one can find an angular region near grazing incidence where optical total reflection no

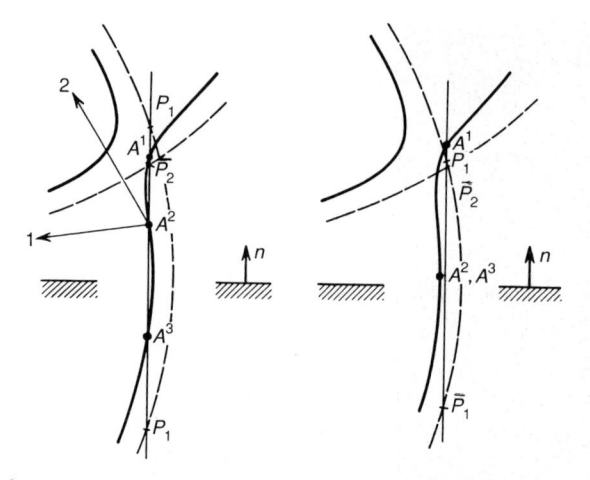

Fig. 6 Combination of X-ray interference and optical reflection. Left: just before occurrence of optical total reflection (the perpendicular to the crystal surface cuts the dispersion surface at three points A^1, A^2, and A^3: the effect of triple Pendellösung!); right: optical total reflection. (Ewald and Schmid 1936.)

longer exists, but interference total reflection has not yet begun, and that in this region a triple Pendellösung effect is to be expected (Ewald and Schmid 1936) (see Fig. 6).

In the same year, Ewald discussed the origin of the use of the reciprocal lattice (Ewald 1936). Apparently Max von Laue had complained that in the recent literature his own contribution had found less attention than Ewald's. Ewald stated that he introduced the reciprocal lattice in the particular case of orthogonal lattices early in 1913, whereas later in the same year von Laue contributed the extension to oblique-angled lattices, thus re-inventing reciprocal vector systems which had not been known to him, although they had been introduced by Gibbs in 1881. (Von Laue evidently soon came across Gibbs, for he cited Gibbs as the source for reciprocal vectors in his publications of 1914; see also the chapter by Cruickshank in this volume). So both Ewald and von Laue share in the credit for the introduction of the reciprocal lattice.

Ewald's other important contributions of the 1930s—the *Handbuch* article of 1933 and the fourth part of his *Crystal Optics* of 1937—have already been mentioned. However, a further contribution, not resulting in a paper of his own, but serving the whole community of crystallographers, should not be forgotten. At a meeting at the Royal Institution in 1929, Ewald convinced Sir William Bragg of the need to prepare tables for the determination of crystal structures. The final decisions about their content were made at a 1930 conference in Zürich, with Ewald as chairman, and, under the editorship of C. Hermann, the resulting two-volume book *Internationale Tabellen zur Bestimmung von Kristallstrukturen* appeared (Hermann 1935), the forerunner of the well-known *International Tables*.

In summary, it can be said that, in his German period, Ewald not only founded the dynamical theory of X-ray diffraction, but also elaborated it in more and more detail, and made important contributions to structure research and other nearby fields. Following his method, Hans Bethe established the dynamical theory of electron diffraction (Bethe 1928). Max von Laue's reformulation of X-ray theory, in terms of Maxwell's theory applied to a medium with spatially periodic dielectric constant (Laue 1931), provided an easier understanding of the dynamical X-ray theory (but arrived at the same final results if the same crystal model is chosen (Wagenfeld 1968)).

In 1932 Ewald became rector of the Technische Hochschule Stuttgart. He resigned this post soon after the Nazi regime had enacted the 'Reichsgesetze zur Wiederherstellung des Berufsbeamtentums' (April 1933; 'Laws for the re-establishment of the professionalism of civil servants', which provided the basis for the elimination of Jews from all higher positions). He still continued in his position as professor. In 1936 he walked out of an obligatory faculty meeting when the speaker, a young assistant recently named leader of the non-professorial faculty organization, read a ministerial directive stating that objectivity was no longer an acceptable concept in science, and shortly thereafter he was pensioned. Not entirely unhappy, he left for the Ammersee, where he found the leisure to complete Part IV of his papers on crystal optics (Ewald 1937). Then an invitation from England changed his life; he left Germany for Cambridge, and half a year later his family followed him.

After the Second World War, a number of scientific and social events brought Ewald back to his native country. It is appropriate to recall some of them here.

The first one occurred on the occasion of von Laue's eightieth birthday in 1959. Ewald came to

Fig. 7 Celebration of Max von Laue's eightieth birthday (9 October 1959) at the Harnack-Haus der Max-Planck-Gesellschaft in Berlin-Dahlem: P.P. Ewald offers greetings from the IUCr (and a medal from the Société française de minéralogie et cristallographie). (Courtesy Dr. G. Kik, Heidelberg.)

Berlin, where von Laue had resigned from his position as director of the Fritz-Haber-Institut a half year earlier, to deliver greetings to his old friend from crystallographers all over the world (Fig. 7).

The next important event was the meeting 'Fifty years of X-ray diffraction', which took place in Munich in 1962. To commemorate this fascinating conference, Ewald edited a book of the same title (Ewald 1962b).

Sixteen years later, and also about sixty years after his founding of crystal optics, Ewald was awarded the Max Planck Medal, the highest honour of the Physical Society of Germany, in recognition of all his meritorious work; this occasion in 1978 was combined with a festive colloquium in Munich (Fig. 8).

A year later, it was again Berlin's turn; the event was the celebration of the one-hundredth birthdays of Albert Einstein, Otto Hahn, Lise Meitner, and Max von Laue. The most vivid contribution was that of the then 91-year-old Paul Ewald, in remembrance of his friend Max von Laue. His speech was published in 1979 in *Physikalische Blätter* under the title (in translation) 'Max von Laue—the man and his work' (Ewald 1979b). It was Ewald's last paper written in the German language.

Fig. 8 Presentation of the Max Planck Medal (Munich, summer 1978). Ewald has just received the document; at his left his wife Ella. (Courtesy LMU München, Sektion Physik.)

Acknowledgements

I am grateful to Dr H. Albrecht (University of Stuttgart, Department of History of Natural Sciences) for making available the transcript of an interview given by Ewald in 1979 on the occasion of the one-hundred-and-fiftieth anniversary of the University of Stuttgart, and to Professor H. Juretschke (Polytechnic University, Brooklyn) for many valuable comments and discussions.

References

Note: These references do not include a complete compilation of Ewald's papers in his German period. A complete listing of these papers, together with their titles, is found in the Bibliography.

Bethe, H. (1928). *Ann. Phys. (Leipzig)*, **87**, 55–129.
Bragg, W.L. (1968). *Acta Cryst.*, **24** (A & B), 4.
Dropkin, J.J. and Post, B. (1986). *Acta Cryst.*, A**42**, 1–5.
Ehrenberg, W., Ewald, P.P., and Mark, H. (1928). *Z. Kristallogr.*, **66**, 547–84.
Ewald, P.P. (1913). *Phys. Z.*, **14**, 465–72.
Ewald, P.P. (1916*a*). *Ann. Phys. (Leipzig)*, **49**, 1–38.
Ewald, P.P. (1916*b*). *Ann. Phys. (Leipzig)*, **49**, 117–43.
Ewald, P.P. (1917). *Ann. Phys. (Leipzig)*, **54**, 519–97.
Ewald, P.P. (1920*a*). *Phys. Z.*, **21**, 617–19.
Ewald, P.P. (1920*b*). *Z. Phys.*, **2**, 332–42.
Ewald, P.P. (1921*a*). *Z. Kristallogr.*, **56**, 129–56.
Ewald, P.P. (1921*b*). *Ann. Phys. (Leipzig)*, **64**, 253–87.
Ewald, P.P. (1923). *Kristalle und Röntgenstrahlen*. Springer, Berlin.
Ewald, P.P. (1924). *Z. Phys.*, **30**, 1–13.
Ewald, P.P. (1925). *Phys. Z.*, **26**, 29–32.
Ewald, P.P. (1927). In *Handbuch der Physik*, Vol. XXIV, pp. 191–369. Springer, Berlin.
Ewald, P.P. (1928). In *Probleme der Modernen Physik* (ed. P. Debye), pp. 134–42. S. Hirschel, Leipzig.
Ewald, P.P. (1933). In *Handbuch der Physik*, Vol. XXIII/2, pp. 207–476. Springer, Berlin.
Ewald, P.P. (1936). *Z. Kristallogr.*, **93**, 396–8.
Ewald, P.P. (1937). *Z. Kristallogr.*, **97**, 1–27.
Ewald, P.P. (1962*a*). *J. Phys. Soc. Jpn.*, **17**, Supplement B-II, 48–52.
Ewald, P.P. (ed.) (1962*b*). *Fifty years of X-ray diffraction*. Oosthoek, Utrecht.
Ewald, P.P. (1965). *Rev. Mod. Phys.*, **37**, 46–56.
Ewald, P.P. (1968). *Phys. Bl.*, **12**, 538–42.
Ewald, P.P. (1974). *Phys. Today*, **27** (September), 42–7.
Ewald, P.P. (1979*a*). *Acta Cryst.*, A**35**, 1–9.
Ewald, P.P. (1979*b*). *Phys. Bl.*, **35**, 337–49.
Ewald, P.P. and Friedrich, W. (1914). *Ann. Phys. (Leipzig)*, **44**, 1183–96.
Ewald, P.P. and Hermann, C. (1927). *Z. Kristallogr.*, **65**, 251–60.
Ewald, P.P. and Hermann, C. (1931). *Strukturbericht 1913–1928*. Akadem. Verlagsges., Leipzig.
Ewald, P.P. and Hönl, H. (1936). *Ann. Phys. (Leipzig)*, **25**, 281–308.
Ewald, P.P. and Renninger, M. (1934). In *Solid state of matter*, pp. 57–61. Cambridge University Press.
Ewald, P.P and Schmid, E. (1936). *Z. Kristallogr.*, **94**, 150–64.
Hermann, C. (ed.) (1935). *Internationale Tabellen zur Bestimmung von Kristallstrukturen*, 2 volumes Borntraeger, Berlin.
Laue, M. von (1931). *Ergebn. Exakt. Naturwiss.*, **10**, 133–58.
Wagenfeld, H.K. (1968). *Acta Cryst.*, A**24**, 170–4.

3

Paul Ewald in Cambridge and Belfast

H.A. Bethe and G. Hildebrandt
Max F. Perutz
Dorothy C. Hodgkin

[*Editors' note*: An obituary memoir for Paul Ewald by H.A. Bethe (son-in-law and Nobel Laureate in Physics) and G. Hildebrandt was published in *Biographical Memoirs of Fellows of the Royal Society* (1988), **34**, 133–76. The following is their description (pp. 147–9) of his period in Cambridge and Belfast.]

H.A. Bethe and G. Hildebrandt

His friend, Sir Lawrence Bragg, secured him a research grant at Cambridge, and Paul left Germany in the autumn of 1937. His wife and the two younger children followed in spring 1938, and his mother soon afterwards. Lux, the older son, emigrated independently. He had a good job in Stuttgart with Bosch, the electrical firm, and the directors, knowing his situation, sent him abroad to a subsidiary with good wishes for his future. So as not to draw attention to their emigration, which would have entailed permits and delays, the family travelled to England separately and with little luggage, as for an extended visit only. Lux, having the task of dissolving the apartment, put the furniture in storage but managed to ship the Ewalds' beloved books, wrapped in bedsheets and other linen, all of which were sorely needed in England. Old Mrs Ewald followed her son's invitation to celebrate her birthday in England, expecting to return soon.

On his research grant of £400 for one year the family rented a house on Guest Road and furnished it with then unfashionable Victorian furniture acquired at auctions, 10 shillings for a table, etc. Professor and Mrs Hutton offered their garden to sit in, their flowers, their vegetables; there were gifts of fruit. Soon the Ewald house was full of visitors and Ella managed somehow always to have enough to offer a good meal along with good talk and comfort. Doter, as Paul's mother soon was called by closer friends, set up her easel and painted several portraits, among them one of Paul Dirac. The two younger children, Linde and Arnold, were enrolled in the Perse schools and Linde achieved her A-levels and enrolled the following year in Queen's University in Belfast to start her medical training.

As happened with many refugees from Hitler's Germany, the relief to be out of Germany and the absence of constant fear created a cheerful willingness to put up with a vastly reduced standard of living, and to accept the friendly charity so generously offered by colleagues and old and new friends.

In late August 1939 the Ewalds—Paul and Ella, Lux, Linde, Arnold, and Paul's mother—moved to Belfast where Paul had been appointed lecturer at Queen's University with a promise of promotion to professor. Owing to a general freeze on promotions during the war, this promise was not fulfilled until May 1945. In October of that year he became a British subject.

Paul and Ella found the people in Belfast congenial and very friendly. There was a very active refugee committee whose members found the Ewalds to be happy and adaptable people, and a great help in consoling other refugees. Paul's mother charmed whomever she met and ushered whomever she could into her studio. This included the chief of police who was officially charged to keep track of refugees and called at the Ewald house when only his mother was at home. They became friends immediately, and later she painted his portrait.

The new house, on Rugby Road, had a bedroom for each, including a large front room on the second floor that could serve 'Doter' as a studio as well as her bedroom. Some furniture came from Cambridge, but most important were the crates of books. The books filled the living-room walls, the hallway, Paul's study, and still there were some for every bedroom wall. The crates became furniture, and when the Ewalds moved from Belfast to Brooklyn in 1949 the crates were filled again, and again were turned into furniture. Paul remained ready to move, ready to accept what fate might bring and willing to adapt, but he never felt really permanent again. He insisted on 'making do' in material things, which was a very good attitude in the years in Belfast, where money remained scarce.

In 1939 he and his family settled into the new existence. Paul for the first time gave introductory courses—judged excellent by good students, very difficult by not so good ones—and had to devise examination questions, which he found a troublesome task. He thought easy questions would be perceived as condescending, and yet the problem must have a simple solution. Also, for the first time, he was teaching engineering students, not future physicists.

This fact was most fortunate later, when the British Government, after the fall of France, found itself sorely pressed and arrested all enemy aliens, whether or not they were refugees. But because engineers were needed for the war effort and most other physicists had departed for wartime laboratories, Paul was essential to keep the department going. These arguments were presented by the Belfast Refugee Committee and through the chief of police directly to the central authorities in London, without going through the local authorities. Had Paul been teaching only physicists and doing pure research, this would probably not have been a sufficient reason, but the training of engineers for the war effort was recognized, and Paul was not arrested in 1940.

Lux, Ewald's elder son, was not arrested because he had an immigration visa for the United States and was waiting for a berth, which came through in July 1940. But the younger son, Arnold, was interned, and for a long time the family did not hear what had happened to him. One of the ships taking many internees to Canada sank, being torpedoed by a German submarine. Arnold's ship fortunately was not sunk, but the six-week journey to Australia was most unpleasant. After many weeks the Ewalds heard that Arnold had safely arrived in an Australian internment camp. In the end it all turned out very well. After the Japanese war broke out Arnold volunteered to become a 'working soldier' in the Australian army who, for example, loaded goods on to trains. Thereby he became an Australian citizen and, after the war, was entitled to government support for his undergraduate studies. His graduate work in physical chemistry was done in Manchester, supported by Petrocarbon Industries.

For the rest of the war the family consisted of Paul, Ella, their daughter Linde, and Paul's mother, with Ella cooking, cleaning, shopping, and keeping everybody's body and soul together, as Paul lectured, Linde studied, and the painter painted. They had the luck of not getting a direct hit during the two air raids on Belfast, though bombs fell on the houses to right and left and one landed in their neighbour's backyard. One day Lux, who had in the mean time been drafted into the US army, appeared at their door: his troop transport from the United States had just landed in Belfast.

Paul had to get used to British methods of teaching, which were very different from the German ones. He was much helped in this by his first assistant. His second assistant was James Hamilton, who later became well known for his work in nuclear theory and became professor at Cambridge University and at Nordita in Copenhagen.

Strange and unexpected insights into different ways of life came through one assistant who grew up on the Isle of Wight and, for instance, had never climbed stairs. A student explained his sleepiness in class as owing to his nightly vigils in attempts to photograph leprechauns. Paul's last assistant was Jose Moyal, who was especially interested in statistical mechanics and is now professor at Canberra, Australia.

Paul could not do much research in the early years of the war, but towards the end he could do so again, including some research for the war effort. After the war he was soon busily engaged with organization of international science.

[*Editors' note*: The memoir by Bethe and Hildebrandt concluded with tributes by Max Perutz and Dorothy Hodgkin, Nobel Laureates in Chemistry. We reprint these below—that of Max Perutz being slightly expanded from the original. The portrait of Dirac by Ewald's mother has been presented by Ella Ewald to the Royal Society.]

Max Perutz

In September 1936 I left my native Austria to become a research student at the Crystallographic Laboratory in Cambridge. I liked the English and hated the Germans. About six months after my arrival there appeared in the laboratory a tall, fair-haired, bemused, attractive, middle-aged German with a red snub nose, who introduced himself as Paul Ewald. He personified all that had been best in the German academic tradition: profound scholarship, integrity, love of literature and the arts, courtesy, and ease of manners. To these qualities he added wit, charm, tolerance, and a sense of humour. Here in Cambridge he gave much of his time to young scientists wanting to profit from his lucid understanding of the interaction between radiation and matter.

Ewald introduced the custom of regular seminars into our laboratory; he also asked me to give my first one. It was on Liesegang Rings, which interested me because I thought that they might explain some features of the radioactive nodules found on certain beaches and cliffs in Devonshire on which I was then working. I talked much too long, but Ewald was very patient. He took a lively interest in everyone's work, took the lead in scientific discussions, and generally enlivened our laboratory. One day Ewald himself was to give a lecture, but in the morning his mother walked in and said to me: '*My boy* cannot lecture today, because he has a sore throat'. The 'boy' was over 50 then.

Having had to leave their beautiful home and most of their possessions in Germany, he and his family moved into a dismal house in a dilapidated street and furnished it with oddments picked up at the weekly auctions at the Corn Exchange. In these drab surroundings the Ewalds lived as naturally as if they had never known anything better. Ewald's mother, a gifted artist, painted portraits of several of his scientific friends, including an excellent one of the great theoretician Paul Dirac. Ewald's upright and lovable character cured me of my blinkered prejudices and made me realize, like Pasternak's Zhivago, that there are no nations, only people.

The Ewalds welcomed me at their home as though I belonged to the family. I was desolate when they all left for Belfast and loneliness descended once more.

Dorothy Hodgkin

Paul Ewald was a legendary figure of whom I first heard in the stories J.D. Bernal told us over lunch in the laboratory. He was the initiator of the study of crystals by X-ray diffraction through his talk with Laue about his thesis. His was the first complete theory of the geometry and intensities of X-ray diffraction effects from crystals, of the reciprocal lattice, of structure factors, and the dynamical theory of X-ray scattering. Others had contributed —Darwin had derived a working theory; only, when Bragg took him to the meeting on the Ammersee which Ewald had organized in 1925, Darwin found he had forgotten his derivation and Ewald explained it for him. At our then present times, 1932–34, Bernal added current stories from his talks with Ewald over the design of the first International Tables.

In 1935 I first corresponded with Ewald. He accepted the first scientific paper I wrote on my own for the *Zeitschrift für Kristallographie* (on the interpretation of Weissenberg photographs) and then discovered his US sub-editor had already accepted a paper on the same subject by Martin Buerger. Ewald wrote very kindly to me suggesting modifications necessary to make something of my contribution still publishable. He was a wonderfully considerate as well as careful and critical Editor.

I must, I think, first have met him in Cambridge in 1938 or 1939. I have a letter I prize on a possible interpretation of the character of the X-ray diffraction effects I had observed for insulin. But I remember him best from the Oxford X-ray analysis group meeting of 1944 when he gave his own account of the history of X-ray analysis in a plenary lecture, and became an abiding influence on our lives.

The Oxford meeting is followed in my memories by the meeting after the war in 1946 in the Royal Institution to discuss the founding of the International Union of Crystallography. Leaders in the field from every country where they were known were invited. The war had hardly ended, and feelings ran high between individuals so recently fighting one another. But Ewald was a strong friend of all. He, in particular, knew Laue's record in saving many threatened by the Nazis, and could make peace between Laue and his warring 'children'. Gradually, the meeting settled to listening to accounts of research over the war years in the different countries and the building of new links for the future.

From then on we could count on seeing him at most of the general congresses of the Union, at first on its executive committee as editor of *Acta*, then as President of the Union 1960–63, then as a constant participant. It was wonderful seeing him at the age of 87 listening to Warner Love on the crystal structure of sickle cell haemoglobin, a recent success far from his own field. One change others had made in his constitution for the Union he hated: the requirement that non-delegates should leave the assembly when applications for membership of the Union were considered, a mark of old conflicts. I was glad we managed to change this in time to keep him with us in his last assembly—to speak freely to us against limiting the tenure of the editors of the Union. I well believe he himself would have been a good editor into his 80s. But luckily he had wholly approved of his chosen successor.

It was a great happiness to visit him in his last beautiful home, with Ella beside him and Rose near at hand, surrounded by his books, still retaining his clear insight and deep interest in new scientific discoveries and international friendships.

4

Paul P. Ewald and the building of the crystallographic community

Harmke Kamminga

Besides contributing in fundamental ways to the theory of propagation of X-rays in crystals, Paul Ewald was instrumental in consolidating the social cohesion of the international crystallographic community. This has been a diverse community since the beginnings of X-ray crystallography, comprising those concerned with the physical foundations of X-ray diffraction phenomena, those involved in developing methods of crystal structure analysis, and those engaged in crystal structure determination for the sake of its value to a range of established disciplines, from mineralogy and chemistry to metallurgy and biology. A specialized area that has close links with such a variety of fields of scientific enquiry might well have fragmented, were it not for the efforts of a number of leading figures within X-ray crystallography to stress the common problems and interests of this diverse community as a whole.

Ewald, in particular, played a seminal role in fostering communication and co-operation between all those connected with X-ray diffraction analysis, most notably by pressing strongly for the foundation of the International Union of Crystallography and its publication programme. The Union, in which Ewald played such an active role, co-ordinated the activities of crystallographers world-wide in a formal way. However, the fact that the Union was founded at all implies that there was a reasonably cohesive group to be represented by such an organization in the first place and, given the heterogeneity of modern crystallography, the question arises how this cohesion emerged. This chapter will focus on Ewald's early contributions to the growth of an international network of crystallographers, set against a sketch of changes in the nature and organization of the science of crystallography.

The study of crystal form was traditionally a part of mineralogy. The foundations of crystal geometry, based on the study of the symmetry characteristics of crystals, were laid in the nineteenth century, with Haüy's law of rational indices, Bravais' lattice theory, and the theory of the 230 space groups. Studies of crystal chemistry and crystal physics also developed in the nineteenth century, and an attempt to bring together these various aspects of crystallography was made by Paul Groth with the foundation of the *Zeitschrift für Kristallographie und Mineralogie* in 1877, which he edited from that date until 1920.

Until 1912, however, no physical or chemical means were available to determine the inner structure of crystals. Space group theory expressed the exact number of different three-dimensional arrangements that can be generated by the operation of crystallographically permissible symmetry elements on the structural constituents of the crystal. The formulation of space group theory was a significant mathematical achievement, but it had limited practical value to crystallographers in the absence of experimental methods of distinguishing between the 230 space groups in real crystals. The precise relation between the geometrical theory and the structure of real crystals therefore remained a matter of uncertainty. (And, of course, knowledge of a particular crystal's space group does not automatically yield its detailed internal structure!)

The subject of crystallography was transformed by the exploitation of the phenomenon of X-ray diffraction by crystals, first demonstrated by Laue, Friedrich, and Knipping in 1912. (Ewald's theoretical contribution to this experiment is described in Chapter 2 of this volume by Hildebrandt.) This experiment was not motivated by crystallographic concerns, but by a controversy in physics regarding the nature of X-rays, which appeared to exhibit both wave-like properties and properties analogous to those of α- and β-particles. The experiment of Laue *et al.* seemed to settle the matter in favour of the wave theory of X-radiation. (See Glusker (1981) for reprints of this and other classical papers in the field.) In the same year, Lawrence Bragg established that the X-ray diffraction pattern provides information about the internal structure of the crystal.

Lawrence Bragg's keen interest in Laue's experiment was also inspired by physical rather than crystallographic concerns. (For a personal account, see Bragg (1961).) His father, William Bragg, was actively engaged in the controversy on the nature of X-rays, and defended the particle theory. In the course of an (unsuccessful) attempt to reconcile Laue's results with his father's particulate theory of X-radiation, Lawrence Bragg reinterpreted these results in terms of reflection of white radiation from planes in the crystal used in the experiment. He derived a simple relation between the wavelength of the X-rays for different orders of reflection and the lattice spacing of the crystal (Bragg's law), and showed how this relation could be used to obtain information about the three dimensional arrangement of the crystal's constituents. With this insight, W.L. and W.H. Bragg began a series of structure determinations which inaugurated the field of X-ray crystallography. The new method was soon applied by mineralogists and chemists, and by the early 1920s the majority of papers in Groth's *Zeitschrift* incorporated X-ray diffraction analysis in one form or another.

The reason for repeating this rather well-known history here is to emphasize that the scientists involved directly in the transformation of crystallography were all physicists who were in the first instance concerned with physical questions and had no prior crystallographic training or interests. Once the power of X-ray crystal analysis was recognized, in terms of the novel information it yielded about the internal structure of crystals, mineralogists and structural chemists might simply have absorbed the new technique into their own respective fields and the physicists might have returned to supposedly more fundamental problems in physics. What happened in practice was that the development of the mathematical formalism of modern diffraction theory, the growth in understanding of the physical principles underlying diffraction phenomena, and developments in methods for crystal structure determination remained closely interlinked. This interlinkage was coupled to the growth of a crystallographic community that to this day represents different disciplines and yet recognizes enough common interests to remain cohesive.

This social cohesion of the crystallographic community was not given, but was built up systematically through the initiatives of leading scientists in the X-ray diffraction field, especially the Braggs and Ewald. Training and teaching, the promotion of collaborative research, the organization of meetings and conferences, the establishment of societies, and the publication of research in specialized journals and reference works all played a role in this process. Ewald took a leading part in many respects and, as witnessed by his own retrospective accounts, was well aware of the importance of his organizational role in the shaping of the subject of modern crystallography (see, e.g. Ewald 1962 (Part VIII), 1968, 1977). The remainder of this chapter will focus on this social dimension of Ewald's contributions, beginning in the 1920s.

It was Ewald who arranged the first specialized X-ray diffraction meeting with international representation, which took place at his mother's house at Holzhausen on the Ammersee, not far from Munich, in 1925 (see Bethe and Hildebrandt, 1988, p. 167, and this volume p. 15 and p. 30). This informal, but by all accounts intense meeting served to establish or re-establish contacts that had been disrupted during the First World War. It was attended by W.L. Bragg, C.G. Darwin, and R.W. James from England; L. Brillouin from France; P.P. Ewald, M. von Laue, H. Mark, H. Ott, and K. Herzfeld from Germany; A.D. Fokker from Holland; I. Waller from Sweden; and R.W.G. Wyckoff from the USA. P.J.W. Debye and K. Fajans participated in some of the sessions. The discussions were focused on diffraction theory, especially Ewald's dynamical theory and Darwin's theory of diffraction by the mosaic crystal, and thus reflected the theoretical concerns of Ewald and those he had invited. The presence of Bragg and Wyckoff especially ensured that these theoretical discussions were linked explicitly to issues concerning crystal structure determination.

In his book *Kristalle und Röntgenstrahlen*, published two years before the Ammersee meeting, Ewald had not only described the theory of X-ray diffraction by crystals, but also included a 20-page list of all crystal structures that had then been determined. He thus illustrated the significance of the physical theory that was his prime area of scientific interest in terms of its application in structure determination. Although Ewald himself was not engaged in crystal structure analysis, his emphasis on the broader crystallographic relevance of a specialized area of physics is likely to have been an important factor in his appointment to the new Board of Editors of the *Zeitschrift für Kristallographie* in 1924.

Paul Groth had handed over the editorship of the *Zeitschrift* to Paul Niggli in 1920, and in 1924 Niggli was joined by Ewald, Fajans, and von Laue (Ewald 1968, 1977). At the same time, it became editorial policy to provide a forum for studies of the crystalline state in general, regardless of the disciplinary origin of published research. In particular, the close connections between the *Zeitschrift* and mineralogy were loosened, as indicated by the omission of 'Mineralogie' from the title of the journal. There was at this time a rapid increase in the number of papers reporting the results of X-ray crystal analysis, comprising a range of crystalline compounds much wider than the natural minerals. Ewald edited a large proportion of these papers until 1939, when his links with the journal were severed as a result of the war.

It was his involvement with the *Zeitschrift* that especially stimulated Ewald's efforts to strengthen contacts between crystallographers at an international level. The journal itself became more international in its authorship and readership after the editors and publisher decided in 1927 to publish papers in English and in French as well as in German. As a result, the *Zeitschrift* was not only the first journal of general crystallography, but it came to occupy a leading role as the first international journal of general crystallography. It maintained its leading position until the Second World War, during the course of which it ceased publication. (When *Acta Crystallographica*, which began publication in 1948, was being planned, it was not foreseen that the *Zeitschrift* would resume publication. It did so in 1955, but *Acta* had already become the central international forum for crystallographic publications and has maintained this status.)

Meanwhile, Ewald had begun to keep an index of newly solved crystal structures, initially with the aim of including an updated list of structures in a second edition of his book *Kristalle und Röntgenstrahlen*. In view of the enormous increase in structure determinations around this time, Ewald sought assistance with his index. Carl Hermann was appointed to this task in 1925, with financial support from the Notgemeinschaft der Deutschen Wissenschaft. In the event, Ewald did not prepare a second edition of his book. (Instead he wrote two substantial and comprehensive chapters for the *Handbuch der Physik*, in 1927 and 1933). It was decided that the structure index should be published regularly as a supplement to the *Zeitschrift für Kristallographie*, but in the form of abstracts of the original papers. The first such supplement, a volume of over 800 pages covering structures determined in the period 1913–1928 and co-edited by Ewald and Hermann, appeared under the title *Strukturbericht* in 1931. Thus, structural work published in a great variety of chemical, physical, and other technical journals was brought together for a general crystallographic readership.

Besides the matters of publication and abstracting in crystallography, there were questions of notation, nomenclature, and pictorial representation connected with space group theory that called for standardization. Also in this area, Ewald played a significant role in involving an international group of crystallographers to promote a broad consensus.

Space group theory was first applied systematically to X-ray analysis of crystal structure by Paul Niggli in his *Geometrische Kristallographie des Diskontinuums*, published in 1919. Niggli's aim of turning abstract space group theory into a practical tool for structure analysis was followed by Ralph Wyckoff in *The analytical expression of the results of the theory of space groups* (1922), and by William Astbury and Kathleen Yardley in their 'Tabulated data for the examination of the 230 space groups by homogeneous X-rays', published in the *Transactions of the Royal Society* in 1924. Each of these works adopted different symbols and conventions for origins and axes of the space groups. Ewald's editorial activities made him acutely aware of the desirability of introducing standard conventions in the tabulation of space groups for use by crystallographers.

After a conference on crystal structure organized by the Faraday Society in London in 1929, Sir William Bragg convened a special meeting of the many crystallographers present to discuss the issues that required closer co-operation among crystallographers at an international level. Ewald and Bernal here pressed for the preparation of standardized space group tables, and a committee was set up for this purpose. (Two other committees, concerned with the standardization of crystallographic nomenclature and with a co-ordinated abstracting scheme, respectively, were established at the same meeting.) The Tables Committee consolidated its plans at a 12-day meeting organized by Ewald and Bernal and held in Niggli's institute in Zürich in 1930 (Ewald 1930). The eventual outcome was the publication, in 1935, of the two volumes of *Internationale Tabellen zur Bestimmung von Kristallstrukturen*.

The *Internationale Tabellen* were prepared by Carl Hermann (with W.H. Bragg and M. von Laue as honorary editors), with the collaboration of 18 crystallographers from Britain, France, Germany, the Netherlands, Switzerland, and the USA. As such, the tables represent the first major publication in X-ray crystallography that involved active collaboration on an international scale. Their authority and value for the crystallographic community is undisputed and encouraged standardization, especially as regards the acceptance of the Hermann–Mauguin notation that had been recommended by the Nomenclature Committee and was incorporated in the *Tabellen*.

Ewald's activities as one of the editors of the *Zeitschrift*, as co-editor of the *Strukturbericht*, and as one of the prime movers behind the *Internationale Tabellen* contributed greatly to the growth of an autonomous international crystallographic community. He went on to build on these achievements in the 1940s, through his role in the creation and shaping of the International Union of Crystallography.

During the course of the 1930s, the political situation in Germany became intolerable for Ewald, and in 1937 he moved to Cambridge to take up a research position secured for him by Lawrence Bragg. In 1939 he was appointed to a lectureship at Queen's University, Belfast, where he taught physics to engineering students (amongst them the late John Bell, whose fundamental work in quantum mechanics—witness Bell's theorem—inspired the famous Aspect experiment). While in Belfast, where he had only limited facilities for research, Ewald took the opportunity to take an active part in the X-ray Analysis Group (XRAG) of the Institute of Physics, which was set up in 1943 under the Chairmanship of Sir Lawrence Bragg and held its first annual meeting in Oxford in March 1944. It was at this meeting that Ewald presented an evening lecture which ended with a strong plea for the foundation of an international union of crystallography and a detailed vision of the responsibilities appropriate to such an organization.

A text based on this lecture was subsequently published in *Nature* (Ewald 1944). The paper is well worth reading for the skilful way in which the common interests of all those concerned with the structure and properties of ordered and semi-ordered materials are presented. In addition, it is striking in retrospect how closely the central activities of the International Union of Crystallography, when it was founded in 1947, conformed to Ewald's vision.

In the text published in *Nature*, Ewald stressed the interdisciplinary nature of crystallography in its new form of X-ray crystallography, including its origins in physics and its applications in chemistry, metallurgy, and engineering and industrial production. He also emphasized the close links between pure and applied crystallography in terms of the similarity in underlying laws and methods of observation and interpretation. He then went on to discuss in detail the role that had been played internationally by the *Zeitschrift für Kristallographie* and the *Strukturberichte* in bringing together much of this multi-faceted work, and the role of the *Internationale Tabellen* in promoting standardization for the convenience of crystallographers in general.

Ewald acknowledged that the international co-operation that had been encouraged by these projects, as well as by various meetings, and training and research programmes, had been disrupted severely by the war, but he was convinced that the rapid development of crystallography would make closer international co-operation even more pressing after the war. He stated:

The desirability of forming, besides regional crystallographic societies, an International Union of Crystallography, should be clear, I believe, from past experience. Co-operation of all authorities is necessary to develop a subject with so many ramifications and applications to optimum efficiency (Ewald 1944, p. 630).

He then listed the tasks he envisaged for an international union, including the publication of a crystallographic journal owned by and edited on behalf of the union, the collection of structural work in abstract form, and the preparation of a second edition of *International Tables*, to be owned by the union. (His experience with the *Zeitschrift*, which was owned by a publisher who had placed commercial interests above the interests of the crystallographic community, made Ewald keenly aware of the importance of property rights.) In the event, Ewald's wishes were fulfilled when the International Union of Crystallography, from the beginning, undertook the publication of *Acta Crystallographica*, of *Structure Reports,* and of the *International Tables* under the Union's ownership (see Kamminga (1989) for a detailed account).

Ewald's text ends with a careful discussion of the term 'crystallography'. He acknowledged that the term in its traditional meaning, associated with the notions of perfect three-dimensional symmetry, homogeneity, and periodicity, does not include the entire field of application of diffraction methods, which, moreover, has very fluid boundaries. He felt, nevertheless, that 'crystallography', broadly interpreted, would be the most appropriate term for the field to be represented by the union he advocated.

The detailed negotiations that led to the foundation of the International Union of Crystallography, and Ewald's role in these negotiations and in the activities of the Union in its early stages, have been described elsewhere (Kamminga 1989; see also the chapters by Evans, Ewald, and McLachlan in McLachlan and Glusker (1983)). Briefly, initiatives, at first directed primarily at founding a new journal, were taken by the American Society for X-ray and Electron Diffraction (ASXRED), by XRAG (most notably by its very active Publications Subcommittee, set up in 1945), and by crystallographers in the Soviet Union. These initiatives rapidly became co-ordinated internationally and merged fully in 1946. In July of that year, XRAG held an international crystallographic conference in London, which was followed immediately by meetings to consolidate plans for an international union as well as a new international journal of crystallography. A Provisional International Crystallographic Committee was formed, and it was its Journal Subcommittee that in fact did most of the preparatory work for the foundation of the Union. When, in 1947, the International Union of Crystallography was formally admitted to the International Council of Scientific Unions (ICSU), the Journal Subcommittee acted as the Union's interim Executive Committee, with Ewald as Chairman and the equally indefatigable Robert Evans as General Secretary.

Clearly, Ewald did not create the Union single-handed, but he played a crucial part in convincing crystallographers of the need for the Union, in delineating the traditions the Union could profitably build on and consolidate, and in characterizing the 'crystallography' to be represented by the Union as a broad and heterogeneous, yet coherent, field. In shaping the International Union, Ewald helped to shape an international crystallographic community that has succeeded in maintaining a separate identity despite its multidisciplinary connections and diverse research interests.

Acknowledgements

The research on which this chapter is based was funded partly by the Leverhulme Trust and partly by the International Union of Crystallography. The Union's archives provided valuable historical source materials. Current support from the Wellcome Trust is gratefully acknowledged.

References

Bethe, H.A. and Hildebrandt, G. (1988). *Biogr. Mem. Fellows Roy. Soc. (London)*, **34**, 133–76.
Bragg, W.L. (1961). *Proc. Roy. Soc. (London)*, A**262**, 145–58.
Ewald, P.P. (1923). *Kristalle und Röntgenstrahlen*. Springer, Berlin.
Ewald, P.P. (1930). *Z. Kristallogr.*, **75**, 159–60.
Ewald, P.P. (1944). *Nature (London)*, **154**, 628–31.
Ewald, P.P. (ed.) (1962). *Fifty years of X-ray diffraction*. Oosthoek, Utrecht.
Ewald, P.P. (1968). *Acta Cryst.*, A**24**, 1–3.
Ewald, P.P. (1977). *Acta Cryst.*, A**33**, 1–3.
Glusker, J.P. (ed.) (1981). *Structural crystallography in chemistry and biology*. Hutchinson Ross, Stroudsburg, PA.
Kamminga, H. (1989). *Acta Cryst.*, A**45**, 581–601.
McLachlan, D., Jr and Glusker, J.P. (eds.) (1983). *Crystallography in North America*. American Crystallographic Association, New York.

5

Paul P. Ewald—a personal appreciation

Hellmut J. Juretschke

Most of the other contributions to this volume deal eloquently with the various primarily scientific ideas and accomplishments of Paul Ewald himself or flowing out of his work. I would like to fill in some more of the background of his numerous broader public contributions to crystallography. At the same time, I would also like to remember him as a friend and colleague, and to touch upon how I saw him in more private roles.

I am not a historian; I am not even a bona fide crystallographer. My main qualifications for this undertaking are that for nearly ten years Paul Ewald was my department head at the then Polytechnic Institute of Brooklyn (still best known as Brooklyn Poly, despite several later changes in name) where I shared a considerable amount of his time and of his ideas, and also was a beneficiary of his deep-felt concerns with the younger research faculty and their families. After his retirement, our friendship, if anything, deepened. Our family was in touch with the Ewalds—Paul and Ella—by mail, phone, or visit, first in New Milford in Connecticut, and later in Ithaca, New York. And reciprocally, we welcomed them from time to time in Brooklyn, and one year even in Grenoble. Clearly, any account I give will not be unbiased.

Let me mention a small dilemma that arose in preparing this account. When I first met Paul in 1950, he was nearly as old as I was at the time of the 1987 meeting in Perth, which to me then seemed *very old*. In fact, before going to see him—about a possible position at the Polytechnic—I made sure that it was the *same* P.P. Ewald whose major contributions in the earlier part of this century were already such a prominent and permanent part of the physics lore and literature. The dilemma is that some readers, especially the younger ones, may now, another forty years later, have the same questioning attitude I had then, and so I will have to remind them that he was then actually a *very young* 62-year-old, whose subsequent career continued to flourish for decades; other readers will have taken this for granted, because they may have known him much longer and better than I have, have followed his career all along, and perhaps even have a much more rounded picture of his personality. It is difficult to meet everybody's expectations, and the shortcomings will be mine.

First a few words about Ewald's background. Some of this is also described in other contributions in this volume, though usually in a slightly different context, and much of it has been the subject of the various obituaries, by now familiar to most readers. So here is my version. A listing of some of the important dates in his life span is given in Table 1. Born into an academic family in 1888, his early education was singularly peripatetic, with frequent and extended travels allowing him to sample schools and languages on the continent and in England. He then studied at the universities of Cambridge, Göttingen, and finally Munich, where Sommerfeld suggested to him the now famous problem of light propagation in anisotropic atomic arrangements on a space lattice. This led, as brought out extensively throughout this volume, to the dynamical theory of molecular optics, and it did so *before* the discovery of X-ray diffraction. Thus, the original intent in its formulation did not extend to applications at very short wavelengths, such as those of X-rays, though the formulation itself was, in fact, valid at all wavelengths. He developed that particular extension in astounding breadth during and after the First World War. The topic of dynamical theory remained one of Ewald's primary concerns for the rest of his life, as he continued to strive for more mathematical ease and formal elegance, as well as for a fuller conceptual grasp of its underlying physics.

His academic career reflected a preference for stability, and, probably just as much, a sense of obligation to involve himself in the affairs of his

Table 1 Paul P. Ewald (1888–1985)

1888	Birth in Berlin
1891–3	Residence in Paris, Cambridge, Montreux
1897	Wilhelm Gymnasium, Berlin
1900	Victoria Gymnasium, Potsdam
1905–6	Cambridge University, Göttingen University
1907–12	Munich University (with A. Sommerfeld)
1912	PhD (Double refraction in crystals)
1912	Assistant to Hilbert in Göttingen
1913	Marriage to Ella Philippson
1913–21	Assistant to Sommerfeld in Munich
1915–18	X-ray technician in German army
1921	Professor, Technical University of Stuttgart
1937	Research Fellow, Cambridge University
1939	Lecturer, later Professor, Queen's University, Belfast
1949	Department Head, Polytechnic Institute of Brooklyn
1958	Fellow of the Royal Society
1959	Retirement to New Milford, Connecticut
1968	Last major papers (with Y. Héno)
1971	Move to Ithaca, New York
1978	Max Planck Medal
1979	Laue Commemoration Address

institution, wherever it was. Despite invitations to other chairs, he remained at Stuttgart, eventually rising to the post of Rektor, until forced into exile. He then went to Belfast as Lecturer and later Professor of Theoretical Physics, where helping to train engineers kept him from being interned during the Second World War. Eventually, he came to the Polytechnic Institute of Brooklyn, where, clearly as a labour of love, he accepted the challenge to develop a strong graduate programme in Physics. He went into the then mandatory retirement before age 72, but continued in research- and science-related activities for more than twenty additional years. Figure 1 shows him at age 89, in a photograph taken by one of his grandsons.

Altogether then, his scientific career spanned an awe-inspiring nearly 70 years. As a young man he saw Lord Rayleigh, and knew his world. He then partook in the various revolutions of modern physics, became colleague and at times teacher, but always a friend, of many of its leaders, and he stayed with it, still eagerly discussing the latest advances in X-ray physics during the 1980s.

A perceptive comment on Ewald's career, in his obituary in the *Soviet Journal of Crystallography*, recalls that the period after the war, and especially that in Brooklyn, saw a remarkable reflowering of his scientific activities. In many respects, this seems a valid observation, though the coincidence of this reflowering with the end of the war and with the move to Brooklyn came for rather different reasons.

By the 1930s, crystallography, and especially its organized community, was mature and in good shape, largely as a result of Ewald's own efforts. The *Zeitschrift für Kristallographie* flourished under his demanding but fair editorship; the *Strukturberichte* had become the systematic repository of tremendous amounts of new data; the *Internationale Tabellen* had set common standards of geometrical classifications and basic physical data; and the theory, together with analyses of various agreed-on standard experimental techniques, received its capstone summary in Ewald's *Handbuch* chapters. It appeared that the still-outstanding problems were well defined, and that the mechanisms for further progress were well in hand.

However, after the Second World War, vehicles for international co-operative ties had to be re-established, the continuity of data repositories had to be ensured, and means for a timely exchange of up-to-date information had to be re-created. Finally, of course, there had been crucial new discoveries during and immediately after the war in X-ray physics itself, as well as in solid-state physics, that put new demands on crystallography. Dynamical effects—most already explicit in Ewald's original formulation of the theory—started to play a practical role. It is unthinkable that he would not have played a leading part in many of these developments.

As to Brooklyn, it provided a modest but stable base of operations for both scientific and organizational activities. It offered an enthusiastically supportive and stimulating setting created by the efforts of I. Fankuchen, H. Mark, B. Post, D. Harker, R. Brill, and their students and associates. At the same time, because of its location near the gateway to Europe, Brooklyn served as a natural stopover point, and therefore as a central information exchange, for most of the crystallographers coming into or out of the United States. Few escaped having to give a lecture on their latest findings and ideas at the famous weekly Point Group seminars at the Polytechnic. Ewald presided, easily

Fig. 1 Paul Ewald in 1977. (Photo Ben Ewald.)

supplying the visitors and the audience with fluent translations, if needed, in a variety of languages, and often ending up as the arbiter of all-too-lively debates. The ashtrays at these meetings, indispensable for the cigars fashionable at that time, were seashells gathered by him on occasional trips to the beaches of Long Island.

Already in 1944 Ewald had argued for an International Union of Crystallography as the most appropriate vehicle for supporting and reinforcing the common interests, methods, and research goals of crystallographers world-wide. Equally, it seems to me, he wanted to avoid a repetition of the state of affairs after the First World War, when it took more than seven years before international collaboration and co-operation could be mobilized—largely on his initiatives—to lay the foundations of an organized development of crystallography. He himself told the story of the beginnings of the Union when he retired from the presidency of the ACA (American Crystallographic Association) in 1953 (Ewald 1953), and a more detailed account is given in this volume by Kamminga in Chapter 4. The idea of a Union was strongly favoured by Bragg and others, and as early as 1946, it became a central topic at the first post-war international conference. Participants at that meeting remember especially Ewald's persuasive roles, including mediating for trust between groups only recently on opposing sides of the war, as well as his stand for fairness and justice in vouching for his close friend von Laue's record in dark times. The creation of the Union itself, just a year later, had him serving first as chairman, and later as president, and on many of its committees. Under his ever-watchful eyes, his encouragement, or his criticisms, where needed, it became the flourishing instrument of international science, one of whose benefits we were able to enjoy at the XIVth Congress in Perth.

The first and foremost task of the Union, as Ewald saw it, was the publication of an international journal of crystallography. It is hardly surprising that this recommendation was not only endorsed, but that he was also named as the first editor of what became *Acta Crystallographica*. Based on his own experiences, he felt strongly that the journal should belong to the crystallographers themselves, through the Union. A subsidy was assembled from UNESCO and from industrial concerns—primarily British—to launch the journal as soon as the Union was established; and further, to do so at a modest price, and without page charges, two aspects close to his heart. In 1953, at the time of his reminiscences cited above, he was still concerned about the journal's future, especially in face of the rapidly increasing volume of worthy manuscripts coming into his office while he was under strict orders from the Union to limit the number of pages of each volume. Carrying out his responsibilities, as he saw them, was not always made easy for him. But in 1987, nearly thirty-five years later, there was no doubt that *Acta* has lived up to his expectations to become the leading journal of record in crystallography world-wide. Probably only the able treasurers of the Union, working behind the scenes, know fully of the never-ceasing efforts to balance the demands of science with the constraints of finance to ensure the continued existence of *Acta*, but they seem to have done very well.

Ewald's prospectus for the first issue of *Acta* (Ewald 1948), wisely anticipated that the growth of diffraction and crystallography in solid-state physics, chemistry, and biology would soon lead to the creation of parallel journals, but he hoped that *Acta* would always continue to play a central role. However, although this was true in general, *Acta* soon fell short of one of the expectations of its first editor, namely, that it would also become the central journal for the static and dynamical properties of solids in relation to atomic forces and structure. This field, crystal dynamics, was Ewald's love only second to X-rays, and he tried to offer all of it a home in *Acta*. But the field grew too rapidly, and too erratically, and on too many fronts, for such a tradition to be established. Although *Acta* certainly published its share of the literature, it never had the unifying influence Ewald envisioned, and which he thought so important for the good growth of any field.

What was he like as an editor? At one level, he is surely remembered for unswervingly questioning and altering points of grammar, or punctuation, or choice of words, of practically every manuscript he examined. His standards for clarity of expression and felicity of style were uniquely demanding. I have seen him spend disproportionate time struggling with the right phrase in someone else's manuscript, or ruthlessly tearing into a paragraph of imprecise logic (including sometimes my own). Surely, this often caused some resentment, or at least led to lengthy correspondence, and even visits by an author 'to meet the editor who thinks he can correct my style'. However, in his defence, Ewald did so only after judging a manuscript's scientific content to merit publication, and then entirely in the spirit of offering his readers the best possible exposition of a new idea or new facts. He certainly disagreed with those who felt that an editor is not responsible for the style, comprehensibility, or even content of a manuscript. At the other end of the scale, he went out of his way to help modify submitted material that overlapped papers just in press or recently published, by working with the author on changes, of emphasis or otherwise, that could best bring out its independently publishable contributions. Equally, he was not afraid to back publication of controversial or out-of-the-way ideas that had a chance of being right, and he fought to obtain a fair hearing for them. All in all, he exemplified an ideally caring, and responsibly critical editor, motivated by the highest regard for both his authors and his readers.

To Ewald's two public roles as organizer and editor, let me add a third, that of historian, or more properly, of keeper of record of the beginnings of crystallography, and of the doers and thinkers involved in it. He was a tireless expositor of the events and personalities of those days, and his enthusiasm surely reflected both the excitement of having been present at the birth of a new branch of science, and the unique opportunity for documenting from personal interactions how it came about. Titles of papers of his own recollections, offered on many occasions over the last thirty years or more, abound in the bibliography of this volume. A central one of these is reprinted in Part D. Even now, traces of many more informal presentations keep turning up; unfortunately, often only partially or not at all recorded. Beyond the reminiscences, he edited and contributed numerous sections to the magnificent *Fifty Years* volume, which gave so many additional perspectives on that period. Diffraction crystallography is probably one of the best documented modern fields in physics.

However, it must be mentioned that for these endeavours he also ran into criticism by a professional historian (Forman 1969) in a paper entitled 'The discovery of the diffraction of X-rays by crystals; a critique of the myths'. Starting with the intriguing question of why crystallographers, alone among the many subgroups of physics and chemistry, managed to create their own Union, and to maintain their separate identity, this historian constructed an anthropological model characterizing crystallographers as a clan needing 'myths' for tribal survival, and naming Ewald as one of the chief mythmakers. He also suggested that, apart from clarifying priorities, scientists should stay away from historical speculations, such as motivations or antecedents of a breakthrough. Ewald's reply (Ewald 1969) confronted arguments based on mythological grounds with factual ones. Here is not the place to enter into the debate; though, taken together, these two papers raise important issues about how scientists and historians answer, or should answer, the same questions. Such dialogues are all to the good, and probably take place too rarely. Within this same context of the beginning of crystallography, let me raise for possible speculation the question of how the development of crystallography might have changed if the first crystals exposed to X-rays had been among the recently discovered quasi-crystals whose sharp diffraction features do not rely on the existence of the long-range translational order that is the traditional point of departure of crystallography.

One may ask 'What made Paul Ewald so successful in these many public roles?' With respect to the Ewald I knew, I would have to answer that first and foremost, he had an aura about him which radiated such self-assurance and confidence that it made his listeners trust and follow him. One of my colleagues at Poly even remarked that 'when Paul Ewald entered a room, everyone was immediately aware of it'. At the same time, his physical stature gave him the advantage of being able to view most other mortals from above, and watch their doings with a certain detachment and bemusement. Yet, underneath this olympian image he was intensely human and informal, full of the warmest interest in his friends and their lives, sparkling with ideas, ready to tease or joke, or to expose pomposity. I will never forget the time when a distinguished European physicist, en route to exploring academic positions in the United States, visited the Ewalds; after dinner he was taken aside by Paul, and admonished in all seriousness that the first thing he would have to learn as a professor in the United States is to help in washing the dishes. And so they both disappeared in the kitchen. Our children loved his dramatically told tales; he himself became transformed into an eager child in the hands-on halls of the Brooklyn Children's Museum, ready to climb into its gigantic model of DNA.

On a more serious level, he was full of deep human concerns. While in Cambridge in the late 1930s, he was intensely involved in helping friends and colleagues emigrate in their times of peril. Later, he recalled as his most memorable contribution to his ACA presidency in 1952 the decision to move the annual meeting away from Florida because black crystallographers were to be treated unequally. He was deeply intolerant of anti-semitism or other prejudices or political oppression, and ready to speak out about them.

I have to admit that this brief look at Ewald's human sides may, in fact, be a composite of that of Paul himself, and of his wife Ella, but if it applies to one, it applies to both. They shared equally in their outgoing warmth for their friends and visitors, in their forthright stands regarding fairness and justice, and in their deep love of knowledge. Being

Fig. 2 Ella Ewald in Ithaca, in 1986, with the author. (Photo Ruth Juretschke.)

with them and talking with them was always an adventure, whether the subject was anecdotes from their past, or the origin of obscure words, some aspect of classical or modern literature, or the latest advances in medicine. Of course, Ella could not share fully in Paul's scientific work—her professional studies were largely in medicine—but she surely was caught up in many excitements of discovery, by Paul and others, and with her extraordinarily acute conceptual grasp she readily partook in what it was all about. Ella explained that she gave up her own career primarily to raise her four children. She may not approve of my saying this, but it appears to me that this decision also enabled her to surround Paul with an environment that allowed him to extend himself as much as he did; and thus in her own way she deeply shared in his success, as well as in the world in which he moved. Some day, biographers of scientists may want to pay much more attention to the role of the wives who—often even nameless in photographic records of the period—stood behind or next to their famous husbands. Figure 2 shows a 1986 snapshot of Ella, still vigorous, charming as ever, and enormously engaged in the world around her.

Back to Paul. His love of words, and of memorizing poetry in half a dozen languages, needs no retelling. But he also enjoyed expressing his own thoughts in little rhymes, and, for example, hardly

PAUL P. EWALD—A PERSONAL APPRECIATION

PAUL P. EWALD

19 FORDYCE ROAD
NEW MILFORD, CONNECTICUT 06776
TEL. (203) ELGIN 4-3582

An Hellmut & Ruth Juretschke   31 Januar 1968
in Brooklyn

von ihrem Quasigrossvater.

Ein Ikosaeder kann nicht ein Jeder
In seinem Geiste sehen
Zwar ist's regulär, doch ist es sehr schwer
Den Aufbau von Grund zu verstehen.
Von fünfzähligen Achsen ist wirres durchwachsen
Und Dreiecke hat's zweimal zehn
Von Plato geboten, Kristallen verboten
Hat Klein es berühmt gemacht.
Nun habt Ihr als Vase zu meiner Extase
Die vollendete Form mir gebracht
So kann ich's studieren und endlich kapieren
Was Plato der Alte erdacht.

                    Vielen Dank!
                              Paul.

Fig. 3 Thank you poem by P.P. Ewald to the Juretschkes on the gift of a vase for his eightieth birthday.

a festive occasion went by without a small verse from him. Let me share one he wrote to us after we gave him for his eightieth birthday a vase in the shape of an icosahedron. The poem is reproduced in Fig. 3, and a rough translation is as follows: 'An icosahedron cannot be readily visualized by just anyone. Though regular, it is very hard to understand its basic structure. Wildly traversed by five-fold axes, it has triangles two-times-ten. Presented by Plato, forbidden in crystals, and made famous by Klein. Now, to my delight, you have brought me the perfect form. I can study it, and finally grasp what old Plato had in mind.' The poem is a beautiful and graceful juxtaposition of the icosahedron's geometrical complexities with the challenges it represented for Plato, and Klein, and for crystallographers.

Poetry, in fact, with its grand visions and deep meanings, seems to me to be close also to Ewald's approach to science, which can be said to have bordered more on an artistic rather than a coldly analytical point of departure. A similar approach came through in his teaching method. Not always the most patient explainer, he tried to lead students to learn by self-discovery, and by experiencing the excitement of suddenly discerning order and grand pattern. Thus, he usually set problems that were challenging, though eventually rewarding, and, in the spirit of his own experience with Sommerfeld, he even posed doctoral thesis problems which he himself had not yet fully seen through. Nevertheless, it nearly always worked out!

His one successful Ph.D. student at Poly wrote about the Borrmann effect using the point dipole model (Mayer 1952)—though the work was never published. Others who began with him shifted to different advisors. Actually he let the junior faculty have most of the research students. But of the grand problems that still occupied him after his formal retirement, the one he felt most satisfied about was the work with Bienenstock on symmetry in reciprocal space (Bienenstock and Ewald 1962). It is also interesting to note that, in a letter in 1966 concerning an English translation of two of his major papers, he lists as his major contributions his five papers on the *Foundations of crystal optics*, of which four are known and 'one is still to come'. One can only speculate that this referred to the later monumental papers with Héno on three-beam interactions. But I know that other challenges he still wanted to face included the phase problem, and the question of optical activity at X-ray frequencies, and he may have had all of these in mind; or something even bigger. Incidentally, the second paragraph of the same letter refers to another of his major and still not fully appreciated contributions, of increasing interest today, namely, the specification of the electromagnetic fields in the transition region at the surface between inside and outside.

It was a joy to see him continue active with his own ideas, and respond with curiosity and enthusiasm to the ideas of others. In his last years he began to complain that his memory was beginning to let him down. In truth, it remained more formidable than that of most persons in their prime, and it continued to be a pleasure to talk with him, about physics or anything else.

His 1983 thank you reply to birthday greetings, shown in Fig. 4, speaks for itself. He was a kind man himself, and a great man, and he left his mark

Thank you for your greetings and kindness to an old man.

Paul P. Ewald.

Fig. 4 Thank you note by P.P. Ewald on the occasion of his ninety-fifth birthday in 1983. (Photo Rose Bethe.)

especially on all those who had the privilege to know him.

Let me close with two short excerpts from a taped message he sent in 1978 to the retirement ceremony of our mutual colleague at the Polytechnic, John Dropkin. To me, it captures perfectly the flavour of his warmth with friends, and his attitude towards work and towards life. He begins:

My dear John Dropkin. I am sad not to be with you and other friends at Poly when your retirement is being marked by a common gathering and feast. But you told me, and wrote to me, that I am too old to come from Ithaca to Brooklyn on this occasion. And although I don't really agree, I am, as so often, following your prudent advice, and thus I am joining only by this tape.

And he ends the tape with a characteristic flourish:

My feeling about your leaving Poly is best expressed like this:

There was John J. Dropkin of Poly,
A physicist learned and jolly.
He was not yet old,
When by law he was told
to quit. What a waste, what a folly.

References

Bienenstock, A. and Ewald, P.P. (1962). *Acta Cryst.*, **15**, 1253–61.
Ewald, P.P. (1948). *Acta Cryst.*, 1–2.
Ewald, P.P. (1953). *Physics Today*, **6** (September), 12–17.
Ewald, P.P. (1969). *Arch. History Exact Sci.*, **6**, 72–81.
Forman, P. (1969). *Arch. History Exact Sci.*, **6**, 38–71.
Mayer, E. (1952). Absorption of X-rays in perfect crystals set at the Bragg angle. Ph.D. Dissertation, Polytechnic Institute of Brooklyn.

C
Aspects of Ewald's work and their legacies

6

The reciprocal lattice imbedded in Fourier space

D.W.J. Cruickshank

1. Ewald's construction

The first point to make is that Ewald's introduction of the reciprocal lattice and the sphere of reflection grew directly out of his dynamical theory, and not from any examination of purely geometrical aspects of diffraction.

The story began in Munich in 1910, when Ewald was offered by Sommerfeld the problem of explaining optical double refraction by considering an anisotropic lattice of dipole oscillators. Ewald treated a primitive orthorhombic lattice with one dipole (oscillating electron) at each lattice point. Although his method has long been called the dynamical theory, it is more easily appreciated today as a self-consistent-field theory. The field acting on each dipole had to correspond to the fields generated by all the other dipoles in response to the fields acting on them.

I will concentrate on the aspects relevant to the reciprocal lattice story. The Hertz potential generated by the oscillating dipoles is given by

$$\Pi(\mathbf{x}) = \mathbf{P}_o \sum_i \exp(2\pi i \mathbf{K} \cdot \mathbf{X}_i) \frac{\exp(2\pi i K_o |\mathbf{x} - \mathbf{X}_i|)}{4\pi |\mathbf{x} - \mathbf{X}_i|}, \quad (1)$$

where $K_o = \nu/c = 1/\lambda$, $K = \nu/q$, and the summation is over all points i of the infinite lattice. In this expression a plane wave $\exp(2\pi i \mathbf{K} \cdot \mathbf{X})$, with wave vector $\mathbf{K}$, is assumed to be travelling through the lattice with frequency ν and phase velocity q. It causes each dipole to generate a spherical wave $\exp(2\pi i K_o r)/4\pi r$, and the total potential at any point is the sum of these spherical waves emitted from every point of the lattice.

By a remarkable piece of mathematics, Ewald managed to convert this set of spherical waves into a set of plane waves—actually a Fourier series with certain coefficients.

It was some astonishing optical aspects of his theory that Ewald went to discuss with Laue in the winter of 1911–12, before Laue's initiation of the Friedrich and Knipping X-ray crystal experiments. Ewald, who had left Munich, heard about these experiments when Sommerfeld lectured in Göttingen in the early summer of 1912. Later that day Ewald realized that the X-ray results could be explained by his theory, as the conversion to plane waves was valid mathematically for waves of any length. He published a short paper on this interpretation in *Physikalische Zeitschrift* in June 1913 (submitted on 8 May). A translation by H.J. Juretschke of this remarkable paper (Ewald 1913a) appears in part D of this volume. Ewald omitted any mention of the details of the brilliant mathematics which transformed the sum of spherical waves into a sum of plane waves. There was just a passing reference to his dissertation.

With some small alterations of notation, his equation for the Hertz potential as a sum of plane waves was

$$\Pi(\mathbf{x}) = \frac{\mathbf{P}_o}{V_c} \sum_{h,k,l} \cdot$$

$$\frac{\exp\left\{-2\pi i \left[\left(\frac{h}{a} - \alpha\right)x + \left(\frac{k}{b} - \beta\right)y + \left(\frac{l}{c} - \gamma\right)z\right]\right\}}{K_o^2 - \left(\frac{h}{a} - \alpha\right)^2 - \left(\frac{k}{b} - \beta\right)^2 - \left(\frac{l}{c} - \gamma\right)^2}, \quad (2)$$

where h,k,l are integers, a,b,c the cell sides, and α,β,γ the components of K. Equation (2) represents a triply infinite set of plane waves characterized by the integers h,k,l and with amplitudes inversely proportional to the denominators. Very small denominators imply large amplitudes, and Ewald noticed that the denominator would be zero whenever

$$K_o^2 = \left(\frac{h}{a} - \alpha\right)^2 + \left(\frac{k}{b} - \beta\right)^2 + \left(\frac{l}{c} - \gamma\right)^2. \quad (3)$$

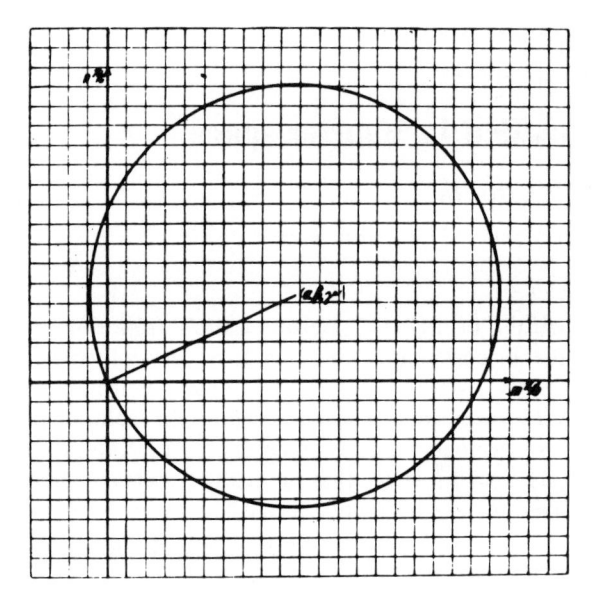

Fig. 1 Ewald's first diagram of 1913.

This is simply the equation of a sphere

$$r^2 = x^2 + y^2 + z^2. \qquad (4)$$

Ewald then enunciated the construction as follows. In a lattice with spacings $1/a, 1/b, 1/c$ (the reciprocal lattice), and around the point (α, β, γ), the sphere which passes through the origin of the lattice is constructed. If other lattice points (h,k,l) lie on the surface of the sphere, there will be waves in the crystal with high intensity which have the same direction as the connecting lines from (h,k,l) to the centre of the sphere.

Figure 1 shows Ewald's first diagram of this construction. The reciprocal lattice and its origin are shown, together with the sphere whose centre is at (α, β, γ). If the sphere passes through no other reciprocal lattice points except the origin, only the one primary wave propagates through the crystal. If the primary wave is in the direction of, say, a three-fold symmetry axis, any secondary waves will necessarily be in symmetrical sets of three. (This idea was developed shortly afterwards in another paper (Ewald 1913b) to explain photographs of cubic crystals.)

Ewald continued by considering the vector potential for the region outside a half-infinite crystal slab with a primary wave either incident upon the slab or emerging from the slab. He showed that condition (3) remained true for the generation of strong interference waves outside the crystal.

Figure 2 shows Ewald's second diagram of the reciprocal lattice and the sphere of reflection. In this diagram, several ideas are combined, and the sense of the wave vectors is reversed, so that waves propagate in directions outwards from the sphere centre. The primary wave has direction cosines proportional to $(-\alpha, -\beta, -\gamma)$, but the sphere centre is retained at (α, β, γ). In Fig. 2 the sphere passes through a number of reciprocal lattice points. The primary wave is incident on a finite crystal with front and back faces, and interference beams are generated on both sides of the crystal.

2. Non-orthogonal axes

The treatment in Ewald's 1913 paper was for an orthorhombic lattice. A little later, von Laue (1914), in a paper first presented to the 1913 Solvay congress, married his geometrical diffraction theory to Ewald's construction, and extended the treatment to non-orthogonal axes by using Gibbs' reciprocal vectors. These were introduced by Willard Gibbs (1881) at Yale in his *Vector analysis*. Though printed privately, this was widely distributed by Gibbs, whose ideas were further disseminated by Wilson (1901). Von Laue quoted Budde (1914) as an example of a German source for Gibbs' vectors.

From three arbitary non-coplanar base vectors $\mathbf{a}_1, \mathbf{a}_2, \mathbf{a}_3$, Gibbs defined reciprocal vectors $\mathbf{b}_1 = \mathbf{a}_2 \wedge \mathbf{a}_3 / V$, etc., where $V = \mathbf{a}_1 \cdot (\mathbf{a}_2 \wedge \mathbf{a}_3)$ is the volume of the parallelepiped (cell). $\mathbf{b}_1$ is perpendicular to the plane defined by $\mathbf{a}_2$ and $\mathbf{a}_3$, and the vectors satisfy the relations

$$\left. \begin{array}{ll} \mathbf{a}_i \cdot \mathbf{b}_j = 1, & i = j \\ = 0, & i \neq j. \end{array} \right\} \qquad (5)$$

Von Laue's (1912, 1913) geometrical diffraction theory had led to equations of the type

$$(\mathbf{s} - \mathbf{s}_o) \cdot \mathbf{a}_i = h_i \lambda \qquad (i = 1,2,3), \qquad (6)$$

where $\mathbf{s}$ and $\mathbf{s}_o$ are unit vectors in the directions of the diffracted and incident rays, and the h_i are integers. Defining a vector $\mathbf{h}$ such that $\mathbf{h} \cdot \mathbf{a}_i = h_i \lambda$, it followed that $\mathbf{h} = \mathbf{s} - \mathbf{s}_o$. In terms of Gibbs' reciprocal vectors, von Laue (1914) now wrote

$$\mathbf{h}/\lambda = h_1 \mathbf{b}_1 + h_2 \mathbf{b}_2 + h_3 \mathbf{b}_3. \qquad (7)$$

(Some remarks on the roles of Ewald and von Laue in the development of the reciprocal lattice were

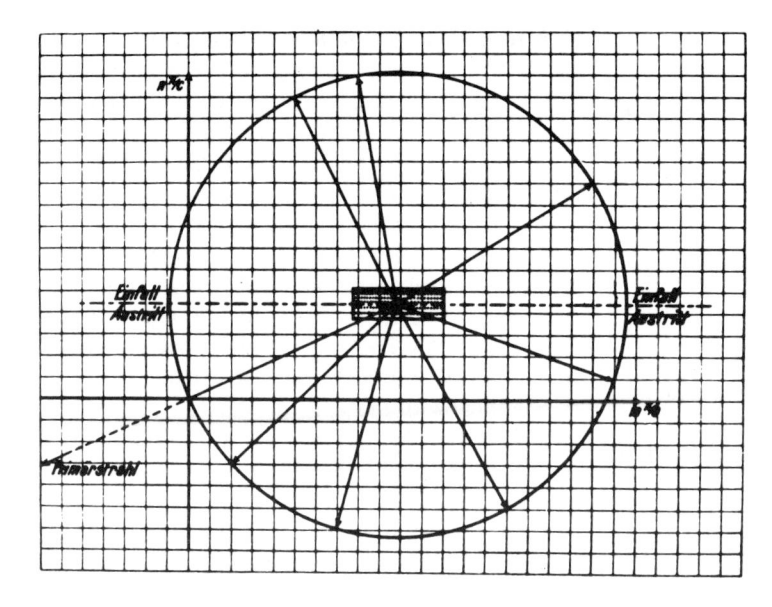

Fig. 2 Ewald's second diagram of 1913.

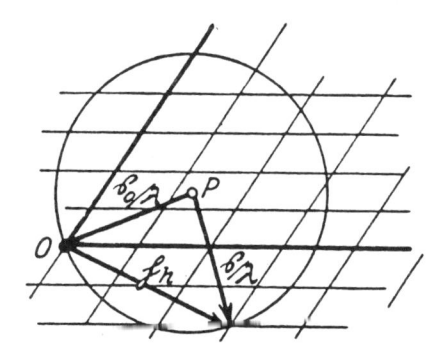

Fig. 3 The two-beam case of 1923.

given later by Ewald (1936); see also Chapter 2 by Hildebrandt in the present volume. Ewald's reciprocal lattice in the paper of 1913 was actually scaled to give $\mathbf{a}_i \cdot \mathbf{b}_i = (2a)(\pi/a) = 2\pi$, rather than the true reciprocal $\mathbf{a}_i \cdot \mathbf{b}_i = 1$ as in (5). This was because he used a cell of side $2a$ and a sphere of radius $K_o = 2\pi/\lambda$ rather than $K_o = 1/\lambda$ as in this chapter.)

Ewald did not present a diagram like Fig. 3 with an oblique lattice and just two waves until 1923 in his book *Kristalle und Röntgenstrahlen* (Ewald 1923). In 1923, the use of white-radiation Laue photographs was at its height, especially by Wyckoff (1924), who made much use of the gnomonic projection. This is a transformation of the diffracted beam directions from 2θ to $90° - \theta$, which transforms sets of spots lying on ellipses in a photograph into sets lying on straight lines in the projection. Figure 4 shows Ewald's (1923) demonstration of the relation between the reciprocal lattice and the gnomonic projection. The projection provides a fish-eye view of the reciprocal lattice, with the fish's eye at the origin.

3. The polar lattice of Bravais

Two years earlier, Ewald (1921) had published a major paper on the reciprocal lattice in structure theory. The early part of the paper offered a systematic treatment of the reciprocal lattice, and contained some historical remarks about J.G. Grassmann (the father) and Bravais. This paper included a demonstration that the reciprocal lattice of a face-centred cubic lattice is body centred, and vice versa. Ewald commented that this result had already been found by Bravais in 1850.

It is perhaps not sufficiently known that a lattice very similar to the reciprocal lattice was extensively explored by Bravais in his famous memoir on lattices in 1850. Von Laue also became aware of this, and in his Historical Introduction to *International Tables*, Vol. I (1952), he even wrote of the 'rediscovered reciprocal lattice theory'.

In Shaler's translation of Bravais' memoir (Bravais 1949), there are some 90 pages on nets and

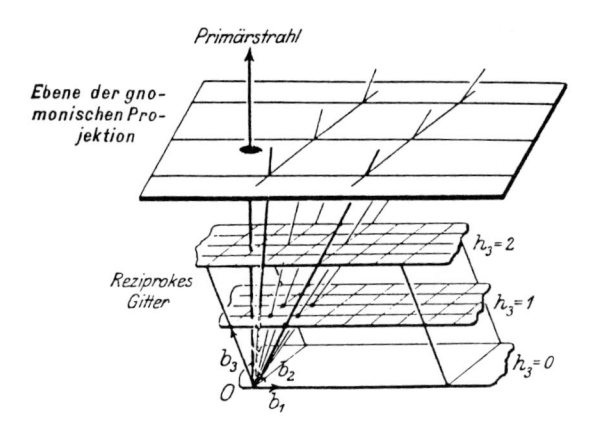

Fig. 4 The reciprocal lattice and the gnomonic projection (Ewald 1923.)

lattices in general, culminating in the enunciation of the 14 possible lattice types, which we call the Bravais lattices. The concluding 20 pages are on the polar lattice. 'Polar' is the term used by Bravais, and is to be understood somewhat in the sense of 'dual', not in any sense of electrical polarity. Bravais defined this polar lattice in terms of axes which might now be written

$$\mathbf{p}_1 = \mathbf{a}_2 \wedge \mathbf{a}_3 / V^{1/3} \quad \text{with } V_p = V, \quad (8)$$

rather than Gibbs'

$$\mathbf{b}_1 = \mathbf{a}_2 \wedge \mathbf{a}_3 / V \quad \text{with } V_b = 1/V. \quad (9)$$

From these polar base vectors Bravais generated the polar lattice. He fully explored the metrical relations between the two lattices. He showed that

(1) the polar lattice of a polar lattice is the original lattice;
(2) the symmetry axes and mirror planes of one lattice are preserved in the other;
(3) distances between points in one lattice are proportional to mesh net areas in the other;
(4) interplanar spacings and the directions of plane normals in one lattice are related to vectors between neighbouring points in the other lattice.

Of course, Bravais did nothing to anticipate the geometry of X-ray diffraction, nor the Ewald sphere construction. Nor had his polar axes, with $\mathbf{a}_i \cdot \mathbf{p}_i = V^{2/3}$, the reciprocal property $\mathbf{a}_i \cdot \mathbf{b}_i = 1$ essential in the analysis of diffraction.

4. Fourier space

I now return to another aspect of Ewald's major paper of 1921.

With one atom at each point of a simple crystal lattice, we have no difficulty in conceiving the points which constitute the reciprocal lattice. What happens when we have a compound or multiple lattice arising from several atoms in each cell of the original lattice? This problem led Ewald to attach 'weights' to the existing reciprocal lattice points. To his astonishment, these weights turned out to be identical with the structure factors, and this led him to realize that the weights were Fourier coefficients.

This development was brought to its summit many years later in papers by Bienenstock and Ewald (1962a,b) on the symmetry of reciprocal space, which they now called Fourier space. Instead of starting with the symmetry elements of direct space, they explored the possible symmetries of Fourier space when there are weights $F(\mathbf{h})\exp[2\pi i\phi(\mathbf{h})]$ at each reciprocal lattice point. Different restrictions on the possible linear transformations from (h,k,l,ϕ) to (h',k',l',ϕ') between symmetry-related points lead successively to the Fourier-space equivalents of the direct-space colour groups, the black-and-white groups and the 230 Schönflies–Fedorov space groups.

Bienenstock and Ewald's work on symmetry in Fourier space has been reformulated and much extended by Mermin and collaborators. The reformulation does not require periodicity in direct space and leads to a treatment of quasiperiodic materials. References can be found in Mermin (1992).

An earlier expression of Ewald's delight in Fourier transforms appeared in a discussion (Ewald 1940) of the shape transforms of finite crystals, a topic considered also by von Laue and Patterson. With monochromatic radiation, if the crystal is very small in some of its dimensions, each diffraction spot will be broadened in a manner depending on the shape and size of the crystal.

Shape transforms may be introduced in the following way. The electron density $\rho_s(\mathbf{x})$ of a finite crystal may be expressed as

$$\rho_s(\mathbf{x}) = \rho_\infty(\mathbf{x})s(\mathbf{x}), \quad (10)$$

where $\rho_\infty(\mathbf{x})$ is the density of the infinite crystal, and the crystal shape function $s(\mathbf{x}) = 1$ inside the

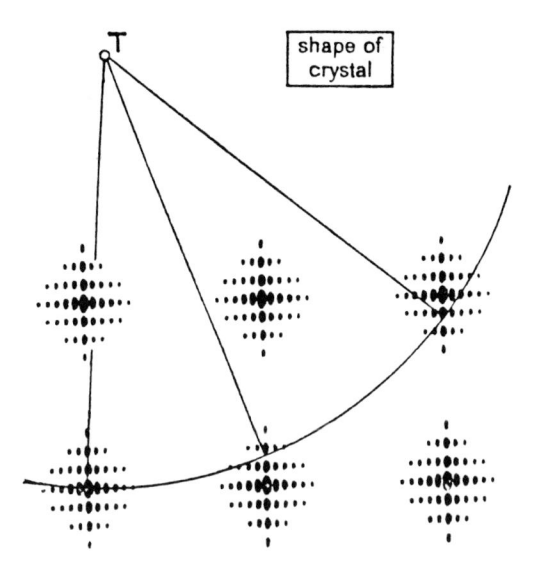

Fig. 5 The effect of shape transforms. (Ewald 1962).

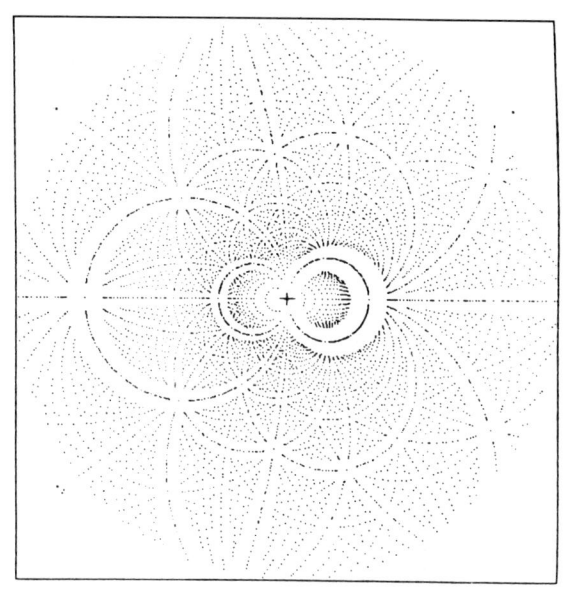

Fig. 6 Laue pattern of a protein. (Helliwell 1985.) A simulation has been chosen for clarity of reproduction.

crystal and $s(\mathbf{x}) = 0$ outside the crystal. Convolution theory then shows that the Fourier transform of the finite crystal is the set of point structure factors $F(\mathbf{h})$ of the infinite crystal each spread out by the Fourier transform $S(\mathbf{h})$ of the crystal shape function.

Figure 5 is an elegant picture of this, presented by Ewald (1962) in Chapter 6 of *Fifty years of X-ray diffraction*. The diffraction from a stationary crystal is given by the intersection of the sphere with the Fourier transform. Ewald went on to remark: 'The space in which the reciprocal lattice is imbedded is best called Fourier Space. This gets rid of the term "reciprocal space", which is a bad term because reciprocity is a symmetrical relation between two things and therefore unsuitable for designating one of them.' We may accept the first of these sentences, but the second is surely dangerous! For it could be applied also to the term 'reciprocal lattice', and no one would wish to rename that.

5. New and old aspects of the Laue method

Ewald's construction of the sphere of reflection and its intersections with the reciprocal lattice has proved enduringly fruitful in the understanding of diffraction geometry and in the development of experimental methods, especially for rotation/oscillation and precession photography, and for diffractometers. Space does not permit any history of these particular developments. Rather, I will gave an example of the current relevance of the reciprocal lattice as manifested in the recent revival of Laue photography.

Figure 6 shows a computer simulation of a typical Laue photograph of a protein crystal obtained with polychromatic synchrotron radiation. Beautiful photographs containing thousands of spots may be obtained with exposures down to the millisecond time range, and appear to have significant potential for dynamic experiments.

One of the difficulties associated with the Laue method is the multiple orders problem; that is, when many orders of a reflection are exactly superimposed. This was already illustrated (Fig. 7) by Ewald in his first 1913 paper. With incident radiation spanning a range ΔK_o of reciprocal wavelengths, if (0,8,16) is an interference spot, then (0,7,14), (0,6,12), (0,5,10), and (0,4,8) are simultaneously interference spots. These are higher harmonics of the inner point (0,1,2). The accessible region of the reciprocal lattice was clearly shown (Fig. 8) by Ewald (1923) as lying between spheres of radii $P'O = 1/\lambda_{max}$ and $P''O = 1/\lambda_{min}$.

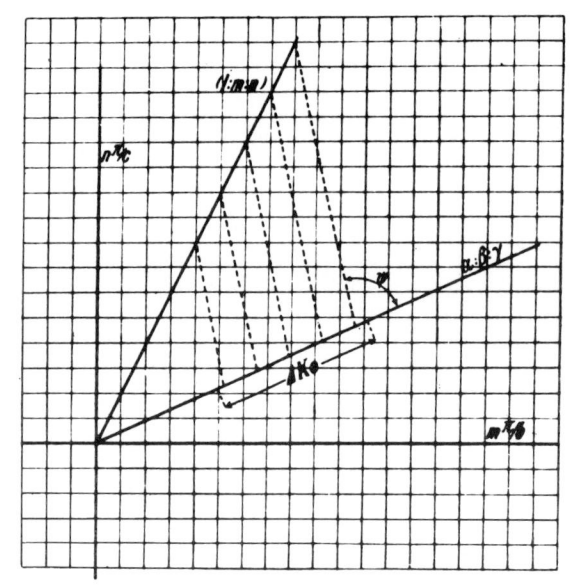

Fig. 7 Multiple orders. (Ewald 1913.)

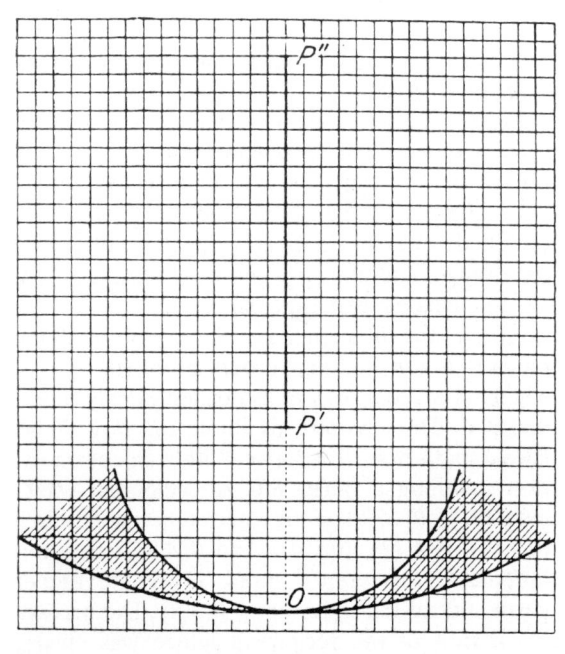

Fig. 8 The accessible region of the reciprocal lattice. (Ewald 1923.)

If the majority of spots were to correspond to multiple orders the utility of the Laue method in protein crystallography would be greatly restricted. Cruickshank *et al.* (1987) have studied the multiplicity distribution in some detail, and their results show that multiple orders occur much less frequently than might be supposed.

The essential reason is found by considering the inner points of a lattice. A point (h,k,l) is an inner point if the highest common divisor of h, k and l is 1. Thus (5,4,3) is inner point, but (8,6,4) is not, as it is the second order of the inner point (4,3,2). The probability that a randomly chosen lattice point is an inner point is

$$(1 - 1/2^3)(1 - 1/3^3)(1 - 1/5^3)(1 - 1/7^3)\ldots = 0.832. \qquad (11)$$

This is an expression of the probability that h, k, and l are not simultaneously divisible by any of the primes 2, 3, 5, 7 … In simple terms, as one travels outwards from the origin of the reciprocal lattice, there is a never-ending constant-density supply of inner points.

The accessible region of the reciprocal lattice may be defined by a slight refinement of Fig. 8 in terms of its two Ewald spheres of radii $1/\lambda_{max}$ and $1/\lambda_{min}$ with another boundary set by a sphere of radius d^*_{max} ($= 1/d_{min}$) centred at the origin. The final results of the multiplicity analysis are shown in Fig. 9. The most remarkable feature is that, provided $d^*_{max} < 2/\lambda_{max}$, the percentage of single-order spots does not depend on d^*_{max} but only on the ratio $M = \lambda_{max}/\lambda_{min}$. It rises from 83 per cent at $M = \infty$ to 88 per cent at $M = 3$, tending of course to 100 per cent as M tends to unity. Thus the great majority of spots seen on a synchrotron Laue photograph of a protein correspond to single orders.

This example is just one illustration of the enormous importance of Ewald's development of the reciprocal lattice for crystal diffraction theory.

6. Ewald's 1914 analysis of Laue photographs

In Chapter 2 G. Hildebrandt drew attention to Ewald's introduction of the concept of the structure factor during the preparations for Sommerfeld's lecture at the 1913 Solvay Conference. This concept was used in the structure analysis of pyrite, FeS_2, by Ewald (1914*a*) and Ewald and Friedrich

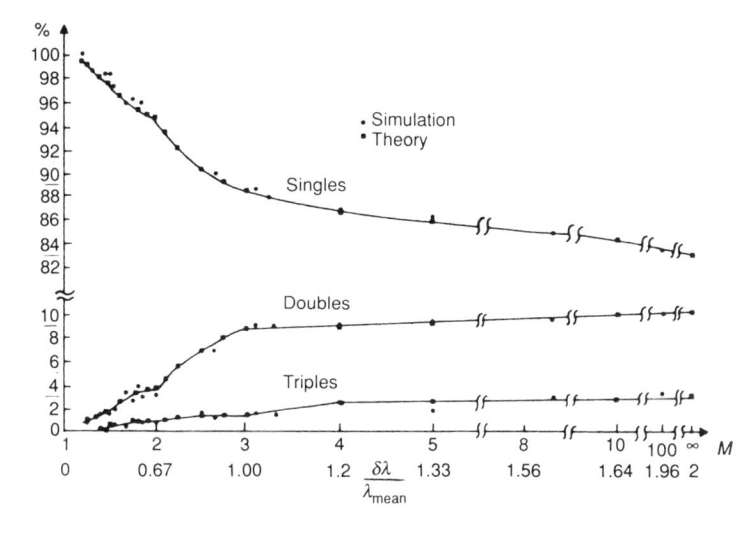

Fig. 9 The variation with M ($= \lambda_{max}/\lambda_{min}$) of the proportions of reciprocal lattice points observed as single-, double-, or triple-order spot. (Cruickshank et al. 1987.)

(1914). Ewald varied a coordinate parameter so as to match the observed intensity ratios for pairs of planes of the same spacing.

In another paper (Ewald 1914b) on the intensities of spots in Laue photographs of zincblende, ZnS, Ewald showed how unsymmetrical photographs could be used to reveal the wavelength dependence of the source. With an unsymmetrical crystal orientation, planes with symmetry-equivalent indices (and hence of the same spacing) reflect at different wavelengths. Thus a comparison of their intensities shows the wavelength dependence of the source.

These two techniques were developed in Japan by Nishikawa (1915), who passed them on to Wyckoff (1920) with 'untiring advice and aid' during a visit to the USA in 1917–18. Ewald's concepts thus had an influence on the vigorous exploitation of the Laue method in the USA during the 1920s by Wyckoff, Dickinson, and Pauling.

Currently, the Laue method in protein crystallography makes use of unsymmetrical photographs, on which symmetry equivalent reflections occur at different wavelengths, to determine a λ-curve for the wavelength dependence of the source, detector response, and other factors. This wavelength normalization curve enables the observed intensities to be placed on a common scale (Campbell et al. 1986). As indicated above, in 1914, Ewald had already realized some of the utility of unsymmetrical Laue photographs.

References

Bienenstock, A. and Ewald, P.P. (1962a). *Acta Cryst.*, **15**, 1253–61.
Bienenstock, A. and Ewald, P.P. (1962b). *Soviet Phys.—Cryst.*, **6**, 665–7.
Bravais, A. (1949). On the systems formed by points regularly distributed on a plane or in space (trans. A.J. Shaler), *J. Ecole Polytech., Cahier* 33, **19**, 1–128 (1850). American Crystallographic Association, Washington, DC.
Budde, E. (1914). *Tensoren und Dyaden im dreidimensionalen Raum*. Braunschweig.
Campbell, J.W., Habash, J., Helliwell, J.R., and Moffat, K. (1986). *Information quarterly for protein crystallography*. No. 18. SERC Daresbury Laboratory, Warrington., UK.
Cruickshank, D.W.J., Helliwell, J.R., and Moffat, K. (1987). *Acta Cryst.*, **A43**, 656–74.
Ewald, P.P. (1913a). *Phys. Z.*, **14**, 465–72.
Ewald, P.P. (1913b). *Phys. Z.*, **14**, 1038–40.
Ewald, P.P. (1914a). *Phys. Z.*, **15**, 399–401.
Ewald, P.P. (1914b). *Ann. Phys. (Leipzig)* **44**, 257–82.
Ewald, P.P. (1921). *Z. Kristallogr.*, **56**, 129–56.
Ewald, P.P. (1923). *Kristalle und Röntgenstrahlen*. Springer, Berlin.
Ewald, P.P. (1936). *Z. Kristallogr.*, **93**, 396–98.
Ewald, P.P. (1940). *Proc. Phys. Soc. (London)*, **52**, 167–73.
Ewald, P.P. (1962) *Fifty years of X-ray diffraction* (ed. P.P. Ewald). Oosthoek, Utrecht.
Ewald, P.P. and Friedrich, W. (1914). *Ann. Phys. (Leipzig)* **44**, 1183–96.

Gibbs, J.W. (1881). *Vector analysis*. Privately printed, New Haven, CT.
Helliwell, J.R. (1985). *J. Mol. Struct.*, **130**, 63–91.
International tables for X-ray crystallography (1952). Vol. I. (ed. N.F.M. Henry and K. Lonsdale) Kynoch Press, Birmingham.
Laue, M. (1912). *Sitz.-Ber. Bayr. Akad. Wiss.*, 303–22.
Laue, M. von (1913). *Phys. Z.*, **14**, 1075–9.
Laue, M. von (1914). *Jahrb. Radioaktivität Electron.*, **11**, 308–45.
Mermin, N.D. (1992). *Rev. Mod. Phys.*, **64**, 3–49.
Nishikawa, S. (1915). *Proc. Tokyo Math.—Phys. Soc. II*, **8**, 199–209.
Wilson, E.B. (1901). *Vector analysis founded upon the lectures of J.W. Gibbs*. Yale University Press, New Haven, CT.
Wyckoff, R.W.G. (1920) *J. Am. Chem. Soc.*, **42**, 1100–16.
Wyckoff, R.W.G. (1924). *The structure of crystals*. Chemical Catalog Co., New York.

7

Lattice sums, the Madelung constant, and Paul Ewald

David H. Templeton

Who was Paul Ewald? Several answers come to mind. Those who knew him as a person will think first of the man himself, whom they loved, respected, and admired. Then they will remember his scientific achievements or his leadership as a teacher, and editor and organizer of international crystallography. 'The Ewald sphere' may be the first response from many crystallographers if asked what they associate with his name. Those more familiar with his work will cite his dynamical theory of diffraction as his monument in science.

But these things are unknown to many chemists and physicists. Among non-crystallographers, his name is more likely to be recognized in association with the 'Ewald method' for lattice sums. One paper on this subject (Ewald 1921) has had an extraordinary influence on the chemistry and physics of the solid state. Written at a time when tedious computations were done by hand, with a desperate need for any shortcut, it retains its value in the context of modern electronic computers. It is an object lesson on the value of good algorithms for some problems no matter how rapid the computer.

The Physics Library at Berkeley includes the old volumes of *Annalen der Physik*, and I went there again to look at this paper. All these volumes are faded and dusty and somewhat fragile, but most show little wear nor even much evidence of having been read at all. But Volume 64 is different. Some pages in the middle, where Ewald's paper is found, are badly frayed and a little loose, and the binding is damaged so that the volume falls open there. A few marginal notes by an unknown hand are further evidence that the paper has been studied carefully.

This paper was chosen by *Current Contents* as a 'Citation Classic', with citations in more than 555 publications since 1955 (Juretschke and Wagenfeld 1985); some comments on its origin and significance are given there. I have noticed it frequently among the references in papers on electrostatic energies of crystals. Not all of these citations suggest that its details are understood. There is a vast literature of invention and reinvention of methods to achieve convergence in calculations of Madelung constants by authors who cite Ewald but seem not to appreciate the advantages of his method.

Ewald studied lattice sums because he needed them for his theory of crystal optics and diffraction, and there are few signs in his subsequent writings that he was interested in crystal energies or the application of his method to them. An exception is a lecture at a symposium (Ewald 1973), in which he discussed the possibility of obtaining this energy from diffraction data. In a later review (Ewald 1979), he mentioned how critical the lattice potential problem was to his crystal optics project and described how Peter Debye provided the key suggestion for its solution. Yet he gives no reference to the 1921 paper!

One of the sums is the electrostatic lattice potential ϕ:

$$\phi_j = \sum_k \frac{\varepsilon_k}{R_{jk}} \quad (j \neq k), \qquad (1)$$

where k includes all the ions in the crystal except one. The electrostatic energy U_c, often expressed as the Madelung constant A with a standard length L, is the dominant term in the binding energy of ionic crystals:

$$U_c = \frac{-e^2 A}{L} = \frac{1}{2} \sum_{j,k} \frac{\varepsilon_j \varepsilon_k}{R_{jk}} \quad (j \neq k); \qquad (2)$$

j includes the ions in a formula unit. These sums are related by

$$U_c = \frac{1}{2} \sum_j \varepsilon_j \phi_j. \qquad (3)$$

The physical problem is well defined, yet the sums (1) and (2) which represent it (somewhat loosely) are meaningless without some qualification, because the positive and negative terms taken separately diverge without limit. The number of terms in a spherical shell at distance r is proportional to r^2, and thus successive shells give larger and larger contributions. Convergence can be achieved by careful grouping of terms, but errors can be made. For example, in the CsCl structure it is easy to group terms in a way which converges to a stationary (but incorrect) limit because the model at each stage has an electric double layer on its surface. A naive idea is that with a fast enough computer one need not worry about convergence if one can continue until each additional term is too small to change the answer. A simple counter-example is the sum

$$\sum_{n=1}^{p} \frac{1}{n} \approx 0.58 + \ln p, \qquad (4)$$

which diverges for infinite p, but when calculated with 32-bit precision on a typical computer converges to about 15.4.

Ewald's solution to this problem is the double sum

$$\phi_j = \phi_j^{(1)} + \phi_j^{(2)}, \qquad (5)$$

where $\phi^{(1)}$ is a sum in reciprocal space and $\phi^{(2)}$ is a sum in direct space. The individual charges in (1) are replaced by a density function ρ represented by a Fourier series; the first sum becomes a sum of integrals of each Fourier component:

$$\phi_j^{(1)} = \int \frac{\rho(\mathbf{R})}{R} dV = \sum_{\mathbf{h}} \left(F(\mathbf{h}) \int \frac{\exp(-2\pi i \mathbf{h} \cdot \mathbf{R})}{R} dV \right). \qquad (6)$$

Here F is a structure factor for the distribution of charge. To achieve convergence, the point charges are replaced by spherical Gaussian distributions of charge. The more extended the distributions, the more rapid is the convergence. Overlap of these distributions creates an error which is corrected for by the direct-space sum. A correction must be made for the effect of the reference ion on itself, but this is simple for a Gaussian distribution. For optimum choice of the size parameter, both series are rapidly and absolutely convergent. Ewald gave the example of sodium chloride (here edited for better numerical precision and to permit direct comparison with the correct Madelung constant 1.74756...):

$$\frac{a}{2e} \phi_{Na}^{(1)} = -2.0000 + 0.3218 + 0.0005 = -1.6777$$

$$\frac{a}{2e} \phi_{Na}^{(2)} = -0.0732 + 0.0034 \qquad = -0.0698$$
$$\overline{-1.7475}$$
$$(7)$$

The sums include only $hkl = \{111\}$ and $\{113\}$ and first and second neighbours, and the answer is correct to the fourth decimal place.

Bertaut (1952, 1953) recognized that, for the Madelung constant, the Patterson function includes all the necessary information about interatomic vectors, and that

$$\frac{-2e^2 A}{L} = \iint \frac{\rho(\mathbf{x}) \rho(\mathbf{x}+\mathbf{R})}{R} dV(\mathbf{x}) dV(\mathbf{R})$$

$$= \sum_{\mathbf{h}} \left(F^2(\mathbf{h}) \int \frac{\exp(2\pi i \mathbf{h} \cdot \mathbf{R})}{R} dV \right), \qquad (8)$$

again with correction terms for self-energy and overlap. This formulation may clarify the physical interpretation for crystallographers not familiar with the theta functions used by Ewald. It has the advantage, for complicated structures, that it includes all the ions in a single calculation. The method using (3) requires a separate calculation for every distinct kind of atomic site. Bertaut also invented truncated distributions of charge which removed the overlap, so that only one series is needed.

My contribution to this subject was to estimate the rate of convergence with variations of Bertaut's shape functions by using Wilson's statistics to approximate F^2 in (8). With the best functions the terms decrease as $|\mathbf{h}|^{-10}$ (Jones and Templeton 1956).

Realistic calculations of crystal energies require consideration of various other effects, some of which fall off with distance faster than $1/R$ but not quickly enough for rapid convergence of sums in direct space. Bertaut (1953) extended his method to dipole and multipole sums. A more general extension uses convergence functions (Nijboer and De Wette 1957; Williams 1971) and is very effective for R^{-n} sums:

$$S_n = \frac{1}{2} \sum q_j q_k R_{jk}^{-n} \qquad (j \neq k). \qquad (9)$$

It too uses sums in direct and reciprocal spaces:

$$S_n = \Phi(R) \, S_n + [1 - \Phi(R)] \, S_n. \qquad (10)$$

The convergence function Φ is unity at $R = 0$ and falls off rapidly enough to make the first sum on the right of (10) converge rapidly in direct space, yet it falls slowly enough to preserve convergence of the second sum in reciprocal space, using Fourier transforms.

Convergence to the correct answer is always a requirement, but it may seem that the speed of computation with modern computers makes the rate of convergence unimportant. This is true for the calculation with specified parameters of the energy of a single structure. But the trend in contemporary work is to investigate the energy as a function of many variables; for example, to predict a crystal structure starting with a model of interatomic forces. These problems are increasingly demanding if symmetry is relaxed to permit a more nearly global search of parameter space, or to investigate dynamic processes. One quickly learns that many configurations of a structure have nearly equal energies, and therefore high precision is required in highly repetitive computations. Ewald's method, with the improvements mentioned here, continues to be the choice for such work.

References

Bertaut, F. (1952). *J. Phys. Radium*, **13**, 499–505.

Bertaut, E.F. (1953). Ferroelectric and high dielectric crystals: contribution to the theory of fields, potentials and energies in periodic lattices. U.S. Air Force Report AD-22696.

Ewald, P.P. (1921). *Ann. Phys. (Leipzig)*, **64**, 253–87.

Ewald, P.P. (1973). Diffraction data and electrostatic energy of a crystal. Am. Cryst. Assoc. Meeting, Storrs, CT, Abstr. B1.

Ewald, P.P. (1979). *Acta Cryst.*, A**35**, 1–9.

Jones, R.E. and Templeton, D.H. (1956). *J. Chem. Phys.*, **25**, 1062–3.

Juretschke, H.J. and Wagenfeld, H.K. (1985). *Curr. Contents*, **25** (49) 20.

Nijboer, B.R.A. and De Wette, F.W. (1957). *Physica*, **23**, 309–21.

Williams, D.E. (1971). *Acta Cryst.*, A**27**, 452–5.

8

X-ray topography

A. Authier and B. Capelle

In the theory of dispersion [...] an important role is played by the reciprocity theorem. That is to say, if you have a ray falling in one direction and generating a secondary ray in another direction, then, if you let the incident ray take the direction of the secondary one, the primary ray will also be generated again. So this showed that you cannot consider a single wave but you have to consider a whole bundle of waves or, as we now say, a wavefield propagating in the crystal, instead of a single wave which is sufficient in Optics. P.P. Ewald (from a lecture given at the International School on X-ray Topography and Dynamical Theory, Limoges, August 1975).

1. Introduction

One of the major applications of dynamical theory is X-ray topography, which enables one to visualize and characterize defects in a crystal. After pioneer works by Berg (1931), Ramachandran (1944), and Barrett (1945), the first topographic techniques were developed independently by Lang (1957, 1959), Barth and Hosemann (1958), Bonse and Kappler (1958), Borrmann et al. (1958), and Newkirk (1958). There are several possible origins for the formation of images on a topograph: misorientation or variation of lattice spacing between one region and another (the so-called *orientation contrast*), or variation of the diffraction and propagation properties of X-rays in regions of different degrees of perfection (the so-called *extinction* or *diffraction contrast*). The latter is the more important effect to be considered in interpreting the contrast of images of isolated defects such as dislocations or planar defects, and it requires the use of dynamical theory and of its extension to deformed crystals. Conversely, the need to interpret these images stimulated the development of the dynamical theory.

Until the 1950s, the applications of dynamical theory were concerned essentially with reflected intensities and rocking curves, but it was with the advent of perfect or nearly perfect crystals and the development of X-ray topography that the notion of *wavefield*, introduced by Ewald (1917), as recalled by the above excerpt from a lecture given by him, became of practical importance. The reason is that the interpretation of topographs required one to follow the propagation of energy within the crystal in the vicinity of defects, and this is precisely what the notion of wavefields helps in doing.

2. Principle of X-ray topographs

The aim of X-ray topographs is to give a one-to-one correspondence between images on a photographic plate or a direct viewing system and the distribution of defects or distorted areas within the crystal. There are many possible set-ups. The geometry may be for reflection or transmission. The incident beam may be monochromatic or white (using synchrotron radiation), pseudo-plane parallel or divergent. The choice depends on the desired exposure time, and spatial and angular resolutions.

2.1 Reflection, or Bragg case

There are two families of set-ups. In the first (Newkirk 1958), a divergent beam falls on the crystal at the Bragg angle at a grazing incidence, and a photographic plate is placed at an angle as small as possible to the crystal surface (Fig. 1). If there are large distortions or variations in the degree of perfection along the surface, a small area A may diffract more, or less, than the neighbouring areas (orientation or extinction contrast, respectively). This will be recorded on area A' on the plate, giving rise to an image. In the second type of set-up (Bonse and Kappler 1958), a nearly plane wave coming from a beam conditioner falls on the crystal at an incidence angle corresponding to a

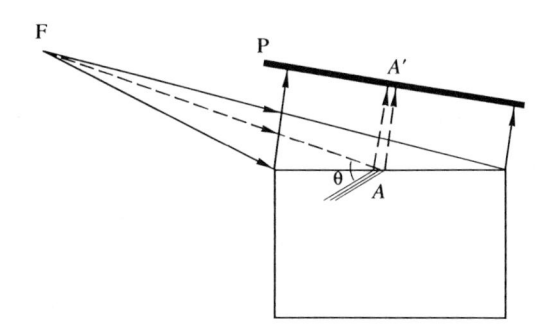

Fig. 1 Principle of reflection X-ray topography. F: X-ray source; P: photographic plate; θ: Bragg angle.

reflected intensity of about half the maximum value. Any small distortion at the surface of the crystal will give rise to a fluctuation of the reflected intensity which will be recorded as an image on the plate.

2.2 Transmission, or Laue case

Again the various topographic techniques can be divided in two classes: those for which the incident beam is either collimated characteristic radiation or nearly parallel synchrotron white radiation, and those for which the incident beam is a nearly plane wave produced by a beam conditioner using radiation coming either from a tube or from a synchrotron source.

The former situation is schematically described in Fig. 2 for the case of collimated characteristic radiation. The angular and spectral widths of the beam are, in general, so large that only a small part of the incident beam is reflected through the crystal. The remaining part, or direct beam, propagates undergoing only normal photoelectric absorption. The part which is reflected can be assimilated to a spherical wave. It generates a fan of wavefields propagating within the so-called Borrmann triangle. The fanning out of wavefield paths within a triangle whose sides make an angle equal to twice the Bragg angle was proposed by Borrmann (Borrmann 1959a; von Laue 1960), but the calculation of their intensities and of their interferences (**Pendellösung**) using dynamical diffraction of X-ray spherical waves is due to Kato (1960).

The images of defects can be due to:

(1) either the kinematical diffraction of the direct beam AB by the distorted areas, giving rise to the so-called **direct image**, which can be used to obtain a general picture of the defect distribution in the crystal;

(2) or the perturbation by even slightly deformed areas of the paths and phases of the wavefields propagating within the Borrmann triangle ABC. New wavefields may also be created in the most heavily distorted regions. These modified and new wavefields interfere among themselves and with wavefields unaffected by the presence of the defect, giving rise to more or less complicated fringes which can be used to find the strain distribution associated with the defect by comparison between computer simulated and experimental images (see, e.g. Epelboin 1987).

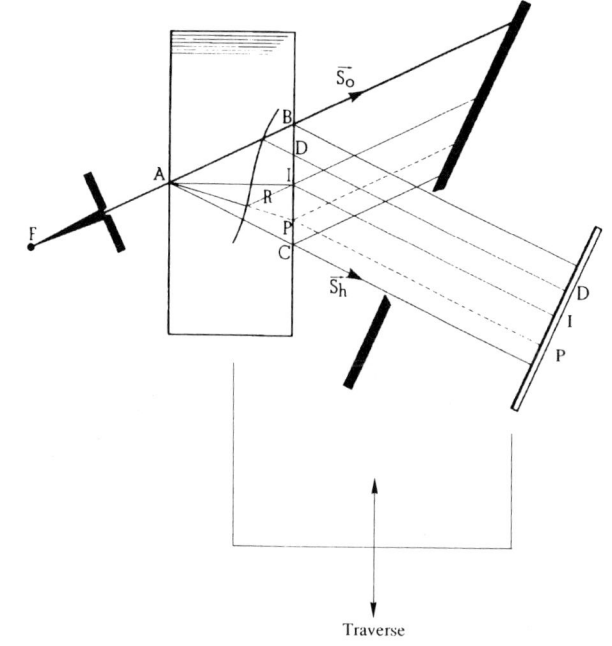

Fig. 2 Principle of transmission X-ray topography. The wavy line is the projection on the plane of incidence of a dislocation line. F: X-ray source; D: direct image: this is the projection of the intersection of the dislocation with the direct beam AB; I: intermediary image; P: dynamical image of the dislocation line; this is a shadow cast by the dislocation on the propagation of wavefields from the apex A of the Borrmann triangle, which appears as a line on the topograph. The height of the slit in front of the crystal in the out-of-plane direction can be up to a few centimetres.

As an example, Fig. 2 illustrates the image formation in the case of a dislocation line (Authier 1967):

(1) The volume around the intersection of the dislocation line with the direct beam AB behaves like a small imperfect crystal and diffracts this *direct* beam, giving rise to the direct image (D in Fig. 2).

(2) The paths of wavefields propagating through the Borrmann triangle are refracted in the vicinity of the dislocation line, creating a shadow P along the projection of the dislocation drawn from the apex, A, of the Borrmann triangle (Fig. 2). This shadow is called the **dynamical image**.

(3) At the same time, some new wavefields are generated at the intersection R of wavefields with the most heavily distorted areas around the dislocation line. These wavefields interfere at I at the exit surface with wavefields which have not been affected by the presence of the dislocation line, giving rise to a third part of the image, called the **intermediary image**.

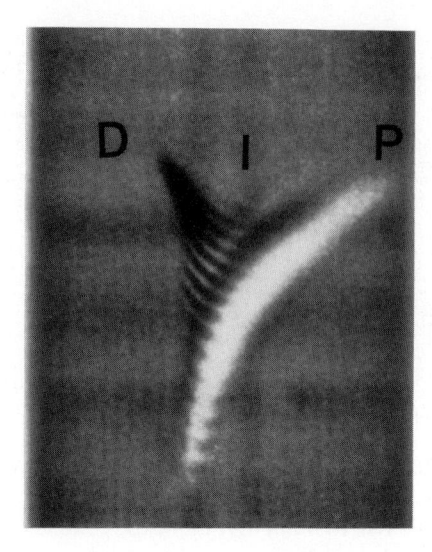

Fig. 3 Section topograph of a silicon crystal (after Authier 1967), 220 reflection, MoK$_\alpha$. D: direct image; I: intermediary image; P: dynamical image. The direction of the fringes (Pendellösung fringes)—that is, the vertical direction—corresponds to the out-of-plane direction in Fig. 2.

There are many variants of the set-ups used in transmission topography, in particular using synchrotron radiation. Only the principles of the most important types are given in the following:

(1) **Section topographs** (Lang 1957). By simply putting a photographic plate after the slit, as shown in Fig. 2, one can record the various components of the image: D, the *direct* image (from the distance of which from the edge of the topograph one can deduce the depth of the defect), P, the *dynamical* image, and I, *intermediary* image. Figure 3 is an example of such a topograph, showing the various parts of the image of a dislocation line in a silicon crystal, and Fig. 9(a) below gives another example of a section topograph in a quartz crystal where the situation is more complicated as a result of the interaction of several strain fields (see Section 3.2). If the defect is a planar defect such as a stacking fault, one can deduce its nature from the analysis of the fringes in its image (Authier 1968).

(2) **Projection topographs**. Using collimated characteristic radiation (Fig. 2), one isolates the reflected beam with a slit and traverses simultaneously the crystal and the photographic plate in a direction parallel to the crystal surface (Lang 1959). The direct images of the defects are then projected on the photographic plate. Figures 4–8 are examples of images obtained in this way. Using white synchrotron radiation which has a narrow natural divergence, one simply puts the photographic plate far enough from the crystal so that the various Laue spots are separated from one another. In these techniques, the *direct* images are predominant for crystals with a small value of μt (μ is the linear absorption coefficient and t is the crystal thickness). The shadows due to the disruption of anomalous transmission along paths ARP are predominant for high values of μt (*dynamical images*—Borrmann 1959b, not illustrated here).

(3) **Double crystal topography** (Renninger 1962). If the incident beam on the crystal is a pseudo-plane wave obtained from either an X-ray tube or synchrotron radiation by means of a beam

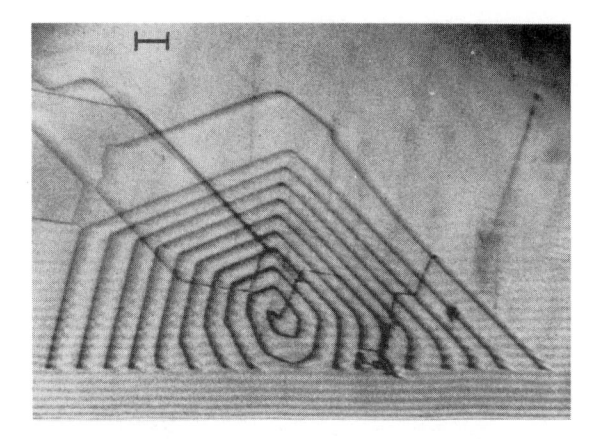

Fig. 4 111 projection topograph of a Frank–Read source in silicon—scale mark: 100 μm. (Authier and Lang 1964.)

conditioner, it is not wide enough for the Bragg condition on the most distorted areas to be fulfilled, and the direct images are eliminated. The reflection width is very narrow because only dynamical diffraction takes place and the observed images are highly strain sensitive.

3. Examples

X-ray topography has been used for the characterization of very many crystals (semiconductors, and metallic, ionic, and molecular crystals), in relation to their growth and their mechanical, electrical, or magnetic properties. Its possibilities have been tremendously increased by the advent of synchrotron radiation which allows photographs to be taken with very short exposure times and dynamic experiments to be performed under unusual pressure or temperature conditions, under an applied stress field, a magnetic or an electric applied field, etc. Some examples are shown for two crystals widely used in modern technology.

3.1 Silicon crystals, Frank–Read source devices

The first example is a historic one (Fig. 4). It shows a Frank–Read source in a silicon crystal which has undergone plastic deformation by twisting at high temperature in a lathe (the experiment was performed by the late W. Dash). The pole of the source

Fig. 5 Synchrotron topograph of misfit dislocations between epilayer and substrate in (111) silicon crystal. (Sauvage and Petroff 1980.)

at the inner end of the spiral is shown distinctly. The outer ring has started to multiply by cross slip (Authier and Lang 1964).

In a second example (Fig. 5), misfit dislocations at the interface between a (111) silicon substrate and a doped silicon epilayer are shown on a topograph taken using synchrotron radiation (Sauvage and Petroff 1980).

It is now well known that there are correlations between the electrical quality of semiconductor devices and the defects present in the crystal. Routine characterization of the defects generated at each step is therefore necessary. Figure 6 shows, for instance, diffused areas at the surface of a silicon crystal at an intermediate stage of the production of a power transistor. A crossgrid of misfit dislocations can be seen in the diffused areas, and emitter edge dislocations propagating in the non-diffused areas from the boundaries with the diffused areas are clearly seen (Sauvage et al. 1981).

Fig. 6 Projection topograph of diffused areas at the surface of a silicon crystal. The diffused areas can be recognized by the cross-grid of straight misfit dislocations inside them. Curvy emitter edge dislocations are formed at the boundaries between diffused and non-diffused areas because of the stress gradient across them. (Sauvage *et al.* 1981.)

3.2 Quartz crystals, resonators

Quartz is still the most widely used material for the fabrication of resonators. It has been shown that their quality is affected by crystalline defects. Synthetic quartz is now mainly used, and X-ray topography is the best way to choose where to cut the resonators in the as-grown crystals. Figure 7 represents, for instance, a slice of a high-quality synthetic quartz crystal. A dashed line shows the position of the seed. The various growth sectors are clearly seen, with growth striations in some of them as well as a few dislocations. Those dislocations which originate inside the seed are refracted when they enter the newly grown areas.

Acoustic vibrations can be made visible using synchrotron stroboscopic topography (Zarka *et al.*

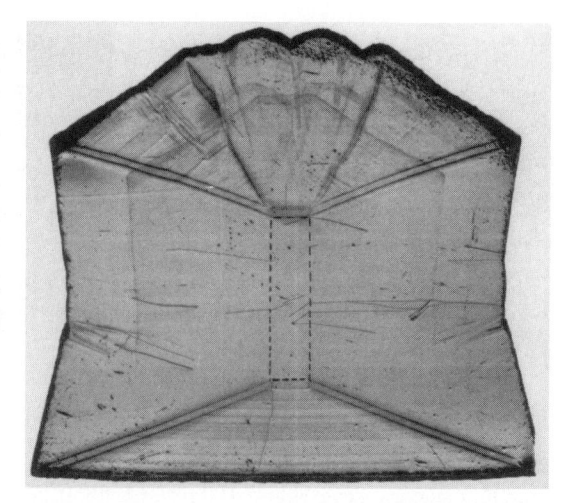

Fig. 7 Projection topograph of a synthetic quartz crystal. (Courtesy A. Zarka.)

1988*a,b*). The principle of the experimental set-up consists in the exact synchronization between the acoustic vibration and the time structure of the synchrotron radiation. Figure 8 shows acoustic vibrations excited in a quartz resonator.

To study the interaction between the strain field of dislocations and the elastic fields of the acoustic vibrations, a comparison has been made between simulated and experimental section topographs taken in an area where both are simultaneously present (Fig. 9). Figures 9(a) and (b) are experimental and simulated images of the dislocation in the absence of acoustic vibrations, respectively. The agreement is very good. The fringes observed in the upper part of the images are due to the normal, spherical wave Pendellösung effect. Figure 9(c) represents the simulated image of acoustic vibrations in the absence of dislocations. Figure 9(d) is the experimental section topograph of a dislocation in the presence of acoustic vibrations, and Fig. 9(e) the corresponding simulated image assuming that the two strain fields can be added up without any interaction term. The relatively poor agreement shows that this hypothesis is not correct.

4. P.P. Ewald and topography

Ewald never did any work on topography. He was, however, very interested in developments in this field and in the possibilities it opened. He was, of

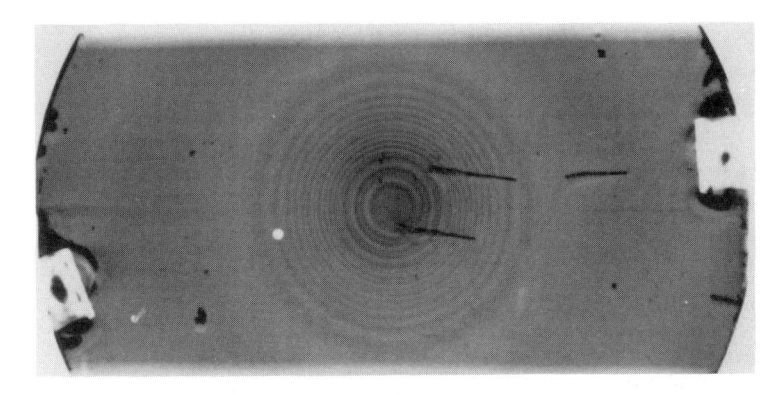

Fig. 8 Stroboscopic synchrotron topograph of a vibrating quartz resonator. (Zarka *et al.* 1988*b*.)

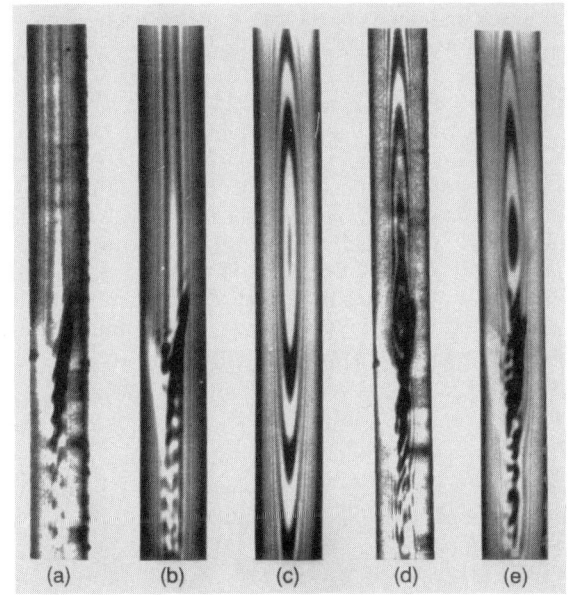

Fig. 9 Experimental and simulated section topographs of a quartz resonator. (Capelle *et al.* 1989.) (a) Experimental image of a dislocation in the absence of acoustic vibrations. (b) Simulated image of the dislocation in the absence of acoustic vibrations. (c) Simulated image of acoustic vibrations in the absence of dislocations. (d) Section topograph of a dislocation in the presence of acoustic vibrations. (e) Simulated image of the dislocation in the presence of the acoustic vibrations, assuming that the two strain fields can be added up without any interaction term.

course, particularly interested in all the extensions of the dynamical theory to strained crystals. As has been shown above, the very basis of topography and of the interpretation of the images of defects is the existence of wavefields, not only as a mathematical concept to solve the fundamental equations of dynamical theory, but also as a physical reality. For this reason, Ewald followed very closely the appearance of the notion that new wavefields may be generated during the diffraction by highly strained areas (see, e.g. Authier and Balibar 1970). During his various stays in France in the 1960s, and the 1970s, he never failed to visit our laboratory and to have long discussions with all of us, and in particular with Françoise Balibar on the generalization of the notion of wavefields. Ewald was a very attentive and active participant (Fig. 10) of the International Summer School on X-ray Topography and Dynamical Theory, which was held in Limoges in August 1975. His wide knowledge, his open-mindedness, and his everlasting good humour were very stimulating for all participants, who included practically everyone active in dynamical theory at the time.

References

Authier, A. (1967). *Adv. X-ray Anal.*, **10**, 9–31.
Authier, A. (1968). *Phys. Stat. Solidi*, **27**, 77–93.
Authier, A. and Balibar, F. (1970). *Acta Cryst.*, **A26**, 647–54.
Authier, A. and Lang, A.R. (1964). *J. Appl. Phys.* **35**, 1956–9.
Barrett, C.S. (1945). *Trans. AIME*, **161**, 15–64.
Barth, H. and Hosemann, R. (1958). *Z. Naturforsch.* (a), **13**, 792–4.
Berg, W.F. (1931). *Naturwissenschaften*, **19**, 391–6.
Bonse, U. and Kappler, E. (1958). *Z. Naturforsch.* (a) **13**, 348–9.
Borrmann, G. (1959*a*). *Beiträge zur Physik und Chemie des 20 Jahrhunderts*, pp. 262–82. Vieweg, Braunschweig.

Fig. 10 P.P. Ewald at the Limoges Summer School (1975).

Borrmann, G. (1959b). *Phys. Bl.*, **15**, 508–9.
Borrmann, G., Hartwig, W., and Irmler, H. (1958). *Z. Naturforsch.* (a), **13**, 423–5.
Capelle, B., Zarka, A., Zheng, Y., Detaint, J., and Schwartzel, J. (1989). *Proc. 43nd Annual Frequency Control Symposium*, pp. 470–6.
Epelboin, Y. (1987). *Prog. Crystal Growth Charact.*, **14**, 465–506.
Ewald, P.P. (1917). *Ann. Phys.* (*Leipzig*), **54**, 519–97.
Kato, N. (1960). *Z. Naturforsch.* (a), **15**, 369–70.
Lang, A.R. (1957). *Acta Metall.*, **5**, 358–64.
Lang, A.R. (1959). *Acta Cryst.*, **12**, 249–50.
Laue, M. von (1960). *Röntgenstrahlinterferenzen.* Akadem. Verlagsges., Frankfurt am Main.
Newkirk, J.B. (1958). *Phys. Rev.*, **110**, 1465–6.
Ramachandran, G.N. (1944). *Proc. Ind. Acad. Sci.*, A**19**, 280.
Renninger, M. (1962). *Phys. Lett.*, **1**, 104–6.
Sauvage, M. and Petroff, J.F. (1980). In *Synchrotron radiation research* (ed. H. Winnick and S. Doniach), pp. 607–38. Plenum, New York.
Sauvage, M., Authier, A., and Petroff, J.F. (1981). *Inst. Phys. Conf. Ser.* No. 60, Section 5, 249–58.
Zarka, A., Capelle, B., Detaint, J., and Schwartzel, J. (1988a). *J. Appl. Cryst.*, **21**, 967–71.
Zarka, A., Capelle, B., Zheng, Y., Detaint, J., and Schwartzel, J. (1988b). *Proc. 42nd Annual Frequency Control Symposium*, pp. 85–92.

9

Multiple diffraction of X-rays and the phase problem

R. Colella

1. Introduction

In this chapter we will discuss a particular aspect of dynamical theory, namely, the use of multiple diffraction of X-rays in crystals for a solution of the phase problem. This subject, multiple Bragg scattering, a situation in which more than one set of crystallographic planes satisfy Bragg's law simultaneously, was an area of diffraction theory in which Paul Ewald applied himself during the last years of his professional life.

In 1968 a special issue of *Acta Crystallographica* was published in honour of Ewald's eightieth birthday, with contributions in dynamical theory from many crystallographers all over the world. The first two papers in this issue (Ewald and Héno 1968; Héno and Ewald 1968) were his own contributions, dealing with the three-beam diffraction case, without and with absorption. The mathematics in those papers is complex, because Ewald, used to solving problems in a radical way, was seeking analytical solutions, which are very hard to come by in the multi-beam case.

Needless to say, those two papers were a great stimulus in my turning to this problem in later years, and represent one of the many examples in which Ewald's legacy is still alive in today's developments of dynamical theory.

2. General statement of the phase problem in diffraction

The structures of very small objects, such as, for example, the lattice cells of a crystal, are normally determined by means of scattering experiments.

Formally, scattering theory gives for the amplitude of an electromagnetic wave S, scattered by a charge distribution $\rho(\mathbf{r})$, in the case of weak interaction,

$$S(\mathbf{K}) = \int \rho(\mathbf{r}) e^{i\mathbf{K}\cdot\mathbf{r}} \, d\mathbf{r} \tag{1}$$

where $\mathbf{K}$ is the momentum exchanged by the electromagnetic wave. The amplitude $S(\mathbf{K})$ is called the 'Fourier transform' of the distribution $\rho(\mathbf{r})$. If S is known for many values of $\mathbf{K}$, the distribution $\rho(\mathbf{r})$ may be retrieved by means of a Fourier synthesis. $S(\mathbf{K})$ is a complex number, characterized by magnitude and phase, but the usual experimental measurements provide only the intensity S^2. The missing piece of information, the phase of $S(\mathbf{K})$, corresponds to a phase shift the electromagnetic wave undergoes while being scattered by the charge distribution $\rho(\mathbf{r})$, and it is an essential ingredient in the process of reconstructing $\rho(\mathbf{r})$.

This does not mean, of course, that there are no methods to circumvent the problem. After all, structural crystallography is a highly developed science, and the determination of atomic sites within a crystal cell is an almost routine operation, in most cases. The approach universally adopted by today's crystallographers is called the 'direct method', and was invented by J. Karle and H. Hauptman, who were awarded the 1985 Nobel prize for this invention. It is a statistical–mathematical method, in which the phases of all reflections are determined as the 'most probable' ones according to the principles of probability theory applied to crystallography. It is a very ingenious method, and is applicable with success to a great number of cases.

It would be desirable, however, to make use of a 'physical' method, in which phases could be determined through the effects produced on a physical property susceptible of 'direct' measurements.

A method of this kind, which for several years has been the subject of considerable investigation and development, is based on **anomalous dispersion** of X-rays; this occurs when the X-ray photons have energies close to the atomic energy levels in the crystal, and resonance effects arise. This method is especially feasible with synchrotron light, because of the tunability feature offered by such a source. It is a promising method, but limited to relatively heavy atoms, whose characteristic frequencies correspond to wavelengths around 1 Å, a region that is suitable for crystallographic work. It is not applicable to organic crystals, consisting of carbon, hydrogen, and oxygen atoms, because in this case the characteristic frequencies are much lower. These are, however, the new compounds produced every day by research laboratories to satisfy the needs of pharmaceutical and biochemical industries.

Another method of a physical nature makes use of simultaneous Bragg scattering, and is based on interference effects between the different diffracted beams coexisting within the crystal. In this technique the 'reference beam', so to speak, is inside the crystal, and is generated by the mechanism of multiple diffraction.

This is a method we consider very promising, and it will be described in detail in this chapter.

The idea was suggested many years ago (Lipscomb 1949; Fankuchen and Ekstein 1949, personal communication quoted by Lipscomb) but could not be profitably exploited before the advent of electronic computers. An attempt to correlate the distortions of the Pendellösung fringes due to multiple diffraction (Hart and Lang 1961) with phases did not go very far for lack of an adequate theory.

3. General theory

The phenomenon of multiple Bragg diffraction was discovered in 1937 by the late M. Renninger, while working in Ewald's institute in Germany, and for this reason it is also called the 'Renninger effect'.

To understand correctly the essence of this phenomenon, we must visualize the diffraction process in the reciprocal space of the crystal, by making full use of the power afforded by the Ewald construction. Let us consider, for example, the case illustrated in Fig. 1(a), in which a monochromatic beam with wave vector $\mathbf{K}_o$ is incident on the (001) surface of a gallium arsenide crystal, at an angle θ satisfying Bragg's law for the (002) reflection, generated in second order by the atomic planes parallel to the surface. The dotted wave vector $\mathbf{K}_{002}$ corresponds to the diffracted beam. In Fig. 1(b), the same situation is described in reciprocal space by means of the Ewald construction. The sphere

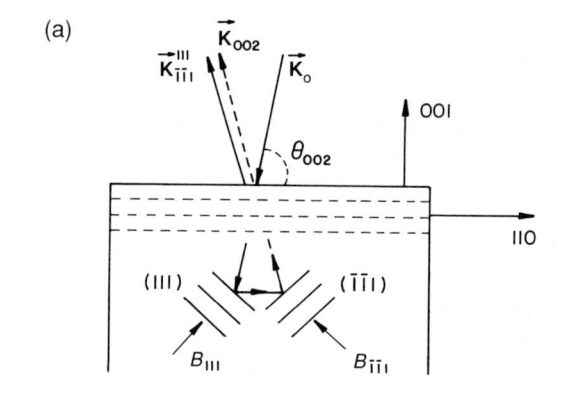

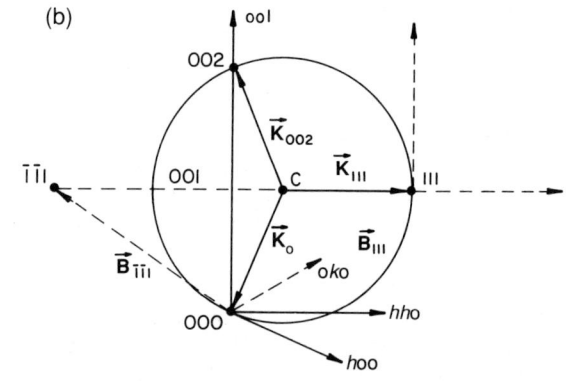

Fig. 1 Simultaneous Bragg diffraction. In (a) the real-space situation is represented. The incident beam $\mathbf{K}_o$ satisfies Bragg's law for two sets of planes simultaneously, the (002) and the (111). For convenience, the 'coplanar' case is considered, in which the normal to the (111) planes lies in the plane defined by $\mathbf{K}_o$ and the normal to the (002) planes. In (b) the same situation is represented in reciprocal space.

of reflection, with radius $1/\lambda$ (where λ is the wavelength of X-rays), passes through the origin and the (002) node. By suitable orientation of the crystal, obtained by rotation around the normal [001] to the crystal surface, it is possible to bring a third node, the (111) for example, on to the Ewald sphere. In this way a third reflection is generated, corresponding to the beam that propagates horizontally inside the crystal. In Fig. 1(b), this beam corresponds to the horizontal wave vector $\mathbf{K}_{111}$. The essence of the Renninger effect lies in the fact that, at this point, *by necessity*, there will always be a third set of crystallographic planes (the $(\bar{1}\bar{1}1)$ in our case) *whose orientation and spacing will be such as to produce a third Bragg diffracted beam, in the direction of* $\mathbf{K}_{002}$, *for which the horizontally diffracted beam acts as an incident beam.*

Now it becomes clear that the two beams, that is, the beam directly diffracted by the (002) planes and that doubly diffracted by the (111) and ($\bar{1}\bar{1}1$) planes, are *coherent* and *exactly superimposed*, inside and outside the crystal, and therefore they will be able to interfere, and affect the intensity of the beam associated with the wave vector $\mathbf{K}_{002}$. The effect can be easily observed while the crystal is slowly rotated around the [001] normal, which keeps θ constant, and it will be manifested by peaks and dips for particular values of the azimuthal angle of rotation. In gallium arsenide the (002) is very weak, and the (111) rather strong. In this situation, a pronounced peak of intensity is normally observed as the (111) node goes through the sphere of reflections. This effect was called by Renninger **Umweganregung (Detour Radiation)**, a term still widely used today. The existence of a third set of lattice planes such as the ($\bar{1}\bar{1}1$) becomes obvious from inspection of Fig. 1(b). When the (111) reflection is excited in the crystal, a strong X-ray beam propagates inside the crystal in the direction of $\mathbf{K}_{111}$. It is useful to think, at this point, that an incident X-ray photon has two choices. The first one, with small probability, is to be directly scattered as (002). The second one, with higher probability, is to be scattered as (111), giving rise, then, to *a new incident beam inside the crystal*. Using the Ewald construction in the usual way, we can then transfer the origin of reciprocal space to the tip of $\mathbf{K}_{111}$, and *use the same sphere of reflection*. The (111) node now becomes the new origin. The other two nodes, (002) and (000), represent two beams propagating inside the crystal, with new indices, as a result of the new choice for the origin. The beam that interferes with that *directly* scattered by the (002) planes will have Miller indices given by the difference 002 – 111 = $\bar{1}\bar{1}1$. This is why, together with the 'main' reflection, (002), and the simultaneous reflection, (111), a third 'coupling' reflection is always involved, the ($\bar{1}\bar{1}1$).

It is clear that this 'sequential' view of the diffraction process inside the crystal does not correspond to reality, even though it helps our intuition to understand what is going on. The three beams mentioned above are never physically separated. They exist simultaneously within each crystal domain, however small it may be.

It is also clear that, in the process of multiple diffraction, the phases of the various reflections are involved in an essential way. In the example of Fig. 1 the beam diffracted by the ($\bar{1}\bar{1}1$) planes interferes with that 'directly' diffracted by the (002) planes, and the intensity emerging from the crystal depends on the difference between the phase shifts associated with these two lattice planes.

An adequate theory is needed to understand the experimental results. The theory of electromagnetic waves propagating in a crystalline medium is called the **dynamical theory** of diffraction, and was developed by Ewald in the period from 1916 onward. The first nucleus of such a theory was in fact the topic of his Ph.D. dissertation. Professor Ewald once said jokingly, in a seminar given at Cornell University in autumn 1967, that he had been working all his life at finishing his thesis, and the job was not yet completed at that time! Dynamical theory was re-elaborated by Max von Laue a few years later [1931–36] in a form that is currently used today. Dynamical diffraction theory is entirely analogous to band theory of electrons in crystals. The electromagnetic wavefield is expanded in a series of plane waves (called Ewald waves), the analogue of Bloch waves for electrons, whose wave vectors differ by reciprocal lattice vectors. The electric displacement is given by

$$\mathbf{D}(\mathbf{r}, t) = \sum_{\mathbf{H}} \mathbf{D}_{\mathbf{H}} \exp\{2\pi i[(\boldsymbol{\beta}_o + \mathbf{H}) \cdot \mathbf{r} + \nu t]\}, \quad (?)$$

where the sum is extended to all nodes $\mathbf{H}$ of the reciprocal lattice, $\boldsymbol{\beta}_o$ is the wave vector of the refracted beam inside the crystal generated by the outside incident beam $\mathbf{K}_o$, and ν is the frequency of X-rays. As in the case of band theory for electrons in crystals, (2) gives rise to a secular determinantal equation.

The most common version of dynamical theory is the so-called 'two-beam case', the most usual situation in which only one node, besides the origin, is excited. In this case, there are analytical solutions which allow one to calculate the intensities of all diffracted beams.

The most general case, with $n > 2$, is not, in general, amenable to analytical formulations. A number of years ago I developed the general formalism (Colella 1974) which is now codified in a FORTRAN program (called NBEAM) for electronic computers of average size.

4. Some applications

A convenient method for observing phase effects is to measure the diffracted intensity of the 'main beam' (see Fig. 1(a)) as a function of the azimuthal angle ϕ as the crystal is rotated around the scattering vector $\mathbf{K}_{002} - \mathbf{K}_o$, and then compare the observed variations with theory. The plot obtained in this way will be called here a **Renninger plot**. This is a widely adopted approach (Chang 1982, 1987), although the photographic method has also been used with some degree of success (Post 1977, 1979).

The ideal situation would be one in which we would be able to change the phase of a structure factor, without changing its magnitude, by simply turning a knob on the front panel of the apparatus. It turns out that such an ideal situation can be practically realized. It is offered by the 'forbidden' reflection (442) in silicon, caused by anisotropies in the charge density associated with covalent bonds (Trucano and Batterman 1972). It has been observed, as the temperature is increased above room temperature, that the (442) intensity decreases, goes to zero, and then increases again with temperature. This phenomenon is interpreted as a competition between two effects with opposed signs. The interatomic charge build-up is temperature independent, whereas the effect of anharmonic vibrations is strongly temperature dependent, and tends to decrease, in the time average, the electron density between nearest neighbours by increasing the interatomic distance as temperature is increased. By a suitable choice of two temperatures at which (442) has the same intensity, 300 and 700 K, we have a situation in which the magnitude F_{442} is the same, but the phases differ by π. It will be recalled here that in centrosymmetric structures, such as silicon, there are only two possible phase values: zero or π.

The experiment has been performed using synchrotron light at Cornell University (Tischler et al., 1985) and Fig. 2 shows the effect clearly. In the proximity of the azimuthal angle $\phi \approx 3.5°$ there is a strong peak in the Renninger plot, as a result of a simultaneous reflection, the $(\bar{1}1\bar{1})$.

What matters here are the details *on the sides* of the Renninger peak. A noticeable asymmetry is clearly apparent, and the character of the asymmetry is reversed at the two temperatures. In both profiles the continuous line is the theoretical re-

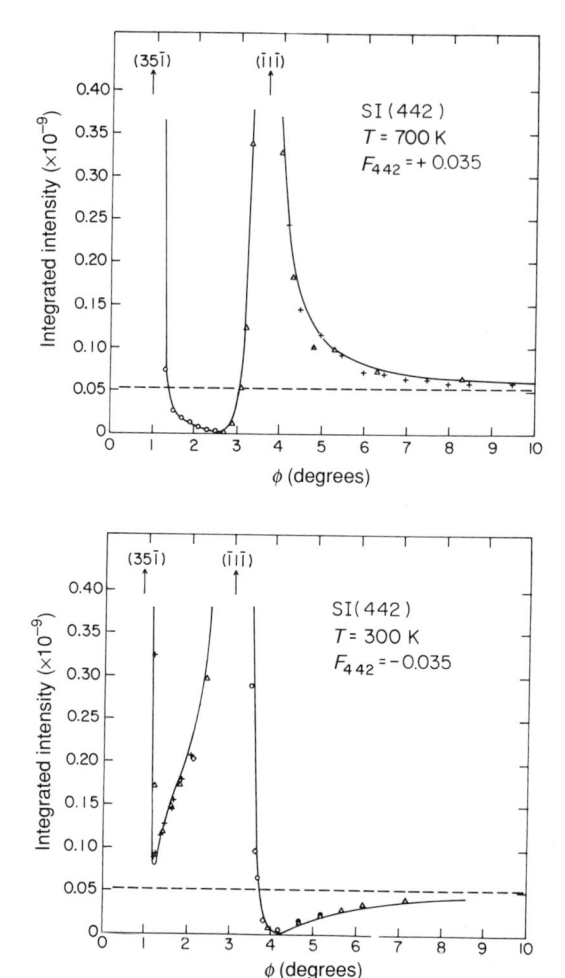

Fig. 2 Intensity of the (442) reflection in silicon as a function of the azimuthal angle ϕ, at two temperatures, in proximity of a strong Umweganregung peak, the $(\bar{1}1\bar{1})$. The dotted horizontal line corresponds to the 'two-beam value', obtained when no other node except for the (442) lies on the Ewald sphere.

sult, when the value of F_{442} is fed in with opposite signs at the two temperatures. The agreement between theory and experiment is perfect.

At this point we felt encouraged to try a more difficult case. The experiment described above was performed using silicon, a very perfect crystal, with no lattice defects of any sort. The theory we used, the *n*-beam dynamical theory, can only be applied, strictly speaking, to perfect crystals, such as silicon, germanium, and a few others. The problem is how

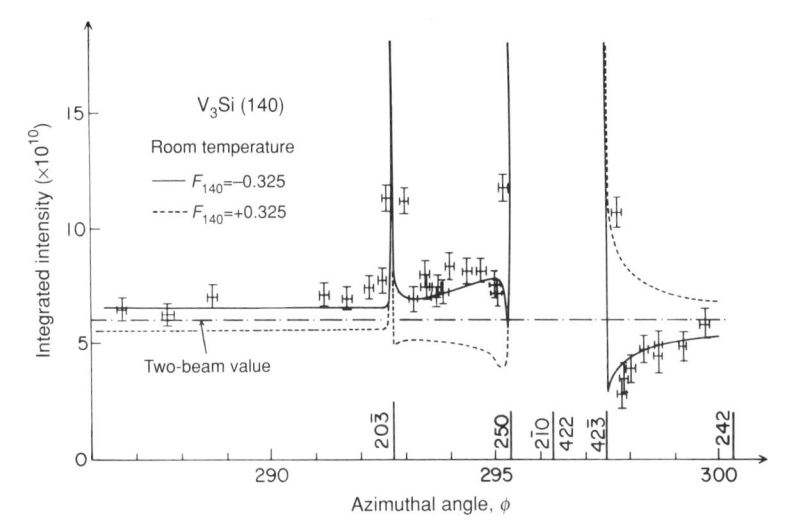

Fig. 3 Intensity of the (140) reflection in V$_3$Si as a function of the azimuthal angle. The smooth solid and dotted lines represent calculated values for negative and positive F_{140}, respectively. The horizontal dot–dashed line corresponds to the two-beam value. The area between the Umweganregung peaks (250) and (42$\bar{3}$) has a complex structure, and has not been explored.

to use this method in the most general situations, when dealing with new organic and inorganic compounds, which inevitably fall in the same class of ordinary not-so-perfect crystals as copper, aluminium, sodium chloride, etc.

That the asymmetry effect should be visible in mosaic crystals was apparent, for example, in the data taken by Gong and Post (Gong and Post, 1983; Post 1983) but it was not clear how the available theory could be applied to such situations.

This problem was a great concern for us, and for some time stopped any further progress because we did not see any way out of it. Further investigations, however, allowed us to conclude in 1981 (Chapman et al., 1981) that if we confine our attention to the neighbourhood of a strong Renninger peak superimposed on a weak 'main' reflection, dynamical theory should be applicable irrespective of crystal perfection (within certain limitations), and that the phase-related asymmetry effects should be visible with imperfect (or mosaic) crystals as well.

The important point to realize here is that, when the 'main' reflection is weak, and a strong Renninger peak is weakly excited because the crystal is purposely mis-set from full excitation, the global interaction between photons and crystal is *weak*. In such a situation each photon is scattered only once, and both dynamical and kinematic theory converge to the same intensity values.

The next step was an attempt to see phase effects in a mosaic crystal, V$_3$Si, a centrosymmetric structure whose Bravais lattice is a simple cube (Schmidt and Colella 1985). The reflection under consideration was the (140), a very weak 'forbidden' reflection due to anisotropies in the thermal vibrations. Contrary to the case of silicon (442), in this case the sign of the structure factor was not known *a priori*. Figure 3 shows that, in this case too, noticeable asymmetries are visible in the Renninger plot, and that between the two possibilities indicated by the dotted and continuous lines, the experimental data unquestionably agree with the curve corresponding to a negative structure factor.

The experiment of Fig. 3 was very important, because it enabled us to conclude that even in a mosaic crystal phase effects are visible, which was not at all clear initially.

As a next step we tried an organic non-centrosymmetric crystal that could be considered a typical representative of a wide class of compounds for which solution of the phase problem plays a substantial part in the process of determining the crystal structure. We considered benzil, C$_{14}$H$_{10}$O$_2$, a crystal with the same space group as quartz, a well-known structure.

Organic crystals often have a large number of atoms in the unit cell—benzil has 78 atoms—and as a consequence the dimensions of the unit cell are large. A problem immediately apparent in dealing with organic crystals is that their reciprocal space is densely populated, and the number of simultaneous reflections per degree of azimuthal rotation is large. In benzil, for example, a Renninger peak is expected on average every 4 minutes of

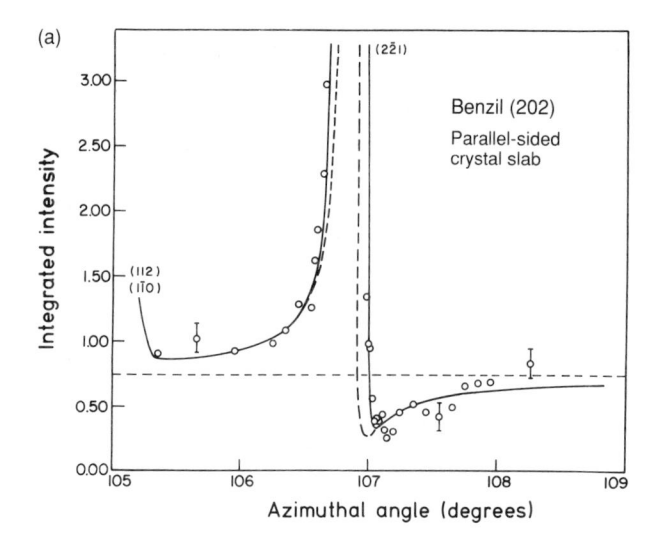

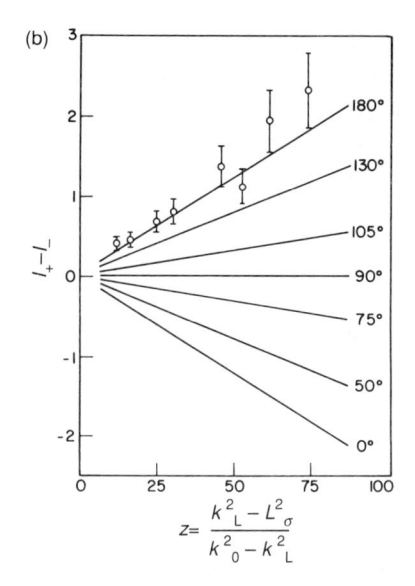

Fig. 4 (a) Integrated intensity of the (202) reflection in benzil, as a function of the azimuthal angle. The continuous curve is calculated from theory, on the basis of the commonly accepted structure. The horizontal dashed line represents the two-beam value. (b) Plot used to obtain the phase of the (202). On the ordinate axis the plotted values are the differences between experimental points symmetrically located on the left and on the right of the value ϕ_0 corresponding to the maximum excitation of the Umweganregung peak ($2\bar{2}1$). The abscissa z is approximately inversely proportional to the azimuthal angle ϕ. (See Shen and Colella 1988.)

arc, which is just barely more than the mosaic spread, about 2–3 minutes of arc.

An additional difficulty with organic crystals is that the asymmetry effect is not visible at ordinary wavelengths, because it is obliterated by the mosaic spread of the crystal.

A physical explanation for this fact is that the ratio between structure factors and unit cell is small compared with that for standard inorganic crystals, with the consequence that all rocking curves are extremely narrow. In other words, the 'size' of nodes in reciprocal space is, on average, very small for benzil and other similar organic crystals.

It turns out that using long-wavelength X-rays has the effect of increasing that size. All rocking curves become wider, and the asymmetry effect extends over a larger angular range of the azimuthal angle. In the case of benzil, an experiment performed with synchrotron radiation at $\lambda = 1.5$ Å failed to show any asymmetry effect, but when the wavelength was increased to $\lambda = 3.5$ Å, the effect became clearly visible (Shen and Colella 1988; see Fig. 4(a)).

The asymmetry effect can be described analytically by means of perturbation theory (Shen 1986). The method is similar to the Born approximation in quantum scattering theory at optical frequencies. Maxwell's equations are solved using a Green function method. As this method deals with wavefields directly, rather than seeking first the equation of the dispersion surface, it eliminates the need for component equations and diffraction geometries to be specified before obtaining the final results. The final expression for the wavefields is in a simple vector form that contains a summation over all the perturbing reciprocal nodes, which is analogous to the expression for a perturbed wave function in quantum mechanics.

Previous treatments, based on different methods, have provided approximate solutions to the dynamical eigen-equations for a three-or-more beam case, in which the additional reciprocal nodes were treated as perturbations (Juretschke 1982, 1984; Høier and Marthinsen 1983; Hümmer and Billy 1986).

Using this theory, which was developed by Q. Shen in the course of his Ph.D. work at Purdue University (Shen 1986), it is possible to construct a graph using all experimental points, from which the phase of the simultaneous reflection, referred to that of the 'main' reflection, can be immediately obtained. Figure 4(b) shows how this can be done. The value we obtain here, 180°, is expected on the basis of the known structure of benzil.

Up to this point, all experiments mentioned in this chapter were performed using large flat-sided crystals, with the surface cut parallel to a well-defined set of crystallographic planes (hkl). As crystallographers always use small crystalline fragments of spherical, cylindrical, or sometimes completely irregular shape, it was important to verify whether the shape of the crystal had any bearing on the asymmetry effect. The perturbation theory developed by Q. Shen has no boundary conditions, which led us to believe that crystal shape was immaterial.

However, an experimental proof was needed. A small spherical benzil crystal (diameter 0.3 mm) was prepared and disposed on a very intense X-ray beam provided by the Stanford Synchrotron Radiation Laboratory. The beam was monochromatic and tuned to $\lambda = 3.5$ Å.

The results obtained were identical to those of Fig. 4 (Shen and Colella 1987).

Recent computer simulations (Colella 1989 and unpublished) of n-beam experiments with protein crystals have indicated that phase effects are indeed visible through the asymmetry effect, on selected combinations of main and simultaneous reflections. As phases in proteins are determined by means of isomorphous replacement, a phase determination based on multiple Bragg scattering would help in assessing the degree of strain introduced by heavy atoms in protein structures. Phase effects in proteins have indeed been observed recently (Chang et al. 1991; Hümmer et al. 1991).

We can then conclude that, under suitable conditions, and irrespective of crystal shape, multiple Bragg scattering can provide useful phase information on perfect or mosaic crystals, and for organic or inorganic structures.

5. Conclusions

When Paul Ewald invented dynamical theory, back in 1916–17, there were not many cases in which his theory could be usefully applied, because it was immediately recognized that ordinary crystals were not sufficiently perfect for effects such as anomalous transmission or Pendellösung fringes to be observable. The best specimens available were rare samples of calcite crystals that exhibited very narrow rocking curves, but in general they were not good enough to allow a widespread verification of all dynamical effects.

We had to wait until the late 1950s, when the semiconductor industry provided us with a wealth of highly perfect specimens of silicon and germanium, to appreciate the power and the beauty of dynamical theory. The Pendellösung effect was promptly observed (Kato and Lang 1959), and the theory was further developed and applied to many specific situations.

Despite the numerous and ever-increasing applications of dynamical theory, perfect crystals are still a rather limited class of the crystals used in solid-state research.

I believe that we are entering now a new age of diffraction physics, in which dynamical theory will be increasingly applied to real, i.e. mosaic crystals.

In this chapter I have shown how phases of X-ray reflections can be retrieved by experiment when this is performed under conditions of kinematic scattering. We know that, under these conditions, kinematic and dynamical theory converge to the same result, but dynamical theory is the only one that preserves phase information.

Recently, a standing wave pattern has been observed under conditions of grazing incidence, using a metal alloy crystal, with non-negligible mosaic spread (Dosch et al. 1986). In another case, a standing wave pattern was observed using the opposite concept, namely, 90° diffraction (Woodruff et al. 1987). The crystal was copper, not silicon or germanium.

In all these cases, and I am sure that we will see more in the near future, we see dynamical effects, such as standing waves, observed in real crystals of average degree of perfection, with some mosaic spread. We see then that, under special circumstances, the language and concepts of dynamical theory can be used in situations that initially did not seem to be relevant to it.

It is fitting to conclude this chapter by recognizing how much diffraction physics has benefited from Ewald's dynamical theory during the last thirty years in applications to perfect crystals. We are looking forward to the new insights waiting for us when further applications will be found in the wider area of real crystals.

References

Chang, S.L. (1982). *Phys. Rev. Lett.*, **48**, 163–6.
Chang, S.L. (1987). *Cryst. Rev.*, **1**, 87–184.
Chang, S.L., King, Jr., H.E. Huang, M.T., and Gao, Y. (1991). *Phys. Rev. Lett.*, **67**, 3113–16
Chapman, L.D., Yoder, D., and Colella, R. (1981). *Phys. Rev. Lett.*, **46**, 1578–81.
Colella, R. (1974). *Acta Cryst.*, A**30**, 413–23.
Colella, R. (1989). *Bull. Amer. Phys. Soc.*, **34**, 418, N. 3, paper A 16–5.
Dosch, H., Batterman, B.W., and Wack, D.C. (1986). *Phys. Rev. Lett.*, **56**, 1144–7.
Ewald, P.P. and Heno, Y. (1968). *Acta Cryst.*, A**24**, 5–15.
Gong, P.P. and Post, B. (1983). *Acta Cryst.*, A**39**, 719–24.
Hart, M. and Lang, A.R. (1961). *Phys. Rev. Lett.*, **7**, 120–1.
Héno, Y. and Ewald, P.P. (1968). *Acta Cryst.*, A**24**, 16–42.
Høier, R. and Marthinsen, K. (1983). *Acta Cryst.*, A**39**, 854–60.
Hümmer, K. and Billy, H. (1986). *Acta Cryst.*, A**42**, 127–33.
Hümmer, K., Schwegle, W., and Weckert, E. (1991). *Acta Cryst.*, A **47**, 60–2.
Juretschke, H.J. (1982). *Phys. Rev. Lett.*, **48**, 1487–9.
Juretschke, H.J. (1984). *Acta Cryst.*, A**40**, 379–89.
Kato, N. and Lang, A.R. (1959). *Acta Cryst.*, **12**, 787–94.
Lipscomb, W.N. (1949). *Acta Cryst.*, **2**, 193–4.
Post, B. (1977). *Phys. Rev. Lett.*, **39**, 760–3.
Post, B. (1979). *Acta Cryst.*, A**35**, 17–21.
Post, B. (1983). *Acta Cryst.*, A**39**, 711–18.
Schmidt, M.C. and Colella, R. (1985). *Phys. Rev. Lett.*, **55**, 715–17.
Shen, Q. (1986). *Acta Cryst.*, A**42**, 525–33.
Shen, Q. and Colella, R. (1987). *Nature* (London), **329**, 232–3.
Shen, Q. and Colella, R. (1988). *Acta Cryst.*, A**44**, 17–21.
Tischler, J.Z., Shen, Q., and Colella, R. (1985). *Acta Cryst.*, A**41**, 451–3.
Trucano, P. and Batterman, B.W. (1972). *Phys. Rev.*, B**6**, 3659–66.
Woodruff, D.P., Seymour, D.L., McConville, C.F., Riley, C.E., Crapper, M.D., Prince, N.P., and Jones, R.G. (1987). *Phys. Rev. Lett.*, **58**, 1460–2.

10

Paul Ewald and the dynamical theory of electron scattering

J.M. Cowley and A.F. Moodie

Introduction

It is one of the marks of the great scientist that with the passage of time specific aspects of his work extend and branch and, not infrequently, emerge in fields unknown at the time of the original discoveries. Examples of this can be found throughout the work of Paul Ewald, but few of these can be more striking than that afforded by his two-beam approximation with its celebrated coupled pendulum solution and its vivid geometry.

Ewald, of course, devised the approximation in the course of his description of the propagation of X-rays through crystals, and here the approximation holds to very high accuracy. In the propagation of fast electrons through crystals, on the other hand, the approximation, at best, is only semi-quantitative, and, at worst, has no range of validity whatsoever. Nevertheless, the approximation found its first applications in fast electron diffraction and microscopy, the topic of this chapter, and still retains its utility over a great part of the field. Indeed, the range of application can be seen to increase with the passage of time.

There are many reasons for this apparently paradoxical circumstance, and a few, it is hoped the more important of these, will be explored in this outline.

Here it might be advantageous to sketch the relationship between the two-beam approximation and the full dynamical theory as it applies to the scattering of fast electrons. Bethe (1928), of course, established the foundations for the theory of the scattering of electrons by crystals, and his work, after the passage of more than sixty years, still illuminates the subject. Ewald's influence in the earliest days of electron diffraction has been well documented by Bethe himself (Bethe 1981), and, over the subsequent years, this influence increased; as often through informal discussion as through formal publication.

As Ewald's key papers on the dynamical theory have now been translated and published in convenient form (Ewald 1991), his elegant procedures will not be duplicated. Instead, an attempt will be made to recall his interest in alternative derivations, particularly those that lay emphasis on underlying symmetry.

To well within the experimental accuracies so far achieved, the Dirac equation can be replaced by a Klein–Gordon equation for the electron wave function ψ, written

$$\left(\frac{\partial^2}{\partial z^2} + H\right)\psi = 0,$$

$$H \equiv \Delta + k^2[1 + \phi/W],$$

with z normal to the crystal surface, $\Delta \equiv (\partial^2/\partial x^2) + (\partial^2/\partial y^2)$, W the accelerating potential, ϕ the potential distribution within the crystal, and k the scalar wave number.

Putting

$$\begin{bmatrix} \mu^+ \exp(ikz) \\ \mu^- \exp(-ikz) \end{bmatrix} = G \begin{pmatrix} \psi \\ \frac{\partial \psi}{\partial z} \end{pmatrix},$$

$$\frac{\partial}{\partial z}\begin{pmatrix} \mu^+ \\ \mu^- \end{pmatrix} \equiv \frac{\partial}{\partial z}|\mu^\pm\rangle,$$

and choosing

$$G = \begin{bmatrix} 1 & -ik^{-1} \\ 1 & ik^{-1} \end{bmatrix},$$

$$\frac{\partial}{\partial z}|\mu^\pm\rangle$$

$$= i\left(\frac{1}{2k}\Delta + \frac{k}{2}\frac{\phi}{W}\right)\begin{bmatrix} 1 & -\exp(-2ikz) \\ -\exp(2ikz) & -1 \end{bmatrix}|\mu^\pm\rangle.$$

For fast electrons, k is large enough to reduce the coupling of forward and backward scattering to negligible levels, leading to the parabolic approximation

$$\frac{\partial}{\partial z}|\mu^{\pm}\rangle = \pm i\left(\frac{1}{2k}\Delta + \frac{k}{2}\frac{\phi}{W}\right)|\mu^{\pm}\rangle. \quad (1)$$

These equations describe the elastic scattering of fast electrons to high accuracy, solutions have been obtained using a variety of techniques, and efficient programs have made numerical evaluation a routine process. In a typical calculation, and for an accelerating voltage of 200 kV, several hundred coefficients will normally be included in the Fourier expansion of the wave function, although there is little difficulty in extending the number to several thousand, and, with a large unit cell or with elements of high atomic number, this option will often be preferred. At higher accelerating voltages, and near a zone axis, an even greater number will usually be required to preserve unitarity.

The two-beam approximation

In those circumstances, it may seem surprising that an approximation involving only two diffracted beams should retain any currency whatsoever, but it does, and the reasons for this are to be found in the fundamental structure of the approximation as, indeed, Bethe effectively demonstrated in his foundation paper. The simplest form of the two-beam approximation, and the one that concerns us here, can be obtained immediately by obtaining the Fourier transform of the parabolic equation with respect to x and y, ignoring the z dependence of the potential, and imposing the boundary conditions appropriate to a plate infinite in the x,y-plane. This leads to

$$\frac{d|u\rangle}{dz} = iM|u\rangle, \quad |u\rangle_0 \equiv \begin{pmatrix} 1 \\ 0 \\ 0 \\ \vdots \end{pmatrix},$$

with the solution (e.g. Bellman 1960),

$$|u\rangle = \exp(iMT)|u\rangle_0,$$

$$\equiv \left[\sum_{n=0}^{\infty} \frac{(iMT)^n}{n!}\right]|u\rangle_0.$$

Here T is the thickness of the crystal measured along z, and, for N interacting beams, M is an $N \times N$ matrix with diagonal elements $a_{ii} = \pi\zeta_i$ deriving from the kinetic energy, and off-diagonal elements $a_{ij} = \sigma V(\mathbf{i} - \mathbf{j})$ deriving from the potential energy. The

$$\zeta_i \equiv \zeta_g = (k^2 - |\mathbf{k} + 2\pi\mathbf{g}|^2)/(4\pi k_z)$$

are the excitation errors, quantities introduced by Ewald to describe the deviations from the Bragg conditions, and $\sigma = 2\pi me\lambda/h^2$ is the interaction constant.

When it is assumed that there are only two beams interacting strongly, that is, the central beam and one diffracted beam, the solution can be written

$$\begin{pmatrix} u(0) \\ u(g) \end{pmatrix} = \exp\left\{i\begin{pmatrix} 0 & \sigma V(\bar{g}) \\ \sigma V(g) & \pi\zeta \end{pmatrix}T\right\}\begin{pmatrix} 1 \\ 0 \end{pmatrix}.$$

The exponential can be evaluated using the anti-commuting properties of the Pauli bases,

$$\sigma_1 = \begin{pmatrix} 0 & 1 \\ 1 & 0 \end{pmatrix}, \sigma_2 = \begin{pmatrix} 0 & i \\ -i & 0 \end{pmatrix}, \sigma_3 = \begin{pmatrix} -1 & 0 \\ 0 & 1 \end{pmatrix}.$$

Putting

$$V(g) = V^R + iV^I, \quad E = \begin{pmatrix} 1 & 0 \\ 0 & 1 \end{pmatrix},$$

then

$$\exp\left\{i\begin{bmatrix} 0 & \sigma V(\bar{g}) \\ \sigma V(g) & \pi\zeta \end{bmatrix}T\right\}\exp\left(\frac{-i\pi\zeta}{2}T\right)E$$

$$= \exp\left[i\left(\frac{\pi\zeta}{2}\sigma_3 + V^R\sigma_1 - V^I\sigma_2\right)T\right],$$

$$= (\cos \Omega^{\frac{1}{2}}T)E +$$

$$i(\sin \Omega^{\frac{1}{2}}T/\Omega^{\frac{1}{2}})\cdot\left(\frac{\pi\zeta}{2}\sigma_3 + V^R\sigma_1 - V^I\sigma_2\right),$$

with

$$\Omega \equiv [(\pi\zeta/2)^2 + \sigma^2 V(g)V(\bar{g})].$$

Explicitly,

$$\begin{pmatrix} u(0) \\ u(g) \end{pmatrix}$$

$$= \exp\left(i\frac{\pi\zeta}{2}T\right)\begin{bmatrix} \cos \Omega^{\frac{1}{2}}T - i(\pi\zeta/2)\sin \Omega^{\frac{1}{2}}T/\Omega^{\frac{1}{2}} \\ i\sigma V(g)\sin \Omega^{\frac{1}{2}}T/\Omega^{\frac{1}{2}} \end{bmatrix}. \quad (2)$$

This result, the pendulum solution, was first obtained for electrons by Blackman (1939) using Bethe's formulation, Bethe himself having been concerned with reflection rather than transmission diffraction. The equations predict that, with changing thickness, intensity should flow back and forth between the beams, and at the Bragg angle, with $\zeta = 0$, the frequency of oscillation with thickness should be minimized and the transfer of intensity should be maximized. Specifically, when the Bragg condition is satisfied, the intensities of the central and diffracted beams are predicted to be given by

$$I(0) = \cos^2(\sigma|V(g)|T),$$
$$I(g) = \sin^2(\sigma|V(g)|T).$$

Alternatively, when the angle of incidence is varied on a crystal of constant thickness, that is, ζ is varied at constant T, distinctive intensity distributions of 'shape function' type are predicted; namely,

$$I(0) = 1 - I(g)$$
$$I(g) = \sigma^2 V(g)V(\bar{g}) \sin^2[(\pi\zeta/2)^2 + \sigma^2 V(g)V(\bar{g})]^{\frac{1}{2}} T /$$
$$[(\pi\zeta/2)^2 + \sigma^2 V(g)V(\bar{g})].$$

When $\pi\zeta/2 \gg \sigma V(g)$ the distributions approximate to kinematical, or single scattering form.

Perhaps the most direct experimental method available to explore the validity of these expressions is convergent beam diffraction. Typical patterns are shown in Fig. 1. Such results are obtained by bringing a converging beam of electrons to a small crossover on a crystal. This cone of incident beams generates corresponding cones of diffracted beams which intersect the plane of the photographic film in discs of typically 0.005 radian angular diameter. The variation in intensity of specific beams with angle of incidence is then mapped in the corresponding discs, that is, $I(g)$ is mapped against $\zeta(g)$.

To approximate most closely to two-beam conditions in such experiments, it is convenient to think in terms of the geometric constructs introduced by Ewald. For fast electrons, the wavelength will be small in comparison with atomic spacings, so that, on the scale of the reciprocal lattice, the Ewald sphere will have a large radius, and hence, near a principal zone, will lie close to many reciprocal lattice points. Further, for appreciable penetration, the crystal must be thin in the direction of the incident beam so that the points of the reciprocal lattice will be extended normal to the bounding faces and a large number of beams will be excited simultaneously. It is this vivid picture that is likely to be in the microscopist's mind as he tilts a crystal into a zone axis orientation. A bright arc of reflections will be seen on the viewing screen, closing into the Laue circle, which contracts to the point at which the Ewald sphere is tangential to the densely populated plane. Reflections lying on concentric circles described about the point mark the intersection of the sphere with successive upper layers in the reciprocal lattice—usually known to electron diffractionists as higher-order Laue zones.

Attention is now focused on a low-order reflection defined by **g**, and the Laue circle is expanded till it passes through the end-point of the vector. Although the Bragg condition for this reflection will now be satisfied, it will, in general, be satisfied, or nearly satisfied, for many others. Interaction with these beams may be reduced by a rotation about **g**, although coupling with systematic reflections, that is, those in the row in reciprocal space defined by **g**, will necessarily persist.

Evidently, the more sparsely reciprocal space is populated, that is, the smaller the unit cell of the crystal, and, broadly, the higher its symmetry, the more closely two-beam diffraction conditions can be approached.

In practice, the form of the two-beam solution can be sufficiently well approximated to have utility in, for instance, the simple body-centred, face-centred, and hexagonal structures; the diamond structures; the sodium chloride and fluorite structure types; and in many of the perovskites.

Generalized two-beam description: weak beam pseudopotential approximation

It is, however, the form rather than the actual numerical value that is well approximated. Thus, typically, in Fig. 1(b), if the obvious non-systematic interactions are avoided, the intensity profile of the diffracted beam can be fitted reasonably well to the function $A^2 \sin^2 \gamma^{\frac{1}{2}} T/\gamma$, with A and γ parameters to be determined from the experimental data. Further reduction is possible, leading to

$$A = \sigma\{V(g) + \Delta V\}$$
$$\gamma = [(\pi\zeta/2)^2 + \sigma^2|V(g) + \Delta V|^2],$$

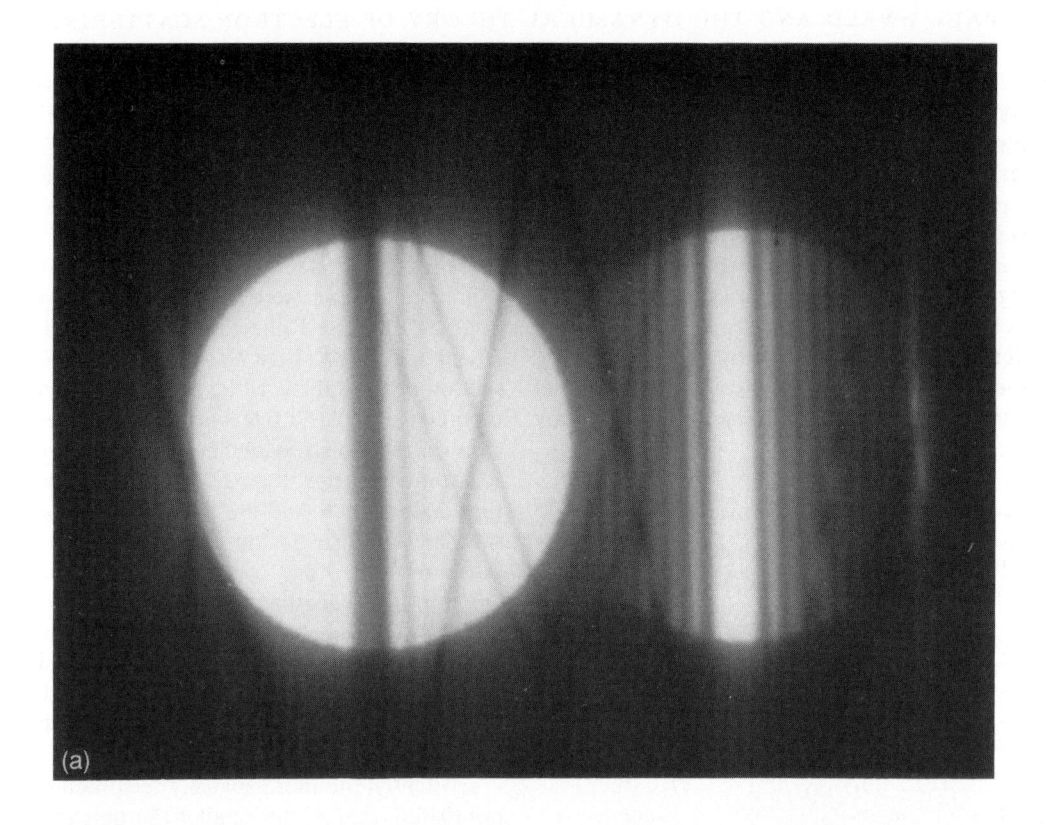

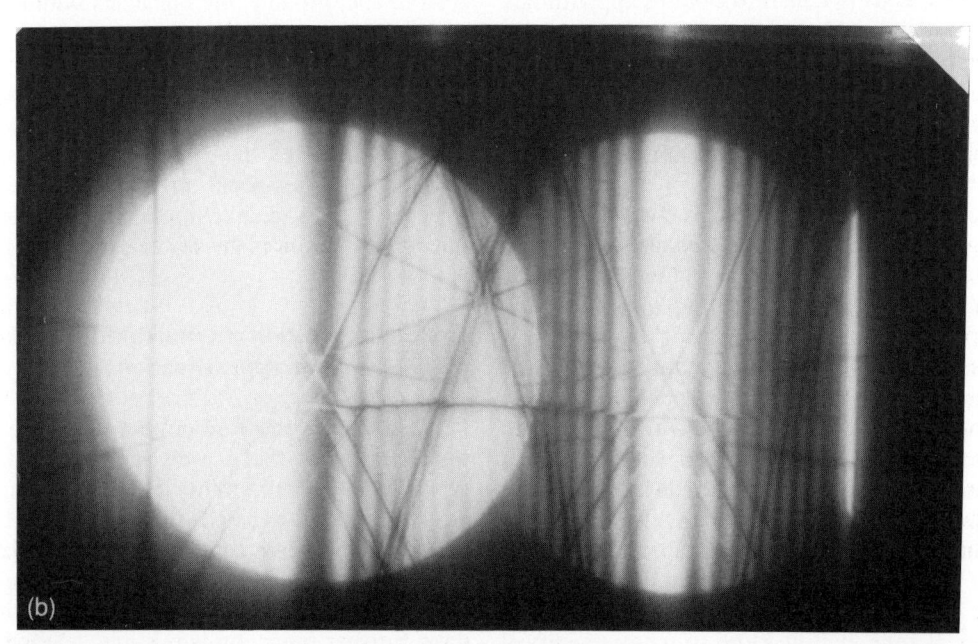

Fig. 1 The two-beam form; the beams 000 and 220 in silicon, (a) taken at an accelerating voltage of 200 kV and remote from an important zone; (b) taken at an accelerating voltage of 400 kV and near the ⟨111⟩ zone. In both (a) and (b), asymmetry in the central beam, in great part due to systematic interactions, is marked, whereas in (b) the orientation is sufficiently close to the ⟨111⟩ zone for the pseudo potential to change appreciably across the disc, with a resulting tapering in the two-beam intensity profile.

where ΔV is now to be determined from the experimental data. Thus, at least for this simple structure and at these specific orientations, the effects of the weak beams can be well approximated by the pseudo potential ΔV, a point made by Bethe (1928) in his foundation paper.

The point is sufficiently important to make closer investigation worth while. Let us suppose that the two-beam approximation in its simplest form holds to indefinite accuracy. Then, for the reflection **g**, rotation about the **g**-axis would leave the scattering unchanged, and the convergent beam diffraction pattern would take the form of parallel bars of intensity with a two-beam profile. If the weak beams involved in interaction are sufficiently remote in reciprocal space, then, over an appropriately restricted rotation about **g**, the bars will remain substantially parallel, but a pseudo potential will have to be adjoined (positive or negative) to $V(g)$ to fit the experimental data. This applies to Fig. 1(a). If the crystal is now rotated towards a principal zone, however, and if the pseudo potential model is valid, it should be possible to find a region in which the two-beam form persists, but within which the magnitude of the pseudo potential changes. The convergent beam pattern in the disc of the diffracted beam should then consist of tapering bars, each with the two-beam profile. This can be seen in Fig. 1(b), which affords a striking confirmation of the essential correctness of Bethe's insights and exemplifies the wide applicability of Ewald's two-beam constructs.

It is true that Bethe's initial treatment suffers from some technical defects (Miyake 1959), and that Bethe (1981) himself has expressed reservations about his approach to this approximation, but the experimental observations are clear. The matter has been resolved by Gjønnes (1962), who, in a powerful and elegant analysis, in terms of Green's functions, has shown that inconsistencies can be removed when the weak beams are made to conform to the boundary conditions, and that this, in turn, demands complex wave vectors, and hence an apparent absorption. Gjønnes put it in these terms:

It may appear surprising that complex wave vectors appear in a problem that is known to be soluble in terms of (an infinite number of) real wave vectors. This apparent contradiction is superficial; in the multiple wave-field picture the imaginary parts of the wave vectors represent the effects of interference between the neglected wave fields and the two wave fields of the two beam theory. This effect can evidently be described by a periodic reduction in amplitude. The effect of the imaginary part of the ΔVs is thus a 'beating' of the two beam solution, not an absorption, except for very thin crystals.

Gjønnes' analysis clarifies the remaining deviation from the classical two-beam distribution. It will be noted in Fig. 1 that, although the diffracted beam retains its symmetry about the Bragg angle, the central beam does not. The effect of the weak beams, in fact, is not only to generate a pseudo potential, but also, through an apparent absorption, to induce a Borrmann effect. In fact, weak beam interactions will often produce greater effects than 'true' phenomenological absorption (Goodman and Lehmpfuhl 1967). Such effects can be conveniently summarized by means of Feynman diagrams. Terms in the series solution for N-beam diffraction are arranged in hierarchies of loops, and as far as possible summed, starting with the single-process g two-beam loops, 2, 4, 6 ... for the central beam, and 1, 3, 5 ... for the diffracted beam, and then continuing through the two-process h, h–g loops to processes of higher order. The loops may be treated as confluences, so that Taylor's series are generated. This is substantially the process outlined by Gjønnes (1962), but set in different language.

Applications

Although the frequent occurrence of generalized two-beam forms in transmission electron microscopy and diffraction can be understood in terms of such arguments, the utility of the phenomenon is not immediately apparent. Indeed, it can lead to misunderstanding in the estimation of structure amplitudes and absorption coefficients.

Defects

There exists, however, a wealth of applications. Foremost among these must be set the analysis of defects by means of the classical electron microscopic techniques due to Hirsch, Howie, and Whelan (e.g. Hirsch et al. 1965). As the study of defects is of particular practical importance in close-packed materials (for instance, simple metals and alloys, high-temperature ceramics, and refractories) and in semiconductors with structures deriving from the diamond lattice, and as these substances tend to have small unit cells, it is plausible that a generalized two-beam description of contrast will provide an adequate basis for analysis. This, indeed, has proved to be the case, and the ramifications extend throughout materials science and metallurgy, and over much of solid-state physics and chemistry. In such work, of course, it is the structure of the defects which is of concern, the structure of the perfect material being sufficiently well known. Hence, in many important cases the effective structure amplitude (i.e. $V(g) + \Delta V$) and the phenomenological absorption coefficients can be treated as, in effect, adjustable parameters, without prejudice to the final results.

With these provisos, a description of the contrast to be expected in the electron microscope from the presence of defects in sufficiently simple structures can be set up in two-beam form, provided, of course, that the resolution is well below that required to provide an image of atomic spacings. The defining differential equations, the Howie–Whelan equations, in their simplest form may be thought of as two-beam equations with an excitation error which is a function of z. With negligible loss in accuracy at the resolutions involved, each element of intensity in the x,y-plane may be considered to be generated by an infinite plate with the excitation error of the element. This is the column approximation, which depends, essentially, on the fact that the diameter of the columns, which is determined by the resolution, is large when expressed in terms of the wavelength of the incident radiation. For isotropic elasticity in particular, simple conditions on contrast can be obtained, most notably $\mathbf{g} \cdot \mathbf{b} = 0$, the diffraction condition that must be satisfied for a screw dislocation with Burger's vector $\mathbf{b}$ to have zero contrast. This condition will, of course, be satisfied, not only for the particular reflection $\mathbf{g}$, but also for the complete line of systematic reflections.

Under conditions of general anisotropy, explicit solutions of the Howie–Whelan equations are not available, and recourse must be made to numerical methods such as those of Head *et al.* (1973). The essential two-beam character remains, however, and the materials scientist analysing a distribution of dislocations in a semiconductor or in a steel will usually start the analysis in a setting devised by Ewald, and not infrequently, use the language introduced by Ewald to describe the results.

Some hint of the wealth of detail that can be interpreted in the generalized two-beam language of Ewald, Bethe, Hirsch, Howie, and Whelan is conveyed in the bright-field images of austenite (Fig. 2) taken by third-year students of the Royal Melbourne Institute of Technology in the course of a project in materials science. The topic, in fact, now forms part of the education of applied scientists at the undergraduate level, as well as continuing to raise delicate problems in parameterization for the algebraist.

Thickness fringes

Thickness fringes are now observed so frequently with the electron microscope, and microscopists, whatever language they use to describe them, think so much in terms of them that it requires a conscious effort to recall that more than quarter of a century separated Ewald's prediction of the phenomenon and the first observations. These observations were made both in real space, where the fringes were observed directly in wedge-shaped crystals, and in reciprocal space, where the fine structure, again associated with wedge-shaped crystals, and described quantitatively by Kato (1952), was observed in cameras of high angular resolution. The reasons for this have been discussed in a previous publication of the International Union of Crystallography (Goodman, 1981); it is the breadth of application that is of interest in the present context.

At the outset, again, the applicability of generalized two-beam descriptions in sufficiently simple structures is notable. For instance, Fig. 3 is a bright-field image of a cube of magnesium oxide that has reacted with palladium at 900°C. The contrast has been reversed to show more clearly the thickness fringes in the metal. Visually, the fringes in the oxide appear to have a cosine-squared form, and hence to be two beam. Microphotometry reveals that this form is approximated reasonably closely apart from a damping factor. A pseudo potential description, in fact, again has useful validity.

The utility of this type of operation is clear. A quantitative description of morphology, apart from a scale factor, is immediately available on

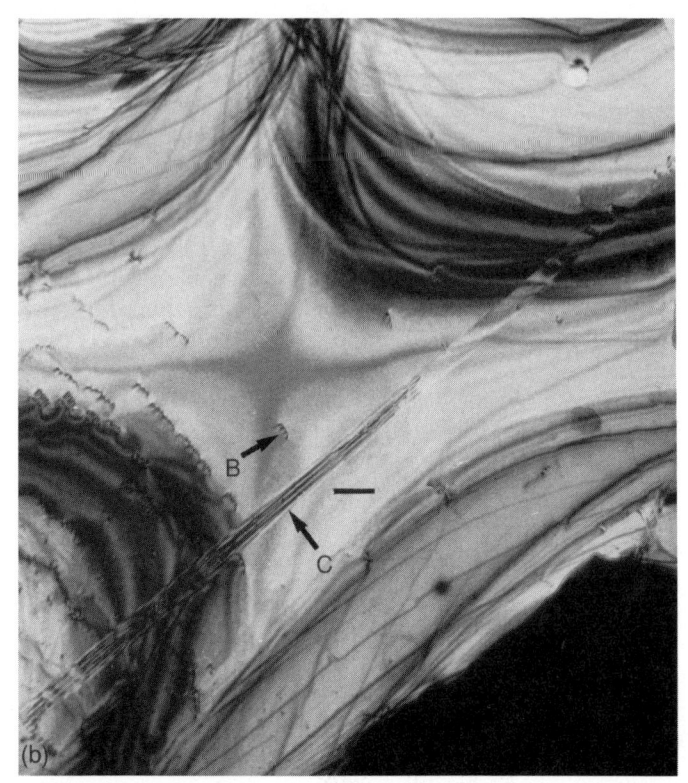

Fig. 2 Two-beam contrast; austenite in ⟨110⟩ orientation, (a) A—the increase in spacing between thickness fringes at the Bragg angle; (b) B—a screw dislocation, C—a stacking fault. The scale bar represents 5000 Å.

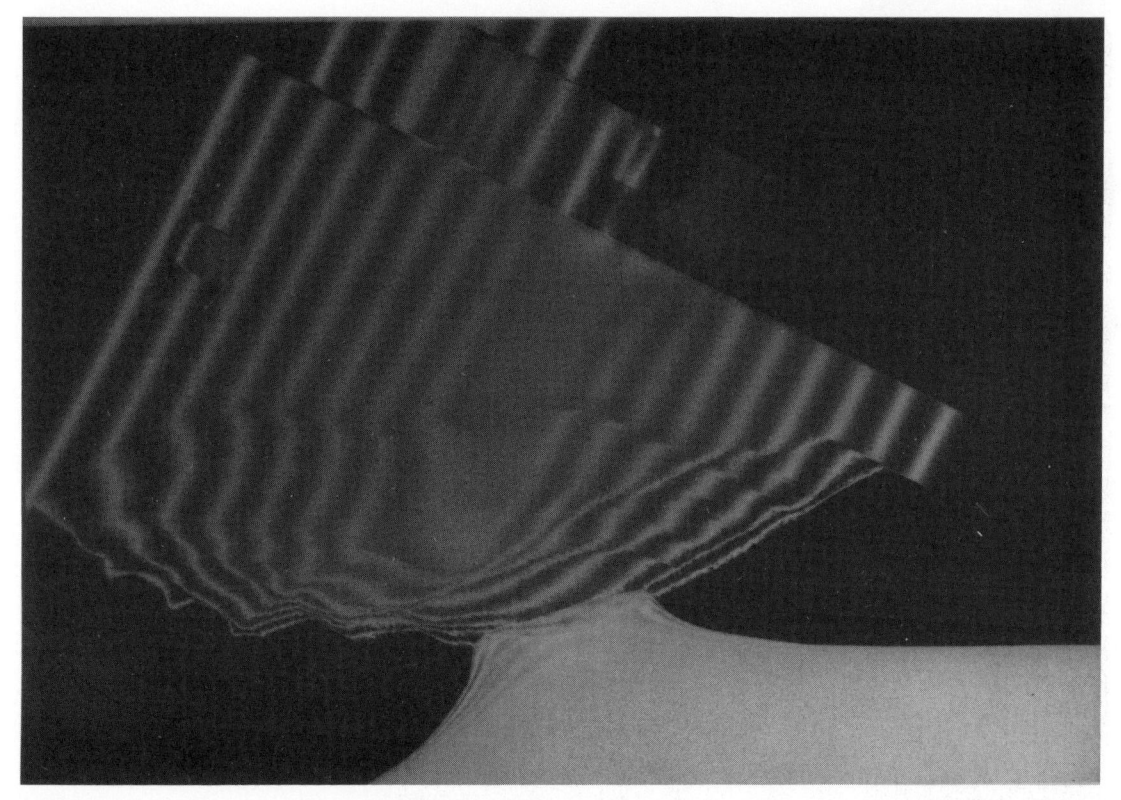

Fig. 3 Thickness fringes—the pendulum solution; thickness fringes in a cube of magnesium oxide in ⟨110⟩ orientation. At 900°C the magnesium oxide has reacted with and been etched by the palladium in the lower part of the micrograph, with which it has now formed an essentially strain-free bond. The contrast has been reversed to make the thickness fringes in the palladium more clearly visible.

interpreting the contrast as being due to fringes of equal increment in thickness, and with little additional work the scale factor can be determined to useful accuracy. In the present instance, it can be seen that the metal, well below its melting point, advanced over the oxide, etched it, and, under the action of forces analogous to those of surface tension, then contracted. The steps on the etched faces can be seen to be of {100} type, and of a few unit cells in height. A small single crystal can be seen to have grown at left centre on one of the main {100} faces, and a group of larger crystals at top centre have grown on a face orthogonal to this. Further observations of this type show that the main etched crystal is free of dislocations. It can also be seen that very little strain is built into the region where the crystals have bonded, a matter of practical importance as well as chemical interest.

Although some information (for instance, the constitution of the Pd/Mg alloy in the immediate vicinity of the interface and its crystallographic relationship to the MgO lattice) can only be obtained by more elaborate experiments involving lattice imaging at atomic resolution, further two-beam information can be obtained on orienting the crystals so that the beam passes through intergrown crystals and down the ⟨100⟩ axis. Interfacial dislocations characteristic of a small-angle twist boundary can then be observed and interpreted in the Hirsch manner.

Two-beam thickness fringes can also be observed in Fig. 2, where it can be seen that, as predicted by Ewald, the spacing is a maximum at the Bragg angle.

Many-beam fringes are normally observed, and, indeed, there are hints of this in Fig. 3, but this constitutes a separate subject. It is clear that two-beam thickness fringes can be used, virtually by

Fig. 4 Dynamical diffraction; large-angle convergent beam diffraction pattern of the forward scattered beam from silicon in the $\langle 111 \rangle$ orientation.

N-beam configurations reducible to two-beam form

Yet another aspect of two-beam work, which has led to further applications, derives from the work of Niehrs (1961), who drew attention to the fact that some typical N-beam configurations reduce to two-beam forms. The simplest example is afforded by the symmetrical three-beam configuration, the wave function for which reduces from considerable complexity (Wagenfeld 1958) to two-beam form as the excitation errors for $V(g)$ and $V(\bar{g})$ approach equality. The seven-beam symmetric case is both striking and of frequent occurrence. Here the six beams must have equal structure amplitudes and equal excitation errors, so that they lie on the vertices of a hexagon with the forward scattered beam as centre.

In practice, these conditions are approximated by $\langle 111 \rangle$ incidence on many centrosymmetric structures based on the diamond lattice. This is illustrated in Fig. 4, the large-angle zero-order convergent beam pattern of silicon taken at an accelerating voltage of 400 kV. Angular ranges over which two-beam forms constitute a reasonable approximation can be identified immediately. The corresponding pattern for the 220 reflection shown in Fig. 5 indicates the very great range over which two-beam forms can persist. Abrupt changes in pseudo potential induced by non-systematic interactions are evident, as are the angular regions within which the two-beam form has no validity whatsoever.

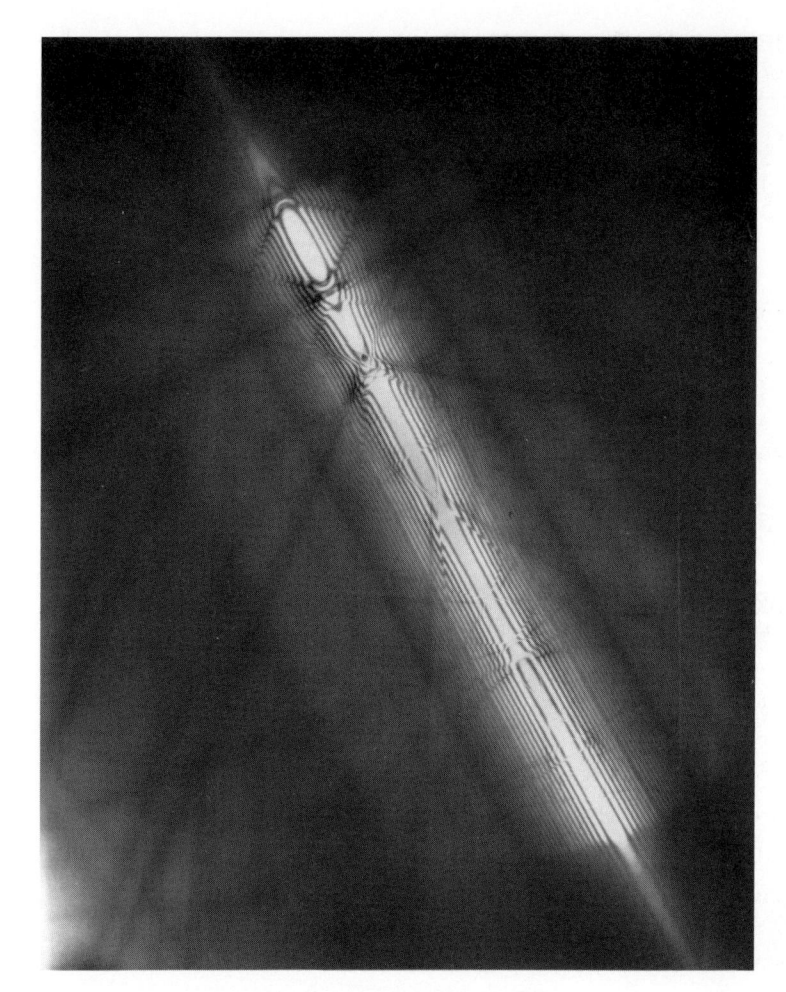

Fig. 5 The two-beam form; large-angle convergent beam pattern of the 220 reflection in silicon taken at an accelerating voltage of 400 kV. The form remains stable over a very large range of scattering angles in spite of marked changes in pseudo potential as a result of both systematic and non-systematic interactions.

This is a particularly favourable case, but approximations to this are observed in many zone axis patterns. Even in structures with large unit cells, prominent sub-lattice reflections, if decoupled from the superlattice by weak first-order structure amplitudes, often exhibit two-beam behaviour.

Inversion of the scattering problem

The main weakness of the simple two-beam intensity distribution is that it contains no phase information, as the structure amplitudes appear in the product $V(g)V(\bar{g})$. However, at certain angles of incidence, the three-beam intensity distribution in a convergent beam diffraction pattern from a centrosymmetric crystal reduces to two-beam form, and the measurement of the parameters associated with this form determines all amplitudes and phases, that is, $|V(\mathbf{h})|$, $|V(\mathbf{g})|$, $|V(\mathbf{g}-\mathbf{h})|$, and the sign of $V(\mathbf{h})V(\mathbf{g})V(\mathbf{g}-\mathbf{h})$ (Moodie 1979). For a non-centrosymmetric crystal no such reduction is possible and another strategy for inversion must be adopted. Again, the two-beam form plays a central role in the classification of symmetries and in the measurement of fundamental parameters in crystallography, in this case the phase, as well as the magnitudes of the coefficients which define the structure.

Summary

While working on this brief sketch, the authors have found that the developments and applications stemming from but one part of Ewald's work have led further and further into science and technology, but an end must be made somewhere, even though the stimulus for further investigation has increased with the compilation. It has, in fact, become increasingly clear that a fundamental reappraisal of the stability of two-beam forms in N-beam scattering is overdue, but here we sorely feel the need for that great figure and gentle teacher, Paul Ewald.

Acknowledgements

The authors are indebted to Dr Y. Bando for his help and generosity in allowing the use of his JEOL 4000 FX electron microscope, and to the third-year students of Applied Physics in the Royal Melbourne Institute of Technology for permission to use their results.

References

Bellman, R. (1960). *Introduction to matrix analysis*. McGraw–Hill, New York.
Bethe, H.A. (1928). *Ann. Phys. (Leipzig)*, **87**, 55–129.
Bethe, H.A. (1981). *Fifty years of electron diffraction* (ed. P. Goodman), pp. 73–6. Reidel, Dordrecht.
Blackman, M. (1939). *Proc. Roy. Soc. (London)*, A**173**, 68.
Ewald, P.P. (1991). *On the foundations of crystal optics*. (ed. H.J. Juretschke) Monograph **10**, American Crystallographic Association, Buffalo, NY.
Gjønnes, J. (1962). *Acta Cryst.*, **15**, 703–7.
Goodman, P. (ed.) (1981). *Fifty years of electron diffraction*. Reidel, Dordrecht.
Goodman, P. and Lehmpfuhl, G. (1967). *Acta Cryst.*, A**22**, 14–24.
Head, A.K., Humble, P., Clarebrough, L.M., Morton, A.J., and Forwood, C.T. (1973). *Computed electron micrographs and defect identification*. North-Holland, Amsterdam.
Hirsch, P., Howie, A., Nicholson, R.B., Pashley, D.W., and Whelan, M.J. (1965). *Electron microscopy of thin crystals*. Krieger, Malabar.
Kato, N. (1952). *J. Phys. Soc. Jpn.*, **7**, 397–405.
Miyake, S. (1959). *J. Phys. Soc. Jpn.*, **14**, 1347.
Moodie, A.F. (1979). *Chemica Scripta*, **14**, 21–2.
Niehrs, H. (1961). International conference on magnetism and crystallography (Kyoto), Paper No. 232.
Wagenfeld, H. (1958). Interferenz Brechung von Elektronenwellen in durchstrahlen Mikrokristallen bei simultaner Anregung mehrerer Interferenzen. Doctoral Thesis, Berlin Free University.

11

Commentary on Ewald's fundamental papers of the dynamical theory of X-ray diffraction

Hellmut J. Juretschke

Introduction

Ewald's dynamical theory of X-rays in crystals is embedded formally in the work of his doctoral thesis (Ewald 1912). Although the immediate goal of the thesis was to develop a microscopic theory of dispersion of visible and near-visible light in anisotropic crystals, its main results were actually valid at all wavelengths. The thesis was submitted a few months before the experimental discovery of X-ray diffraction. Hence the substance of its theoretical formulations predates that event, as well as the initial (kinematical) interpretation of the experimental results.

Once the application of his theory to short-wavelength light became of interest, it took Ewald only a short time—unavoidably prolonged by wartime conditions beginning in 1914—to recast the entire dynamical theory in terms directly suited to the X-ray problem. (Later, he himself felt that wartime service actually gave him the leisurely opportunity to work on it; see Ewald (1979, p. 5).) His three great papers on the subject published in 1916 and 1917 (Ewald 1916a, 1916b, 1917) (later supplemented by a fourth (Ewald 1937)) offered a surprisingly comprehensive first-principles theoretical foundation of the field, including a detailed discussion of the experimentally important two-beam Bragg and Laue cases. In fact, as Ewald has commented, the emphasis on short wavelengths resulted in considerable simplification of the original more general formulation of crystal optics.

These four papers are universally cited as *the* basic reference in the field. Nevertheless, for various reasons they did not become the accepted starting point for interpreting experimental results, for designing new experiments, or for exploring new facets of the theory. It remained for the dynamical formulations by Born for crystal dynamics and optics (beginning in 1912), and by von Laue for X-rays (beginning in 1931), to 'popularize' dynamical theory (Born and Goeppert-Mayer 1933; von Laue 1944).

No doubt one reason for the reluctance to use Ewald's version of the theory has been its austere lattice-dynamical formalism. So familiar today in all of condensed matter theory, at that time it was completely new. Much of it was introduced by Ewald himself, and then applied by him for the first time. Furthermore, the theory was too far ahead of experiment, and often concerned itself with fine structures which would not be observable for another 40 years. Finally, the sparse elegant style, the concern with many questions of primarily theoretical interest, and the sophisticated embedding of the results in mechanical analogies of the behaviour of coupled pendulums—largely for the sake of establishing precedent and continuity with known physics—have never made for an easy reading of Ewald's papers.

An additional minor obstacle, often mentioned in informal discussions among non-German speaking workers in the field, has been the absence of translations to make the original work accessible in English. This is now no longer so. The first two papers were translated by L. Hollingsworth in 1970, and the subsequent two have recently become available in a translation by me (Hollingsworth 1970; Juretschke 1989). Both translations have now appeared together as a monograph of the American Crystallographic Association (Ewald 1991).

However, during the close examination of the text while engaged in the translation, it became apparent to me that even a serviceable rendering into English would still put serious demands on the attentive reader. It therefore seemed worth while to try to develop a summary of the work, as well as commentaries on it, which then could either accompany the translation or serve as an independent overview of its contents. In the context of the present *Memorial volume* for Ewald, I will emphasize those parts dealing with X-rays, and I will concentrate my comments on some of their specific aspects that, in looking back to the time of their development, have struck me as underappreciated or neglected, or perhaps misinterpreted. Obviously, the topics chosen here for discussion represent, to a large extent, personal choices, and the emphases and points of view are entirely mine. (In addition, the occasional references to other work in no way attempt to give balanced or systematic weight to the wider literature of either X-rays or solid-state theory, by Ewald or by others.)

In either of the two above roles, this chapter will serve its intended purpose if it is able to lure the reader to a more detailed study of the text, as translated or, even better, in the original, as some of its subtleties and nuances may not have fully survived translation. This may perhaps also lead to the discovery of other statements or ideas in the text that have not received their due, or to the correction of any misreadings on my part. All of such outcomes, focusing on keeping Ewald's work alive and appreciated, would only be to the good.

Summary of Ewald's dynamical theory

A good part of Ewald's own efforts over the years (Ewald 1962, 1965, 1969, 1979) has been dedicated to explaining and clarifying the content and the methods of the theory laid down in his fundamental papers, and it would not be justified to go over the same ground once more, especially as some of his reviews are reprinted in Part D of this volume. One could hardly improve on the clarity achieved by their author. Other than Ewald himself, many others have also produced excellent summaries; the most recent are those by Bethe and Hildebrandt (Bethe and Hildebrandt 1988) and by Kato in Chapter 1 of this volume.

Rather than retrace the logic of Ewald's equations, I would like to concentrate here on the underlying conceptual structure of his approach because of the pioneering role it seems to me to have played in much subsequent work, not only in X-rays but also in solid-state theory in general. This role has become obscured as his methods and concepts were taken over and spread by others.

The task Ewald set himself was essentially twofold: (1) to establish the modes of propagation of electromagnetic waves in microscopically periodic dielectric media; and (2) to develop the rules for coupling these internal modes to electromagnetic waves in free space either incident on the medium or moving away from it.

To be able to concentrate on the essential features of this problem that are peculiar to a lattice, the model medium was chosen to be a bare-bones description of a crystal, with a set of identical, linearly responding point dipoles located at the lattice points of a simple orthorhombic network. Surfaces, when needed, were taken to be along the faces of the unit cell.

This simple model avoids the problems of the angular dependence and phase shifts of scattering by extended distributions, and of the details of how the crystal is terminated. It can be formulated for all frequencies, once the atomic resonances of the dipoles are included. Also, by breaking up the problem of the interaction of light with solids into the two stages listed above, Ewald was able to simplify the rules for finding particular solutions; they reduced to constructing those special linear combinations of the proper modes of the system—themselves having to be established only once and for all—that will satisfy the 'boundary conditions'.

The first of the tasks listed above was accomplished most elegantly, and along lines that have been followed ever since in condensed matter theory. It must be recalled that a dynamical solution in the crystal is a microscopic many-body problem; the fields propagating without change inside the crystal must be such as to produce excitations of all scatterers that are in full equilibrium with the fields themselves. This condition of self-consistency strictly requires taking into account multiple scatterings to all orders by all dipoles. It also involves the commonly encountered problem of separating out from the total field at any point the field of excitation acting locally on the scatterers. Ewald solved both aspects of the problem by recasting it in reciprocal (Fourier) space. In that fashion the typical dipolar divergence is spread over many **k** values, most of them ignorable at X-ray frequencies. Multiple scattering leads to a 'renormalization', that is, a change in wave velocity as a function of direction through the crystal, as well as to a normal mode description intimately tied to all points of the reciprocal lattice. This description, applicable to all wave phenomena in periodic media, is the same as that later popularized under the name of Bloch waves.

At X-ray frequencies, in particular, the normal modes are found to consist primarily of a superimposition of a finite number of strong plane waves, now increasingly referred to as Ewald modes (corresponding to **k** vectors lying on the faces, edges, or corners of the extended Brillouin zone of solid-state physics). All of these waves are on equal footing (reciprocity theorem), and any one fully determines the others. There is no mention of an 'incident' wave, nor any special emphasis on it. The dispersion relation of n transverse waves with two degrees of polarization becomes an equation of finite degree $2n$. Its solutions give the actual normal modes as constituting $2n$ orthogonal sets of combinations of the $2n$ waves. Each plane wave component in a particular set has both a propagation velocity ('index of refraction') and a direction of propagation slightly differing from those of the corresponding components in the other sets. It is these sets or bundles of plane waves that form the elementary electromagnetic excitations of the crystal (denoted by Ewald as 'elementary interference fields'). Both the number of plane wave components, and their particular choice, are fixed by the group of reciprocal lattice points in proximity of the sphere of propagation, the Ewald sphere. *The most important aspect of the dynamical approach relative to kinematical theory is that, rather than requiring an exact location on this sphere, this proximity suffices to define the propagating modes (even for monochromatic waves).*

The approach to solving the boundary problem was even more original. Ewald found that the purely formal truncation to a half-space of the lattice sum over radiating dipoles produced two additional sets of fields, both emanating from the boundary region, the so-called boundary fields. One set is composed of waves moving from the boundary into the crystal, the other of waves moving away from the crystal. Waves in both sets travel with the velocity of light in vacuum, c. The amplitudes of the $2n$ dominant terms in both boundary fields happened to be related in such a manner that they could be adjusted to eliminate simultaneously all fields in the crystal interior travelling with velocity c (including the 'incident beam'), and produce both incident and reflected beams on the outside. This unorthodox procedure was a necessary ingredient of a microscopic solution where there is no clearcut 'boundary', but at the time it was completely new, and Ewald himself had sufficient qualms about its correctness (in the long-wavelength limit) that he consulted Laue about it in the now famous conversation in the winter of 1911–12. The explicit demonstration of the disappearance of the incident field in the interior of a solid had also been shown independently by Oseen (1915) on a more macroscopic basis, and their common result became known as the Ewald-Oseen extinction theorem. (See also Chapter 12 by Bullough and Hynne in this volume.)

With both the nature of the normal modes and an understanding of the inside–outside connection in place, it became possible to discuss concrete experimental configurations. In his Part III, Ewald (1917) analysed both symmetric Laue and Bragg cases in detail, showing the regions of total reflection and of Pendellösung, the shift from the geometric Bragg angle, the shape of the rocking curves, the differences between the two polarizations, the influence of the lower surface in the case of a crystal slab, absorption, etc. (The Borrmann effect, although not mentioned explicitly here, was evidently well understood by him, as indicated by its appearance as a proposition to be defended in his 1917 preliminary lecture as a prospective member of the faculty at Munich.)

Twenty years later, his Part IV addressed how many of the questions of interest that had arisen in the intervening time could be incorporated in the original formulation; these questions included the generalization to a cell containing more than one atom, and the extension to non-orthogonal lattices (which von Laue had already treated for his own theory, and which Ewald much later (Ewald 1979, p. 4) deemed, through faulty recollection, to have included, for one atom per cell, in his Part III) and to a general continuous distribution of electrons in the cell. This paper also reopened the derivation of the existence of the equation of dispersion under the most general conditions, and relaxed the distinction between strong and weak beams. Various special cases arising from the general equation of dispersion were also treated. Ewald's

Part IV essentially completed the programme of increasingly more comprehensive solutions beginning with the treatment of a single beam, then leading to n strong beams, and finally to an infinite number of beams in arbitrary structures.

Both their wealth of material and the examples of innovative ideas they contain mark these four papers as very special, and many detailed aspects of the material in these papers deserve further discussion. The sections below will be concerned with several of these that have struck me as being of particular relevance to a modern reader.

The boundary conditions in a microscopic model

As already pointed out, the transition from the outside to the inside in a microscopic model of a solid presents features which cannot be handled by the conventional mathematical procedure of demanding continuity of certain quantities across the 'boundary', principally because there is no unique definition of such a boundary. This is particularly true in the open structure of Ewald's model, where, except at discrete points, all of space even *inside* the crystal is empty.

He was therefore confronted with the question of how the incident field, which obviously penetrates such a structure without difficulty, actually disappears. It was known to do so in the optical case (Fresnel equations) and must therefore also surely do so at shorter wavelengths. The answer had to lie in the self-consistent solution of the problem that includes both inside and outside. Throughout his papers, Ewald stressed that knowledge of only the elementary excitations inside the crystal is completely insufficient for making contact with experiment, because one needs to establish how and in what relative proportions these are excited, if at all, by some outside field. In fact, it was not clear at the time that the elementary excitations of the infinite crystal would have any connection with the fields generated in the inside by an outside source.

His elegant answer to these questions rests on generalizing the fields produced by a plane dipolar wave in an infinite crystal to the situation where the dipoles exist only in a half-space. A strictly formal manipulation shows that one obtains the same interior fields as before, but they are now augmented by fields originating in the boundary region. These so-called boundary fields travel away from the boundary in both directions, towards the inside or the outside, but always with velocity c. In particular, strong interior waves are accompanied by strong boundary waves of the same amplitude but slightly different direction. These results show, first of all, that the same elementary excitations as found in an infinite crystal persist inside the crystal even in the presence of a boundary. Furthermore, by choosing a correct combination of the elementary excitations of interior waves, it became possible to nullify exactly all of the strong 'non-physical' waves travelling in the interior with velocity c, including a full cancellation of the incident beam in this region. In parallel with an algebraic solution, the conditions for the correct combinations also lent themselves to an easy graphical visualization in reciprocal space, in terms of the location of the tiepoints on the various sheets of the dispersion surface that characterize the elementary excitations.

This procedure, then, yields a complete solution of the coupling between inside and outside. It leads directly to Snell's law in crystal optics and to its generalization for X-rays.

I cannot help but marvel at the fact that the solution of this formidable microscopic boundary problem took such clearcut elegant form; and I am sure that for Ewald it must have been a personal triumph to obtain it and thereby gain deep physical insights into the microscopic validation of the correctness of Maxwellian boundary conditions. However, nowhere in these papers is there any sign of exultation over this 'miracle'. He only comments drily at one point that if one were to use as interior fields those proposed originally by Laue (kinematical, and moving with velocity c), there would be no excitation of the crystal whatsoever, as then all diffracted interior fields cancel perfectly. That he approached this daring new result in optics, and its consequences, with some trepidation is indicated only by the fact that it formed the topic of his famous consultation with Laue just before Ewald submitted his thesis. Most likely he wanted to have reassurance that he was not getting into trouble with something that may have been thought of, tried, and perhaps rejected before. At that time, however, Laue's preoccupation was elsewhere.

In some respects, however, Ewald's discussion

of the surface problem in these papers has remained incomplete. He repeatedly raised the question of the nature of the X-ray fields in the transition region, and how far into the crystal one would have to go before the interior fields are fully established, that is, the magnitude of the extinction distance for the incident beam. He did provide an answer for long wavelengths, pointing out that there is a typical attenuation of 1000 per lattice distance, and for X-rays he commented that, in the Laue case, the field in the first few layers of atoms is primarily that of the incident beam. However, there is no explicit general construction of the fields in the transition region.

At the same time, he was aware that the laws of reflection and refraction derived by him were independent of the detailed structure of the surface, in all wavelength regions. Therefore they represented a necessary and sufficient condition for properly connecting the propagating waves on both sides of the boundary. Any additional attenuated fields that might exist close to any particular boundary had no bearing on these results. This was also the conclusion of Oseen's work, which was fully within a macroscopic continuum approach.

In connection with another aspect of these surface fields, Ewald did not pursue the physical meaning of the fact that there exists a region between the first two atomic layers where the field representation can be in terms of either external or internal fields. Other than a proof that formally the two representations become equivalent in this region, there is no discussion of how interior and exterior fields and their associated boundary fields interconnect there. Such discussion would be of importance in fully specifying the fields in the transition region, which is certainly a question of current interest.

And yet, as his solution sets are, in principle, complete, they surely contain the answers to these questions. It is of interest to note that the corresponding problem of electromagnetic surface fields in free electron metals—of course, a quantum-mechanical non-local many-body problem—has been solved only a few years ago (Feibelman 1982).

Mean field formulation at X-ray frequencies

Any electromagnetic treatment of solids based on microscopic interactions (ever since Lorentz's original 'electron theory') is faced with the problem of connecting the strictly derived microscopic fields with the average field quantities in Maxwell's equations. In addition, it must disentangle from both types of fields the 'exciting field', that is, the locally acting field that is the prime mover in producing the scattering needed for the self-consistent dynamical equilibrium of fields via multiple interactions. As is therefore only natural, the question of identifying these fields permeates all four Parts of Ewald's papers, and enters into the discussion of the modes in an infinite crystal as well as of the transition at the boundary of a finite crystal.

At long wavelengths (Parts I and II), Ewald included an explicit discussion of the first of these connections (Part II, Section 5), largely to justify that the 'optical' field of Maxwell is obtained by averaging over the periodic part of the Bloch wave, a procedure he had introduced and then used early on (Part I, p. 30 in the translation). The removal of the field of the dipole under discussion was solved by the famous use of the Θ function (Part I, pp. 31–7) (see Ewald's remarks on the Θ function (Ewald 1979, p. 2)). This removal is essential in establishing a phase velocity that differs from c inside the crystal. And, as in this wavelength regime there is only one strong beam, 000, it is evident that dispersion (i.e. the index of refraction) is, from the point of view important for X-rays, effectively due to the contributions of all weak beams. All must be included to establish dynamical equilibrium. The Fresnel formulae for the connections across the crystal boundary follow from the same assumptions, although here they are based on the explicit cancellation of the interior boundary wave, i.e. a field not appearing in standard Maxwellian theory (although that theory could be reformulated in the same manner, as shown by Oseen (1915)).

At X-ray frequencies (Part III), with wavelengths now comparable with atomic sizes, the same averaging problems become more complicated, but in some ways the results are simpler. Here the optical field is identified with the elementary interference field that consists of a bundle of n coupled strong waves (strong in the sense of being related to reciprocal lattice points close to the Ewald sphere). In this formulation, the optical field is, indeed, also the exciting field, as the dipolar divergence is distributed primarily over the many weak terms that have been left out. The strong waves define the dispersion surface in the given neighbourhood, and they are the dynamically active ones. Weak beams, though excitable, do not contribute to the self-consistency. The generalized Fresnel formulae for n beams result from detailed cancellation of the strong interior boundary waves, again invoking fields not present in standard Maxwellian formulation.

It is of interest in this connection to note that von Laue's formulation of X-ray diffraction (von Laue 1931), based entirely on Maxwellian fields and local response of continuous dipole distributions, is formally fully identical with Ewald's equations. Wagenfeld (1968) has given an explicit transformation between the two. Hence von Laue's theory contains both strong and weak beams, on apparently equal footing. These is no explicit acknowledgement of the approximation involved in dealing with the problem of the exciting field, other than a comment that the quantum-mechanical justification of his treatment (Kohler 1935) relies on the Hartree approximation for describing electrons in a solid, itself a procedure already involving averaged fields. In practice, of course, the solution of the dispersion surface automatically favours the contributions of the strong beams, with the weak ones providing little or no back-reaction on the self-consistency of the field.

Hence at X-ray frequencies the standard averaging procedure for deriving optical from microscopic fields used at long wavelengths is replaced by the selection of the strong term elementary excitations, without involving any explicit averaging. (Of course, some average over the weak beams is implied by these procedures.) In this sense, then, both long- and short-wavelength formulations reduce to what are now termed mean field solutions. In the Ewald formulation, at least, these approximations are explicit, and, in principle, it is possible to return to the fundamental equations and derive higher-order corrections to standard theory. To what extent such corrections might play any practical role in current experimental practice remains to be established.

The question of the exciting fields was reopened in Ewald's Part IV, partly to deal with the problem that in composite structures the field is sampled at more than one point in the unit cell, and partly to investigate the validity of the short-wavelength formulation as this wavelength is increased. The first part was handled in terms of reformulating single dipole scattering in terms of 'vectorial structure amplitudes', which represent the effective dipole moment amplitudes of the cell, with appropriate phase factors, that are scattering into any one beam. For the second, it was stated only that a satisfactory extension to longer wavelengths can be formulated, although 'it would lead too far afield to analyze its details here'.

Umweg interactions and degeneracies

The search for rigorous solutions of the n-beam problem in arbitrary structures, in Part IV (1937), required new insights and also had to confront a number of questions that arise only for $n > 2$. First of all, multiple scattering could now be envisioned in terms of all orders of interactions of the n beams. Renninger's work had only recently demonstrated explicitly the transfer of energy between different beams, leading to either enhancement or diminution relative to their strength in two-beam situations. Umweg interaction became an important concept. (Ewald called *all* such back-scattering interactions 'Unweganregung'—excitation via intermediate paths—which they are, of course, although among many theoretical and experimental practitioners there somehow arose the practice, current even today, of treating enhancement or diminution separately, to the extent of giving them separate names.) Thus, Ewald could state that the n-beam elementary excitation is obtained by determining the manner of self-consistent propagation of these n beams 'subject to all possible Umweg interactions between them'. This approach automatically led to an easily implementable ordering of the interactions in terms of the number of intermediate terms they contain, obviously the same kind of foundation for any systematic treatment of many-body interactions as those that appeared in other fields at about the same time. However, Ewald also emphasized that (for point scatterers?) the problem of Umweg excitation becomes decisive only when the cell basis consists of more than one atom.

For two beams, the dispersion equation is the product of two factors, labelled by Ewald Δ and F, such that the product $\Delta \cdot F = 0$ (Δ here is not to be confused with the common differential operator!); one factor vanishes for each polarization. For $n >$

2, this product continues to define all solutions, but the meaning of the two separate factors is changed. All regular solutions are obtainable from $F = 0$. Special solutions are also obtainable from $\Delta = 0$, but only if certain constraints on the Umweg coupling coefficients are obeyed. Furthermore, the nature of these solutions depends on whether there exists degeneracy within the elementary interference field, in the sense that more than two of its beams fall into the same plane. For that case, Ewald discussed the possibility of counter-moving (standing wave?) solutions, of planar fans, and other peculiar field structures. Except in isolated cases (an early one was treated by Ewald's student Lamla, for the case of all beams in one plane) such anomalous solutions of possible modes of X-ray fields have not received systematic attention. It remains to explore whether they lead to experimentally interesting consequences.

Formally, such special solutions arise because the rank of the determinant of the equation of dispersion is reduced below $2n$, or two or more of its roots are equal. It is of interest to note that, at about the same time, similar questions were being investigated with respect to the hanging together or splitting of energy bands of electrons in solids (Herring 1937).

Promises

Throughout these papers, Ewald had to resist the temptation to deviate from his overriding goals by becoming involved in additional aspects that, though important in their own right, would detract from the basic conceptual edifice that he was trying to construct. One therefore finds in various places either brief outlines of what would be involved in following such a branch line, or promises of additional papers in which these topics would be taken up in more detail. I have not fully checked how many of these points actually came to exist as parts of later publications. But a number of major topics, which would certainly have deserved a separate publication, were apparently never followed up by Ewald. Some have in the mean time been covered by others; some, to my knowledge, have never received full systematic treatment. I want to list here a collection of these 'promises', partly to put them on record for the modern reader, but mostly to call attention to the broad overview of the entire field that they project, in what were, after all, the first, ground-breaking, efforts in developing a new field of physics. (Page references are to the English translation.)

Theory of resolving power

In his concern to make contact with actual experiments, Ewald immediately raised the question of the influence of a finite lateral width of experimental incident beams on the fine structure of intensity distributions of diffracted beams that plane wave theory predicts (Part III, p. 103). He saw this as a third part of the overall programme beginning with the mode structure in an infinite crystal and then followed by the boundary problem. However, he did not follow it up as part of the fundamental enquiries, but reserved it to a theory of resolving power similar to that of optical instruments. He only pointed out that a treatment of finite apertures would require a spherical wave approach, which, by introducing angular uncertainties in the diffracted beams, would automatically produce some smoothing of any fine structure. A similar question was raised earlier (Part III, p. 60) in his discussion of the treatment of diffracted waves travelling parallel to the surface. Here Ewald felt that any realistic approach to energy flow would have to give the crystal a finite lateral extent. Of course, both of these points were worked out later, and have become important contributions to the interpretation of experiments.

The general boundary problem

Although the boundary problem had been solved so elegantly for orthorhombic lattices with one atom per cell, the generalization to arbitrary lattices with composite structures was never carried out. It was mentioned in the introduction to Part IV as the topic of a paper to follow (Part IV, p. 112), in the original spirit of Ewald's contention that a solution of the elementary excitations of an infinite crystal lacks any contact with experiment unless the boundary problem is also understood.

It seems to me that for Ewald a major stumbling block in an elegant theory probably was the question of where to locate the boundary. For one atom per cell this boundary can be taken as a mathematical plane, even for arbitrary surface orientations. But with a distributed cell content, such a surface would split cells, surely violating some chemical principles. This question of the true nature of the surface was one of continuing debate between Ewald and von Laue (who preferred the 'building block' approach), and it resurfaced in modified form in 1953 (Ewald and Juretschke 1953). On the other hand, for surface orientations where this question does not arise, the general treatment of composite lattices in Part IV as a superimposition of simple lattices would seem to carry over also to the boundary problem, and presents no difficulties in principle.

Generalized theory of dispersion in optics

The central result of Part IV is the derivation of the general equation of dispersion as a product of two factors, $\Delta \cdot F = 0$. A re-examination of the assumptions involved in this derivation (Part IV, pp. 149–50) shows it to be valid even in the limit $n \to \infty$, and therefore fully applicable at all wavelengths. The only point still requiring closer investigation is the rigour of the notion of the 'exciting field' in the self-consistent formulation. Although not problematical at short wavelengths, it becomes increasingly important as the wavelength increases. Ewald commented that this matter could be resolved, even for distributed structures, and that the long-wavelength theory would lead to the anisotropic dielectric constant tensor, but its details were too lengthy to be included, and 'will be reserved for a further Part'. To my knowledge, such an extension has never appeared. It would have been interesting to see Ewald's approach to what becomes an increasingly non-local problem, and of whose nature he was fully aware.

Papers to follow

The remark on 'papers to follow' in the introductory footnote to Ewald's Part IV suggests an overall plan for a series of papers to deal with the major remaining questions of a complete theory. That he had not given up on the plan is suggested by the comment in a 1966 letter (mentioned in Chapter 5 of this volume) that there is still a fifth paper to come. This may refer to the 1968 papers with Héno on three-beam cases, but Ewald could also have intended to implement the original plan further. Perhaps the answer can be found in any unpublished manuscripts among his papers; perhaps the turmoil of the times made their completion secondary to other concerns?

In any case, however, he clearly saw what had to be done.

References

Bethe, H.A. and Hildebrandt, G. (1988). *Biogr. Mem. Fellows Roy. Soc.* (London), **34**, 135–76.
Born, M. and Goeppert-Mayer, M. (1933). *Handbuch der Physik*, vol. **24**. Springer, Berlin.
Ewald, P.P. (1912). Dispersion und Doppelbrechung von Elektronengittern (Kristallen). Doctoral Thesis, Royal Ludwigs-Maximilian University, Munich.
Ewald, P.P. (1916a). *Ann. Phys. (Leipzig)*, **49**, 1–37.
Ewald, P.P. (1916b). *Ann. Phys. (Leipzig)*, **49**, 117–43.
Ewald, P.P. (1917). *Ann. Phys. (Leipzig)*, **54**, 519–97.
Ewald, P.P. (1937). *Z. Kristallogr.*, A**97**, 1–27.
Ewald, P.P. (1962). In *Fifty years of X-ray diffraction* (ed. P.P. Ewald), Chapter 15. International Union of Crystallography. Oosthoek Utrecht.
Ewald, P.P. (1965). *Rev. Mod. Phys.*, **37**, 46–56.
Ewald, P.P. (1969). *Acta Cryst.*, A**25**, 103–8.
Ewald, P.P. (1979). *Acta Cryst.*, A**35**, 1–9.
Ewald, P.P. (1991). *On the foundations of crystal optics*. (ed. H.J. Juretschke) Monograph **10**, American Crystallographic Association, Buffalo, NY.
Ewald, P.P. and Juretschke, H.J. (1953). In *Structure and properties of solid surfaces* (ed. R. Gomer and C.S. Smith), pp. 82–117. University of Chicago Press.
Feibelman, P.J. (1982). *Prog. Surf. Sci.*, **12**, 287–407.
Herring, C. (1937). *Phys. Rev.*, **52**, 365–73.
Hollingsworth, L. (1970). *Translation of Parts I and II*. AFCRL-70-0580 Translations, No. 84.
Juretschke, H.J. (1989). *Translation of Parts III and IV*. Polytechnic University, Brooklyn.
Kohler, M. (1935). *Sitz.-Ber. Preuss. Akad. Wiss. Phys.-Math. Kl.*, 334–8.
Laue, M. von (1931). *Erg. Exakt. Naturwiss.*, **10**, 133–58.
Laue, M. von (1944). *Röntgenstrahl-Interferenzen*. Akadem. Verlagsges., Leipzig.
Oseen, C.W. (1915). *Ann. Phys. (Leipzig)*, **48**, 1–56.
Wagenfeld, H. (1968). *Acta Cryst.*, A**24**, 170–4.

12

Ewald's optical extinction theorem

R.K. Bullough and F. Hynne

1. Introduction

In his commentary on Ewald's fundamental papers (1916a,b, 1917, 1937) in Chapter 11, Hellmut Juretschke draws attention to the elegant way in which P.P. Ewald solved the boundary problem for the propagation of electromagnetic (e.m.) waves in a crystal. He describes how by using a half-space (with a boundary) Ewald was able to show that the incident e.m. field (the field incident upon the crystal from outside) was *extinguished* inside the crystal and replaced by another e.m. field with different wave number. Similar considerations were independently developed more macroscopically by C.W. Oseen (1915) and the extinction process gained the name of the 'extinction theorem' (or the 'optical extinction theorem' in the sense of 'optics') of Ewald–Oseen. This chapter reanalyses the action of this 'extinction theorem' in an *amorphous optical dielectric* rather than a crystal and traces some of the subsequent history of the extinction theorem and its various applications in this context.

It will be remembered that, in his Doctoral thesis (1912), Ewald solved the problem of e.m. wave propagation in an orthorhombic crystal. It was then a short step to the great papers (1916a,b, 1917) in which he gave his dynamical theory of X-ray diffraction from crystals. The crystal problem offers special difficulties for a theoretical description because of the long-range order of the crystal, namely its strictly periodic character, and because, more particularly, that periodicity is essentially lattice-point like—with large electron density scattering the radiation from only relatively small parts of the whole unit cell. There is, however, a completely opposite physical situation where scattering material is almost *uniformly* distributed throughout some macroscopic region of volume V. At sufficiently long wavelengths λ, such a situation obtains in liquids and glasses, substances which are amorphous rather than crystalline, and it is the application of the extinction theorem to such materials that we shall predominantly describe in this chapter.

2. The smooth dielectric

Ewald's theory was applicable at both short wavelengths (X-rays, $\lambda \approx 10^{-10}$ m) and longer wavelengths including the optical region and near infrared ($\lambda \approx 5 \times 10^{-7}$–$10^{-6}$ m). In principle, it applies at any wavelength. In crystals the atomic structure is both long range (macroscopic $\approx 10^{-2}$ m) *and* on a length scale closely comparable with the wavelength (10^{-10} m) of X-rays. Thus Max von Laue's famous experiment was able to demonstrate the existence of crystalline space lattices through the *diffraction* of e.m. waves at these wavelengths. In a liquid, one is concerned only with *short-range order* on a scale of a few atomic or molecular diameters (10^{-10} m); on the much larger scale of optical wavelengths ($\lambda = 5 \times 10^{-7}$ m) atomic structure plays no part; that is, the scattering material (the atoms) is smoothly and uniformly distributed.

These ideas can be described by a 'correlation function' and particularly by the so-called pair correlation function $g_2(r)$; this is essentially the probability distribution for finding a second atom at a distance r away from a first atom. In a crystal there is a probability of finding the first atom at a particular place inside it (e.g. at primitive lattice or other lattice points) but in a liquid all places at which the first atom may be found must be supposed equally likely (unless there is physical evidence to the contrary) and it is only the *correlation* between atomic positions which matters. Moreover, in such a liquid, atoms are uncorrelated at large distances r (no long-range order) and atomic structure (correlation) arises only at the small (compared with λ) distances where $r \approx 10^{-10}$ m. The lack of correlation at large r means $g_2(r) \to 1$

as r becomes large and 'large' is on the scale of 10^{-10} m. Thus if we make the trivial separation

$$g_2(r) = [g_2(r) - 1] + 1, \qquad (1)$$

the term $[g_2(r) - 1]$ contains all of the structure on this local (10^{-10}m) scale and the '1' is smooth and uniform everywhere. For the '1' the probability of pairs of atoms being in the two volume elements $d\mathbf{x}_1$ (a three-space volume element) and $d\mathbf{x}_2$ (another three-space volume element) is then proportional to $d\mathbf{x}_1 d\mathbf{x}_2$ only.

As we need to know only the influence of other atoms distant r from *one* particular atom at $\mathbf{x}_1$, it is obvious that the expression (1) scales as the number of atoms per unit volume, and this number we shall call n. Thus the '1' in (1) scales, in fact, as n. However, if we are concerned with atoms with this density n inside some definite volume V, n is n everywhere inside V and n is zero outside V. As Juretschke notes, Ewald was very much concerned with the physical problem of when an e.m. wave could be said to be inside the scattering medium. The construction we have just made indicates that there is now a mathematically sharp boundary of V for the smooth distribution of atoms n, and the problem of being inside or outside is transferred to the behaviour of the correlation function contribution $n[g_2(r) - 1]$ at the boundary. As the range of $[g_2(r) - 1]$ is on the scale of 10^{-10} m, it is evident that this problem is isolated as a small (but physically interesting) surface effect. It turns out (e.g. Yvon 1937; Mazur 1958; Bullough 1962; Hynne and Bullough 1984, 1987, 1990, and the references below) that *all* orders of interatomic correlation become involved. Therefore, such corrections must arise at each and every order, $g_3(\mathbf{r}_1, \mathbf{r}_2)$, for triplets of atoms, $g_4(\mathbf{r}_1, \mathbf{r}_2, \mathbf{r}_3)$, etc. It is then possible (though not completely, and it is this which both complicates the multiple scattering—see the scattering theory of Hynne and Bullough (1990)—and introduces effects on the larger scale of λ which must affect the extinction theorem) to make a distinction between local correlation on the 10^{-10} m scale and an otherwise smooth distribution n of atoms everywhere else in the medium V.

These remarks show that it is possible to isolate the problem of the *smooth dielectric*, which has the smooth 'one-body' density of n atoms per unit volume in a finite macroscopic region of volume V, and solve this problem separately first of all. The theory can then be corrected for the much smaller effects of atomic correlation. For example, Hoek (1939) and Rosenfeld (1951, 1965) corrected for the two-body correlations $[g_2(r) - 1]$ in this way, and Hynne and Bullough (1984, 1987, 1990) handled the correlations to all orders, as well as the consequent effects on the optical extinction theorem, in this fashion. We can now see that this problem of the smooth dielectric furnishes a simple, beautiful, and *wholly physical* example of the action of the optical extinction theorem, and we describe this below.

3. Ewald's illuminating crystalline example

Before we describe this problem and trace its history since c. 1916, it is worth discussing a particular example of the crystal considered by Ewald (1925). Ewald assumed throughout that the 'modes' of e.m. excitation in a crystal were governed by dipole waves. Maxwell's famous 'displacement current' exhibits dipole waves generally as the vector current $n\partial \mathbf{P}(\mathbf{x},t)/\partial t$ of dipole density $n\mathbf{P}(\mathbf{x},t)$ ($\mathbf{P}$ is scaled to one atom) at points $\mathbf{x}$ at time t. It is the time derivative of the current which drives the electric field $\mathbf{E}(\mathbf{x},t)$, so it is the second derivative $n\partial^2 \mathbf{P}(\mathbf{x},t)/\partial t^2$ which drives this field. In this way, waves of dipole density $n\mathbf{P}(\mathbf{x},t)$ *create* waves of electric field $\mathbf{E}(\mathbf{x},t)$. Ewald (1925) was concerned with the particular case of dipole waves travelling in a crystal lattice. At atomic level one might need to consider not only dipole allowed transitions but also arbitrary multipole transitions. Such a theory for the smooth dielectric at classical level was developed by C.G. Darwin (1924) and later by one of us (Bullough 1962), and the appropriate forms of the 'extinction theorem' were given in each case. Bullough (1962) also included anisotropic atomic correlation and two kinds of anisotropic 'atoms' (i.e. molecules), for application to the optics of rubber elasticity. Thus his paper contained within it the amorphous counterpart of Ewald's orthorhombic crystal as a special case.[†] Here, we

---

[†] There is a better formulation of the theory than that given in Appendix 2 of Bullough (1962), which shows that the optical extinction theorem acts as simply there as it does in the less complicated context described in this chapter. It should be noted that the 1962 theory is actually an anisotropic dielectric constant theory (see Juretschke, Chapter 11).

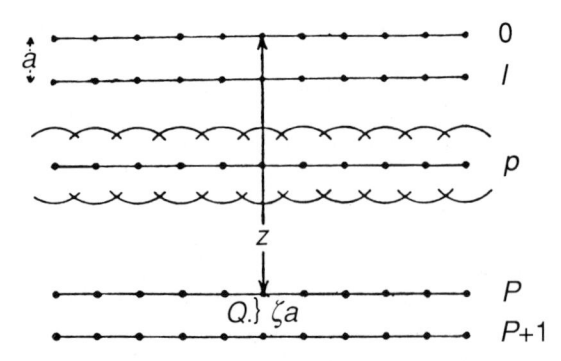

Fig. 1 Dipole wave excitations on crystal planes $p = 0, 1, \ldots, P$ induce electric fields at Q distant $z = (P + \zeta)a$ below plane $p = 0$. (Figure taken from James (1954).)

consider Ewald's dipole waves and briefly summarize his crystal theory of 1925.

Let us consider, following James (1954), a set of lattice planes $p = 0, 1, \ldots, P$, which are equally spaced, as in Fig. 1, a distance a apart. A dipole wave is supposed to pass in a direction normal to the crystal planes (downwards in Fig. 1) and all dipoles in one plane oscillate in phase with an angular frequency ω. Oscillating dipoles are Hertzian dipoles and emit radiation according to Maxwell's equations.[†] The dipoles are vector 'sources' emitting radiation radially with the well-known $\sin^2\theta$ intensity profile (θ is the angle between the radial outgoing direction and the direction of the dipole; this latter direction is vertical out of plane in Fig. 1). The radiation is an e.m. field with electric field **E** and magnetic field **B** built from elementary fields from each separate dipole. Evidently, the oscillating *plane* of dipoles must build a plane e.m. field; the 'front' of this field must be parallel to the 'front' of the dipole field and hence to the parallel crystal planes. These fronts propagate along z, and **E** (like **P**) points vertically in Fig. 1. Equally evidently, there must be *two* such electric fields, one travelling down the crystal and the other travelling up.

The electric field at point Q between planes P and $P + 1$ and at a distance $z = (P + \zeta)a$ below the surface, which is the plane $p = 0$, is the sum of contributions from the dipole fields at each crystal plane. The pth crystal plane contributes an electric field at Q proportional to

$$\exp(ipka) \cdot \exp[ik_o(P - p + \zeta)a]. \quad (2)$$

In this chapter 'wave number' means $2\pi\lambda^{-1}$ not λ^{-1}. Thus in eqn (2) it is assumed that the dipole field has wave number k so the phase relative to $p = 0$ at plane p is $\exp(ipka)$. On the other hand, the field radiated is the sum of dipole radiation fields each radiating *in vacuo* with wave number $k_o = \omega c^{-1}$ (c is the velocity of light *in vacuo*). The total field radiated to Q from planes $p = 0, 1, \ldots, P$ above Q is the sum over $p = 0, 1, \ldots, P$ of the quantity (2). It is easy to check that this *finite* sum over a geometrical progression consists of the *two* terms which can be written as (see Ewald (1925), where even the formula for a geometrical progression was given)

$$\frac{\exp(ik_o z)}{1 - \exp[i(k - k_o)a]} - \frac{\exp[i(k - k_o)(1 - \zeta)a]}{1 - \exp[i(k - k_o)a]} e^{ikz}. \quad (3)$$

It is evidently important that the wave number k is *not* equal to k_o (the terms in (3) individually diverge though their sum is zero if $k = k_o$). We shall see that $k = m(\omega)k_o$, where $m(\omega) \neq 1$ and is the frequency-dependent refractive index. Thus the first term in (3) is a wave travelling down the crystal with wave number k_o, the wave number *in vacuo*. Ewald was able to associate this wave with the boundary at $p = 0$ (in a sense, it is created there) and we shall have much to say about this kind of wave below.

The second term in (3) has the wave number k, not k_o, of the originating dipole wave. It is exactly in phase with the dipole wave at plane $p = P + 1$ where $\zeta = 1$ and, as P is arbitrary, it is exactly in phase with the dipole wave at every plane—at which there is then a phase jump of $(k - k_o)a$, becoming $(k - k_o)(1 - \zeta)a$ as ζ increases from zero to unity. Such a phase jump cannot take place in the smooth dielectric, as planes of scattering points are infinitesimally close to each other.

The conclusion from (3) is that in the smooth dielectric we expect to find *three* waves travelling down the crystal: one is the dipole wave with wave number $k = mk_o$, the second is an electric field (an e.m. field) with wave number $k = mk_o$, and the third is an electric field (an e.m. field) travelling with the vacuum wave number $k_o \neq k$.

However, we have not yet included the waves

[†] See, for example, eqn (1) (p. 53); $\Pi(\mathbf{x})$ is the Hertz potential for the dipoles; the vector potential is $c^{-1}\partial\Pi/\partial t$ in Gaussian units, that is, $-ik_o\Pi$.

travelling up to Q from the planes below Q, planes $P + 1, P + 2, \ldots$ It is important whether there is a second boundary at some plane $P + N$ (say). If there is not, the same analysis as above shows that there is another wave with wave number k, namely

$$(\exp[i(k + k_o)(1 - \zeta)a]/\{1 - \exp[i(k + k_o)a]\})e^{ikz},$$

which actually travels *down* from $p = 0$, and nothing else. This wave is in phase with the dipole wave at plane $P + 1$ (and therefore P), with the jump $(k_o + k)a$ of phase at each plane (where $\zeta = 0$). This line of analysis simply *neglects* any effects of planes p at $p = \infty$ which involve $\exp[i(k + k_o)(p + 1)a]$ with $p = \infty$. We shall see that there is a real problem with this behaviour 'at infinity'; this problem is avoided by creating another boundary at some plane $P + N$, where N can still be very large. If this is done, there is a further wave travelling *up* the crystal but this has wave number k_o. Evidently, we are then dealing with a parallel-sided slab—in this case, a parallel-sided slab of crystal forming a Fabry–Perot interferometer with a crystalline interior. We might expect to reach the smooth dielectric by letting the quantity $a \to 0$, keeping a finite density n of atoms per unit volume. We find when $a \to 0$ appropriately that the two waves travelling down with wave number k contribute an electric field $4\pi n\, \mathbf{P}(\mathbf{x},\omega)\, k_o^2(k^2 - k_o^2)^{-1}$, where $\mathbf{P}(\mathbf{x},\omega)$ is the Fourier transform of $\mathbf{P}(\mathbf{x},t)$ for a single atom and $\mathbf{P}(\mathbf{x},\omega) = \mathbf{P}(0,\omega)e^{ikz}$, a dipole wave travelling down. We shall find this important expression again within the theory of the smooth dielectric.

It is noteworthy that Ewald in his 1925 paper in effect found this expression also. His result was for a crystal when the wavelength λ is large, and we might expect this situation to approximate to the smooth dielectric. He used a parallel-sided slab with planes at $p = 0$ and $p = P + N$ *and* he computed Fresnel reflection and transmission coefficients at the surfaces. He even briefly considered behaviour in electric and magnetic fields. Thus this very simple analysis of the crystal in 1925 is really very complete.

4. Theory of the smooth dielectric: the Lorentz polarizing field

We now investigate this smooth dielectric. We suppose that there is indeed a dipole density wave $n\, \mathbf{P}(\mathbf{x},\omega)$ travelling in it, with

$$\mathbf{P}(\mathbf{x},\omega) = \mathbf{P}_o(\omega)e^{ikz} \qquad (4)$$

scaled to a single atom. The axis of z is chosen as the direction of propagation of this wave. Maxwell's theory says that such a dipole wave must be a *transverse* wave, that is, $\mathbf{P}_o$ is orthogonal to z. Then div $\mathbf{P}(x,\omega)$, the divergence of $\mathbf{P}(\mathbf{x},\omega)$, is zero and there are no longitudinal dipoles.

Each point with the dipole density $\mathbf{P}(\mathbf{x},\omega)$ will now scatter radiation. We could calculate the total field at a field point $\mathbf{x}$ from all dipoles at points $\mathbf{x}'$ in a sheet of width dz' and then sum (integrate) over all elements dz' to obtain the total field at $\mathbf{x}$, much as Ewald did with the crystal planes. In fact, this is how one of us (R.K.B.) learnt about the extinction theorem as a student of D.R. Hartree in 1952.

Here we shall take the more general point of view described by Léon Rosenfeld in his little book *Theory of electrons* (Rosenfeld 1951, 1965). A Hertzian point dipole at $\mathbf{x}'$ radiates a dipole field to $\mathbf{x}$ which has the $\sin^2\theta$ intensity profile mentioned above. Now θ is the angle between the direction of $\mathbf{x} - \mathbf{x}'$ and the direction of the dipole at $\mathbf{x}'$. However, this concerns the 'radiation' or 'far' field of the dipole, which falls as r^{-1}, $r = |\mathbf{x} - \mathbf{x}'|$. It will be recalled there is also the 'near' field and the 'intermediate' field, which fall as r^{-3} and r^{-2}, respectively.

In mathematical terms, there is a *tensor* $\mathbf{F}(\mathbf{x},\mathbf{x}';\omega)$ depending, in fact, only on $\mathbf{x} - \mathbf{x}'$ and ω, which carries the total dipole field from $\mathbf{x}'$ to $\mathbf{x}$; it must be a tensor, as the elementary field from $\mathbf{P}(\mathbf{x}',\omega)$ is $\mathbf{F}(\mathbf{x},\mathbf{x}';\omega) \cdot \mathbf{P}(\mathbf{x}',\omega)$. The dot means 'scalar product' and if (for example) $\mathbf{r} \equiv (\mathbf{x} - \mathbf{x}')$, then, following Willard Gibbs (1881), $\mathbf{rr}$ is an outer product, a dyadic form of a second-rank tensor. Then $\mathbf{rr} \cdot \mathbf{P} = \mathbf{r}(\mathbf{r} \cdot \mathbf{P})$ is a vector in the direction of $\mathbf{r}$ and $(\mathbf{r} \cdot \mathbf{P})$ is the scalar product. In this way, $\mathbf{F} \cdot \mathbf{P}$ is likewise a vector, the vector field at x from the dipole at $\mathbf{x}'$.

The near, intermediate, and far fields are contained in the actual dyadic form for $\mathbf{F}$, which is

$$\mathbf{F}(\mathbf{x},\mathbf{x}';\omega) = (\nabla\nabla + k_o^2 \mathbf{U})\, f(\mathbf{x} - \mathbf{x}'). \qquad (5)$$

Here, ∇ is the gradient operator, a vector operator, so $\nabla\nabla$ is a tensor operator; $\mathbf{U}$ is the unit tensor (often written in its component form δ_{ij}) and is a sum of unit dyadics.

Readers may care to 'play with' (5) to familiarize

themselves with it. For our purposes, it is the function f which is most important:

$$f(\mathbf{r}) = (\exp ik_o r)/r, \qquad (6)$$

where r is the *distance* $|\mathbf{x} - \mathbf{x}'|$ between $\mathbf{x}'$ and $\mathbf{x}$. It is the function $f(\mathbf{r})$ which governs the far field. However, it should be noted that the near field, where $k_o r$ is small, is essentially $\nabla\nabla (1/r) \cdot \mathbf{P}(\mathbf{x}',\omega)$. This is a 'Coulomb' dipole field, which rises as r^{-3} as $r \to 0$.

Evidently, the total field at $\mathbf{x}$ due to all of the dipoles in a region V is given by

$$n \int_V \mathbf{F}(\mathbf{x},\mathbf{x}';\omega) \cdot \mathbf{P}(\mathbf{x}',\omega) d\mathbf{x}', \qquad (7)$$

and $\mathbf{x}'$ is integrated over the three-dimensional region V. There is now the difficulty, however, that as $\mathbf{x}'$ approaches $\mathbf{x}$, which is also inside V, the near field diverges as r^{-3} whereas even in polar coordinates $d\mathbf{x}' \sim r^2 dr$. This would mean a logarithmic divergence in r as $r \to 0$. On the other hand, if spherical polar co-ordinates *were* used, integration on the polar angle first, before the integration on r, produces zero and the integral (7) is mathematically undefined.[†]

The physics of all of this is that if atomic correlation is included we have the correlation function $g_2(r)$ under the integral sign, that is, $n \to n g_2(r)$, and $g_2(r)$ has the property, by the Pauli principle, that two atoms cannot overlap, that is, $g_2(r) \to 0$ as $r \to 0$. To handle this aspect in the total field (7), we remove a small sphere of radius a about the field point $\mathbf{x}$ from the integration region V and let $a \to 0$. Such a step gives the integral (7) a well-defined mathematical value, making it 'conditionally convergent.' However, such rather arbitrary mathematics is actually good physics, for the step is made on the '1' in expression (1), so it must be made also on the '1' in $[g_2(r) - 1]$ in that equation. Thus the result is not an approximation of the theory, it is an identity. The physical test is that the $[g_2(r) - 1]$ should make only a *small* contribution to the smooth dielectric theory. We shall see that it does.

The total field (7) we are calculating is the field acting on an atom at the field point $\mathbf{x}$. This is the 'Lorentz polarizing field' of the title of this section. It is the equivalent of Ewald's exciting field referred to by Juretschke and to which Ewald gave particular attention in his 1937 paper.

5. The total fields in the smooth dielectric: self-consistency and the refractive index

Indeed, as Rosenfeld (1951, 1965) showed, the result of extracting the small sphere is to produce the famous Lorentz field term $(4\pi/3) n \mathbf{P}(\mathbf{x},\omega)$ as the near-field contribution to the total field (7), and in consequence (though this is not obvious) we need to evaluate only the integral

$$n k_o^2 \int_V \frac{\exp(ik_o r)}{r} \mathbf{P}(\mathbf{x}',\omega) d\mathbf{x}', \qquad (8)$$

in which the small sphere plays no role. There is an elegant application of Green's theorem (Green 1828) which converts this volume integral to a surface integral, showing that (8) is exactly

$$\frac{4\pi n k_o^2}{k^2 - k_o^2} \mathbf{P}(\mathbf{x},\omega) + \frac{n k_o^2}{k^2 - k_o^2} \int_\Sigma, \qquad (9)$$

where $\int_\Sigma$ is a surface integral taken over points $\mathbf{x}'$ on the surface Σ of V for a fixed choice of the field point $\mathbf{x}$ inside V. It should be noted how $\mathbf{P}(\mathbf{x},\omega)$ (given by (4)) has moved from the points $\mathbf{x}'$ inside V in (8) to the point $\mathbf{x}$ in V in (9). Though we do not need the details anywhere else, we must now quote here the expression for $\int_\Sigma$ given by Bullough (1962) or in our later papers:

$$\int_\Sigma = \int_\Sigma [\mathbf{P}(\mathbf{x}',\omega) \nabla_{\mathbf{x}'} f \cdot d\mathbf{S} - f d\mathbf{S} \cdot \nabla_{\mathbf{x}'} \mathbf{P}(\mathbf{x}',\omega)]. (10)$$

In this surface integral, $d\mathbf{S}$ is a directed element of the surface Σ of V, so in effect dipoles on the surface of V radiate to the point $\mathbf{x}$ inside V with the vacuum wave number k_o; $\nabla_{\mathbf{x}'}$ operates at the point $\mathbf{x}'$ on the surface Σ. The $\int_\Sigma$ is thus a vector function of $\mathbf{x}$ only and satisfies the 'free-field' wave equation

$$(\nabla_{\mathbf{x}}^2 + k_o^2) \int_\Sigma = \mathbf{0} \qquad (11)$$

because $(\exp ik_o r)/r$ satisfies this equation ($\nabla_{\mathbf{x}}^2$ operates at the field point $\mathbf{x}$). That $\int_\Sigma$ satisfies (11) is all we need to know about $\int_\Sigma$ for our purposes in this chapter.

[†] Darwin (1924) pointed out the unpleasant features of $\mathbf{F}$; namely, that it induces the potential logarithmic divergence from small r *and* a divergence from $(\exp ik_o r)/r$ at large r controlled only by the oscillatory nature of $(\exp ik_o r)$.

On the other hand, the total field at **x** is, from (9) and including the Lorentz term,

$$\left[\frac{4\pi n}{m^2(\omega) - 1} + \frac{4\pi n}{3}\right] \mathbf{P}(\mathbf{x},\omega) + \frac{n}{m^2(\omega) - 1} \int_\Sigma, \quad (12)$$

as $k^2 k_o^{-2} = m^2(\omega)$, the squared refractive index. This field at **x** induces a dipole in the atom at **x**. According to Kramers and Heisenberg (1925), there is a quantum mechanical atomic (or molecular) polarizability $\alpha(\omega)$ of the form

$$\alpha(\omega) = (e^2/m_e) \sum_s f_s (\omega_s^2 - \omega^2)^{-1},$$

where the sum is over all atomic (molecular) excitation states and f_s is an 'oscillator strength' satisfying a sum-rule such as $\sum_s f_s = 1$; therefore, this formula simply generalizes the single *classical* oscillators considered by Ewald (e and m_e are the charge and mass of the electron). In this way, we arrive at the consistency condition

$$\mathbf{P}(\mathbf{x},\omega) = \alpha(\omega) \left[\frac{4\pi n}{m^2(\omega) - 1} + \frac{4\pi n}{3}\right] \mathbf{P}(\mathbf{x},\omega)$$
$$+ \frac{n\alpha(\omega)}{m^2(\omega) - 1} \int_\Sigma. \quad (13)$$

6. The optical extinction theorem

Now, at last, we have arrived at the extinction theorem. It is plain from (4) that $\mathbf{P}(\mathbf{x},\omega)$ satisfies $[\nabla_x^2 + m^2(\omega) k_o^2] \mathbf{P}(\mathbf{x},\omega) = \mathbf{0}$. However, $\int_\Sigma$ satisfies the 'free-field' equation which is (11). This is possible only if terms in $\mathbf{P}(\mathbf{x},\omega)$ and the $\int_\Sigma$ in the consistency condition (13) vanish separately. We thus see that we have the condition on $m(\omega)$ that

$$\frac{m^2(\omega) - 1}{m^2(\omega) + 2} = \frac{4\pi}{3} n\alpha(\omega), \quad (14)$$

the famous relation of Lorentz (1880) and Lorenz (1880), and that

$$\int_\Sigma = \mathbf{0}. \quad (15)$$

However, the condition (15) means that the dipole field amplitude $\mathbf{P}_o$ must vanish (this is the only solution of (14) valid for any Σ) and the theory is empty! Of course, it is clear what is needed: there can be no dipole waves excited unless there is a field incident on V to excite them. This will be a free field $\mathbf{E}_o(\mathbf{x},\omega)$ travelling with wave number k_o, the vacuum wave number, and (15) becomes

$$\mathbf{E}_o(\mathbf{x},\omega) + \int_\Sigma = \mathbf{0}. \quad (16)$$

This *is* the extinction theorem. In the absence of matter inside V only the free field $\mathbf{E}_o(\mathbf{x},\omega)$ propagates. The result (16) shows that when the matter is present inside V this free field is extinguished by the radiation scattered into the interior of V which formally originates on the surface of V. The free field inside V is thus replaced by the dipole field (4) inside V together with the 'polarizing electric field'

$$\left[\frac{4\pi n}{m^2(\omega) - 1} + \frac{4\pi n}{3}\right] \mathbf{P}$$

travelling with the new wave number $k = m(\omega) k_o$. It is also possible to identify a 'Maxwell field'

$$\mathscr{E}(\mathbf{x},\omega) = \frac{4\pi n \mathbf{P}(\mathbf{x},\omega)}{m^2(\omega) - 1},$$

and thus the **D**-field

$$\mathbf{D}(\mathbf{x},\omega) = m^2(\omega) \mathscr{E}(\mathbf{x},\omega) = \mathscr{E}(\mathbf{x},\omega) + 4\pi n \mathbf{P}(\mathbf{x},\omega).$$

We thus have all three of the fields found by Ewald (1925) as described in Section 3, and there are further fields, such as the Maxwell field, if one wishes to identify them. Ewald also identified a Maxwell field. Figure 2 (taken from Hynne and Bullough (1984)) shows just how intricate but elegant the action of the optical extinction theorem can be. The theoretical argument throughout has this intricacy and elegance. It is indeed remarkable that Ewald was able to find it.

The sections which follow outline some additional fundamental questions as well as more recent extensions of the application of the extinction theorem.

7. Associated damping by incoherent scattering

Rosenfeld (1951, 1965) followed closely the analysis of H. Hoek (1939) presented in a Ph.D. thesis at the University of Leiden. Hoek included the two-body correlations $g_2(r)$ and found that the right side of the Lorentz–Lorenz relation (14) became $(4\pi/3) n\alpha(\omega)[1 + G(\omega)]^{-1}$, in which $G(\omega)$ is an integral expression,

$\frac{2}{3}ik_o^3\, n\alpha(\omega) \int h(k_o r)[g_2(r) - 1]\, d\mathbf{r},$

in which $h(k_o r) \approx 1$ when $k_o r$ is small. However, at optical wavelengths $k_o r$ *is* always small, as the range of $[g_2(r) - 1] \approx 10^{-10}$ m and $k_o^{-1} \approx (2\pi)^{-1} \times 5 \times 10^{-7}$ m. The integral is thus

$\frac{2}{3}ik_o^3\, n\alpha(\omega) \int [g_2(r) - 1]\, d\mathbf{r},$

which is a pure imaginary expression, and is very small ($\approx 10^{-9}$). This shows that the refractive index $m(\omega)$ is a complex number and there is small but observable *damping* of the dipole wave $\exp(imk_o z)$. (Strictly speaking, $G(\omega)$ is not small but actually diverges! More completely, it contains the classical radiative level shift of H.A. Lorentz (1909), which becomes the famous Lamb shift in quantum electrodynamics.) This small damping is due to incoherent optical scattering which is not a part of the build-up of the coherent wave fronts. It is possible to develop in this way a rather complete theory of optical scattering from 'molecular fluids', such as liquid or gaseous argon, and, based on the 1984 and 1987 papers, this theory was reported by Hynne and Bullough (1990).

The microscopic theory of optical scattering has a long history, and the reader is referred to the many references in these papers. Mazur (1958) and co-workers, in particular, based their analyses on the extinction theorem as it was described by Rosenfeld. A result of Hynne and Bullough (1990), also ultimately based on the extinction theorem as it was developed by Hynne and Bullough (1984), is a *microscopic* derivation of the phenomenological optical scattering cross-section derived macroscopically first of all by Einstein (1910) in terms of local thermodynamic fluctuations of the dielectric constant. It should be noted that this microscopically scattered radiation always has the wave number k_o. However, Einstein found the wave number $m(\omega)k_o$ for it macroscopically. The problem of this discrepancy was solved by Hynne and Bullough (1990), with a new 'extinction theorem' for the scattered radiation. We mention this again below in Section 9.

8. Conceptual problems of the extinction theorem: need for a real physical volume

We return to Rosenfeld's calculations for the smooth dielectric (Section 4) and the result (16). The extinction theorem in the form (16) is exact and complete. However, if $\mathbf{P}(\mathbf{x},\omega) = \mathbf{P}_o(\omega)\exp(ikz)$, as was assumed, then we find that, for arbitrary surfaces Σ, $\mathbf{E}_o(\mathbf{x},\omega)$ must take on an extremely complicated form even when $(\nabla_\mathbf{x}^2 + k_o^2)\,\mathbf{E}_o(\mathbf{x},\omega) = \mathbf{0}$, whereas if $\mathbf{E}_o$ is made simple, $\mathbf{P}$ almost certainly is not. It is easy to suggest that this problem is avoided if we consider the dielectric to be infinite in all three independent directions (and implicitly this is what Rosenfeld did). Now, however, we have the conceptual difficulty of where to draw the surface Σ. Certainly, we can apparently draw Σ as a closed surface anywhere and with any shape whatsoever (except some mathematically pathological surface such as a fractal, for example). However, we can compute $\int_\Sigma$ for positions $\mathbf{x}$ outside Σ as well as inside it: $\int_\Sigma$ is finite but non-zero outside Σ and still satisfies the free-field equation; however, as it has different values outside Σ (outside V), it can no longer extinguish the incident free field $\mathbf{E}_o$ there, and, to ensure *that*, we must draw another surface Σ' embracing the new field point $\mathbf{x}$ which is now outside Σ.

It is easy to see that the application of the extinction theorem to an infinite smooth dielectric permeated by an incident free field $\mathbf{E}_o$ satisfying $(\nabla_\mathbf{x}^2 + k_o^2)\,\mathbf{E}_o(\mathbf{x},\omega) = \mathbf{0}$ offers conceptual difficulties. Ewald evidently understood these and, as his 1925 paper shows, he understood their resolution. The solution indeed is to take for V a real physical macroscopic volume which is necessarily finite in all three independent directions. This was done for the smooth dielectric by C.G. Darwin (1924); he took a parallel-sided slab for V, of finite thickness L, and one can take a large but finite radius R orthogonal to the direction of L (D.R. Hartree showed one of us how this large radius R simply introduces the diffraction pattern of the large disc by the action of the extinction theorem). Darwin (1924) showed that if $\mathbf{E}_o(\mathbf{x},\omega) = \mathbf{E}_o\exp(ik_o z)$ is incident from outside V then, whether the z-axis coincides with the axis L of the slab or not, refraction occurs exactly as Snell [1621] (see, e.g. p. xxi of Born and Wolf (1980)) originally stated, even though the extinction theorem extinguishes $\mathbf{E}_o(\mathbf{x},\omega)$ everywhere inside V and replaces it by dipole waves. In the case when z and L coincide (that is, the case of normal incidence from outside), there is a pair of dipole waves

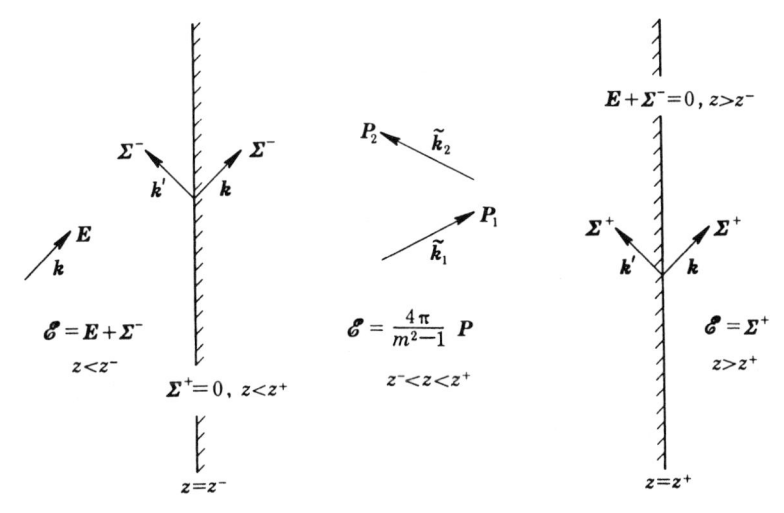

Fig. 2 (Reproduction of Fig. 1 of Hynne and Bullough (1984).) Fields and wave vectors for oblique incidence. The changes of notation in this figure should be noted: **E** is the externally applied field (called **E**$_o$, eqn (16), but now applied obliquely to the surface $z = z^-$ of the parallel-sided slab $z^- < z < z^+$ forming the region V); $\mathscr{E}$ is the macroscopically averaged field of Hynne and Bullough and inside the slab is the total Maxwell field also called $\mathscr{E}$ in this chapter; but **P** is the dipole density called $n\mathbf{P}$ in this chapter. Outside the slab, in $z < z^-$, $\mathscr{E}$ is the total electric field, namely $\mathbf{E} + \Sigma^-$, there: Σ^- is the total electric field reflected from the surface $z = z^-$ through the action of the extinction theorem, and both **E** and Σ^- have the wave number $k_o = \omega c^{-1}$ of this chapter. Likewise, Σ^+ in $z > z^+$ has wave number k_o and is the total electric field transmitted, through the action of the extinction theorem there, into the vacuum from the surface $z = z^+$, and is the actual transmitted wave, in agreement with Maxwell–Fresnel macroscopic theory.

The wave vectors $\mathbf{k}, \mathbf{k}'$ of this figure are oblique to the z-axis of the slab but have the wave number $|\mathbf{k}| = k_o$ of this chapter. The wave vectors, with tildes, $\tilde{\mathbf{k}}_1$ and $\tilde{\mathbf{k}}_2$, have the wave numbers $k = m(\omega)k_o$ of this chapter. It should be noted that $\mathbf{E} + \Sigma^- = 0$ inside the slab and $\Sigma^+ = 0$ there also; these are the *two* identities of the extinction theorem generalizing eqn (16). There are two identities because the dipole waves inside V consist of two waves, one travelling from left to right with wave vector $\tilde{\mathbf{k}}_1$, the other right to left with vector $\tilde{\mathbf{k}}_2$. The condition $\Sigma^+ = 0$ inside V determines the reflection coefficient at $z = z^+$ relating these two dipole waves. This reflection coefficient becomes $\Lambda(\omega)$ as described in Section 8 for normal incidence, and coincides with the Maxwell–Fresnel macroscopic reflection coefficient. Thus the figure shows how all of the macroscopic theory emerges through the action of the extinction theorem in this more complicated case of oblique incidence on a slab.

$$\mathbf{P}(\mathbf{x},\omega) = \mathbf{P}_o(\omega)\exp(ikz) + \mathbf{P}_o(\omega)\Lambda(\omega)\exp(-ikz)$$

running in opposite directions inside V and $\Lambda(\omega) \equiv [m(\omega) - 1]/[m(\omega) + 1]$, exactly the reflection coefficient expected from the analysis by Fresnel (1832), and subsequently, by Maxwell. Oblique incidence is more complicated, and Fig. 2 shows all of the fields for this case as described by Hynne and Bullough (1984).

We return to normal incidence. The dipole waves inside V exhibit the interference features of the Fabry–Perot interferometer (an exact formula is given by eqn (18)). On moving the field point outside V, where there is no scattering matter, the extinction theorem (16) shows that we have a reflected free-field wave proportional to

$$\mathbf{E}_o[(m-1)/(m+1)]\exp(-ik_oz)$$

reflected by the interference pattern in V, and a transmitted wave, also a free field, emerging from the 'back' face of the slab with its proper Fresnel transmission coefficients also modulated by the interference pattern in V. (More exact details for the more general case are given in the caption to Fig. 2.) The boundary conditions imposed on the function f, (eqn (6)), which determine the scattering by the individual atomic dipoles, are 'outgoing at infinity' (the energy flow must be out from the dipole, not into it as the 'ingoing' choice $f = [\exp(-ik_or)]/r$ would imply). We can now see that the action of the extinction theorem, which is a collective action due to all the atomic dipoles induced inside V, is in effect to *replace* these outgoing boundary conditions at infinity by the usual Maxwell boundary conditions on the surface Σ of the 'dielectric' inside V, that is, continuity of the normal component of **D** and of the tangential component of **E**. The 'microscopic' extinction theorem thus builds up *all* of the macroscopic smooth dielectric theory, as Fig. 2 already shows.

9. Scattered light is ordinary light: extinction theorem for the scattered radiation

At the level of additional incoherent scattering, the literature is extensive, as already mentioned. Yvon (1937), and Mazur and Mandel (1956), Mazur (1958), and later co-workers, in particular, built up a microscopic scattering theory of optical scattering involving all the correlation functions $g_2(r)$, $g_3(\mathbf{r}_1, \mathbf{r}_2)$, etc. This theory is quasi-static and elastic. Both Mazur and Fixman (1955) considered scattering processes into a collector essentially within an infinite medium so that the detection process involved

$$\tilde{\mathbf{F}}(\mathbf{x},\mathbf{x}';\omega)$$
$$\equiv \frac{1}{m^2(\omega)}[\nabla\nabla + m^2(\omega)k_o^2\mathbf{U}]\exp(imk_o r)r^{-1} \quad (17)$$

instead of the 'free-field' tensor **F** (eqn(5)). The wave number of the resulting detected radiation is now mk_o, as found phenomenologically by Einstein (1910). The extinction theorem was still effective because the build-up of the refractive index $m(\omega)$ was still described in terms of **F** with wave number k_o.

This theory introduces the problems of the infinite dielectric and, by working consistently with a 'real' finite system in the same quasi-static approximation, Hynne and Bullough (1984, 1987, 1990) showed how either **F** or **F̃** (or, more precisely, yet another propagator $\mathscr{F}$) could be used for every microscopic scattering process provided certain corrections essentially related to the finite boundary for V were included. This is an enormously complicated application of the extinction theorem because of the enormous complication of the multiple scattering processes that occur microscopically at every order. However, as mentioned, Hynne and Bullough (1990) still managed to show that there is an 'extinction theorem' for this scattered light also.

The mathematical form derived for this extinction theorem is exactly that of Ewald's form. Now, however, there is no free field incident upon the system to be extinguished inside it, and this aspect has to be reinterpreted. Even so, the incoherently scattered radiation *is* in effect refracted at the boundary Σ of V according to Snell's law. Thus the conclusion from this new extinction theorem is that the incoherently scattered light does behave exactly like 'ordinary light'. The empirical rules used in actual scattering experiments, such as those used for defining the solid angle for the collection of light scattered from a scattering cell and based on macroscopic optics, are in this way wholly substantiated.

10. Other applications of the extinction theorem especially in 'many-body' theory

We should make reference to other presentations of the usual extinction theorem (16). Born and Wolf (1970, 1980) interpreted (16) as an integral equation for $\mathbf{P}(\mathbf{x}',\omega)$ and its normal component on the surface Σ of V, given $\mathbf{E}_o(\mathbf{x},\omega)$ (and V), as indeed it is. The only tractable V is the slab however, because, if V is a sphere, for example, it necessarily acts as a lens, which enormously complicates the technicality of the analysis. *In this sense, the whole of macroscopic optics rests in the extinction theorem*. Other important developments in dielectric theory concern the $(\mathbf{k},\omega)$-dependent dielectric constants $\varepsilon(\mathbf{k},\omega)$. In this connection, Sein (1969, 1970) and Birman and Sein (1972) made a remarkable extension of the extinction theorem. Work by Agarwal *et al.* (1971) and by Patternayak and Wolf (1972) must also be consulted in this connection.

In a different direction, Bullough (1970) was concerned with the 'many-body theory' of a dielectric and the problem of computing its binding energy. We should like to comment first on the idea of 'modes' in theoretical physics, as Juretschke has already said something about this and it is intrinsic to Ewald's methods. Ewald took as modes of the crystal the plane dipole waves with wave number k (and wave vector **k**). In theoretical physics one tries to calculate the modes for each **k** and deduce their frequencies ω (or their energies $\hbar\omega$) which will satisfy some dispersion relation $\omega = \omega(\mathbf{k})$, as Juretschke has explained. These modes are usually supposed to run as they are, for they are uncoupled and independent, and there is the separate problem of exciting these modes (typically the modes are 'normal modes'). However, in the problems considered by Ewald, and considered again here, his extinction theorem really shows that the dipole waves cannot run *unless they are excited by the free field*. Moreover, this free field excites at the *frequency* of the dipole waves and it is $k = |\mathbf{k}|$ which becomes a function of ω; namely, $k = m(\omega)k_o$. If this is inverted to $\omega = \omega(k)$, there are many branches of *this* dispersion relation. For example, the excitations associated with the 'acoustic branch' running through $k = \omega = 0$ and close to the first 'optical branch' are called 'polaritons' by solid-state physicists. We think that some of these physicists may not have been acquainted with Ewald's extinction theorem(!), which offers a conceptually comprehensive and simpler view. An attempt to marry these two viewpoints for the molecular crystal was given by Bullough and Thompson (1970).

Moreover the papers by Sein (1969, 1970). Birman and Sein (1972), and Agarwal et al. (1971) successfully treat the 'polariton' problem and the problem of spatial dispersion in the context of the extinction theorem. The paper by Patternayak and Wolf (1972) generalizes the Agarwal et al. (1971) paper. Their main point in effect concerns the extinction theorem as being the origin of 'usual' boundary conditions on the surface Σ in agreement with Darwin (1924), Bullough (1962) and with our remarks at the end of Section 8, and with our use of it in the nonlinear context as described in Section 12. We believe De Goede and Mazur (1972), who take up Sein's argument, misunderstand some of our own work, for example Bullough (1970) and papers in 1968–70 referenced in Hynne and Bullough (1984).

For the many-body theory, Bullough (1970) was concerned with the 'linear response' of a dielectric to a probe wave proportional to $\exp[i(\mathbf{k} \cdot \mathbf{x} - \omega t)]$ in which $\mathbf{k}$ and ω are free, not satisfying $\omega = \omega(\mathbf{k})$ or $k = m(\omega)\omega c^{-1}$. In this situation, one can define a formal $\mathbf{k}$, ω-dependent 'dielectric constant' such that $(1/4\pi)[\varepsilon(\mathbf{k},\omega) - 1]$, (or functions of $\mathbf{k}$ and ω like this) is a linear response function whose resonances (poles) satisfy a dispersion relation, and from which a general many-body theory of intermolecular forces inside an infinite volume V can be developed. However, intermolecular forces are electromagnetic and the problem of the infinite volume V re-emerges.

Specifically, if dipole waves $\mathbf{P}(\mathbf{k},\omega)\exp[i(\mathbf{k} \cdot \mathbf{x} - \omega t)]$ run in V they generate light with wave number $k_o = \omega c^{-1}$. Indeed, the 'extinction theorem' (16) shows that dipole waves generate the negative of the free field $\mathbf{E}_o(\mathbf{x},\omega)$ whether k and ω satisfy dispersion relations or not. Bullough (1970) used this light as a new free field inside V and extinguished *this* by a dipole field travelling with wave number $k = m(\omega)k_o$. The extinction theorem is thus used twice. The theory then provides the binding energy of the infinite dielectric in terms of $\varepsilon(\mathbf{k},\omega) - 1$, and there is a contribution from the surface regions and shape Σ of the finite V involving both $\varepsilon(\mathbf{k},\omega)$ and $m^2(\omega)$.

11. Extinction theorem for the quantized field

We now mention yet another form of the optical extinction theorem. Quantum mechanics enters dielectric theory through the Kramer–Heisenberg formula for $\alpha(\omega)$ (although many-body theory shows there are other strictly quantum contributions to dielectric theory also). However, we know from a continuously increasing mass of evidence that the e.m. field itself is quantized—probably in the fashion described first by Dirac (1927) as canonical quantization, although a new and exciting possibility, 'quantum group' quantization, has recently been described (for example, by Chaichian et al. 1990). The dielectric theory described so far is 'semi-classical' in that the atoms, but not the fields, are described by quantum mechanics. It is also a *linear* theory: all of the fields, especially the incident field $\mathbf{E}_o(\mathbf{x},\omega)$, are weak. For such linear theory one can (roughly speaking) quantize by replacing all field quantities by operators in such a way that, provided operator order is maintained, each piece of the theory is preserved formally unchanged. There is then an operator form (i.e. a quantized form) of the extinction theorem (16) (see Jones 1974).

12. Extinction theorem in nonlinear optics

Once the fields are not weak we enter the regime of *nonlinear* optics. Powerful laser sources mean that this regime is now of great technological and fundamental importance. In a very early paper in this field, Bloembergen and Pershan (1962) used the optical extinction theorem for a half-space occupied by a nonlinear dielectric. They were concerned to justify a Maxwellian approach in their preceding paper (Armstrong et al. 1962) to the fields produced at the boundary by an incident intense laser field. They were able to provide that justification.

Ahmad (1973) developed the extinction theorem for the case of intense (gigawatt cm^{-2}) short (nanosecond) resonant pulses entering a dielectric. The intense driving field induces nonlinear behaviour in the resonant dielectric in such a way that optical solitons (e.g. Bullough 1977) are formed. These propagate without loss of energy and the system becomes transparent to the pulse—so-called 'self-induced transparency'.

An exciting recent application is the following (Bullough et al. 1983; Bullough and Hassan 1983; Ibrahim 1989). The extinction theorem provides the complete formula,

$$|E_{out}|^2 = |E_{in}|^2 \left|\exp[i(m-1)k_oL]\right|^2 \left|\frac{4m}{(m+1)^2}\right|^2$$

$$\times \left|1 - \left(\frac{m-1}{m+1}\right)^2 \exp[2imk_oL]\right|^{-2}, \quad (18)$$

for the output of the Fabry-Perot interferometer of width L for a normally incident input intensity $|E_{in}|^2$. However, it is now supposed that an *intense* c.w. (continuously working) laser field of intensity $|E_{in}|^2$ is imposed from outside and this means that m becomes dependent on the value of $|E_{in}|^2$. We can now control the transmission, namely $|E_{out}|^2$, by controlling the intensity $|E_{in}|^2$ imposed. In simplest terms, the system is optically bistable (the relation (18) provides multistability as explained below, although there are some small corrections to this formula involving the atomic inversions at the surfaces—see below) and the plot of outcoming intensity $I_T(\propto |E_{out}|^2)$ against incoming intensity $I_I(\propto |E_{in}|^2)$ is S-shaped, as shown in Fig. 3.

Because the middle branch of the S is unstable there are *two* states for a given I_I in this region. There is evidently an optical *switch*, which is of great potential value in optical information technology (for application in an optical computer perhaps). It is possible to develop the optical extinction theorem

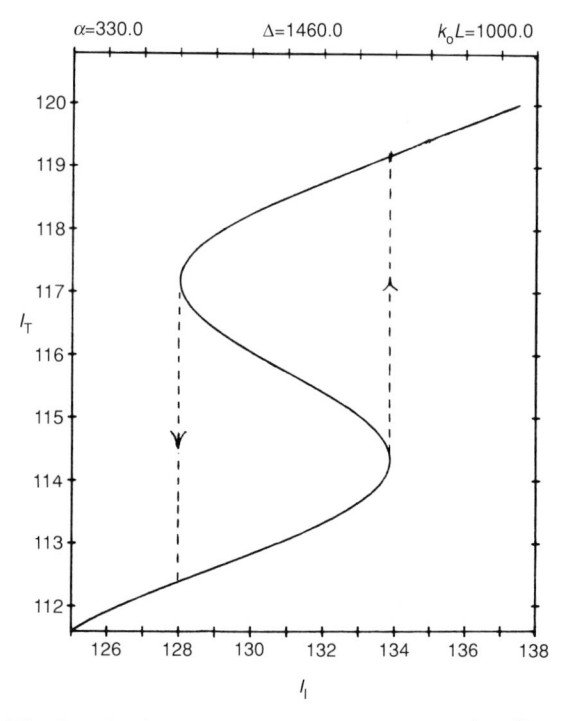

Fig. 3 Optically bistable S-shaped plot of $|E_{out}|^2 = I_T$ against $|E_{in}|^2 \equiv I_I$. The parameters α (= 330) and Δ (= 1460) marked on this figure are α proportional to n, the number density, and Δ to the so-called 'detuning' $\omega_s - \omega$, where ω_s in an atomic resonance frequency: k_oL is as in the text. The middle branch of the S in this figure is unstable, and the system follows the hysteresis loop marked. The hard curve, stable and unstable, continues however—see Fig. 4, which, for a different value of Δ, shows this optically *multi*stable continuation.

for this system and the result is still (18), as explained. The new feature is the dispersion relation for m, which proves to depend on $|E_{in}|^2$. Apparently, there is an immediate theoretical objection to such an argument, as, apparently, there must be to Bloembergen and Pershan's (1962) application and to Ahmad's (1973) application of the extinction theorem. All of the work on the dielectric or the crystal we have described has been *linear* theory, and all of Ewald's work was linear optics. In linear theory, it must be supposed that the electric fields are vanishingly weak. In a nonlinear theory, higher powers of the fields are involved, and, indeed, this shows up in the intensity-dependent refractive index. So how is the extinction theorem still applicable?

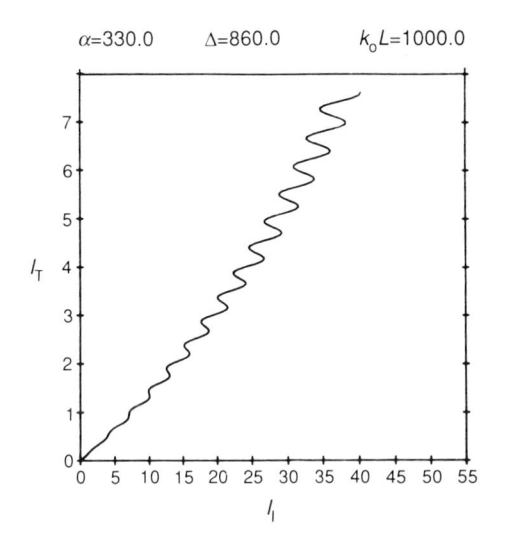

Fig. 4 Plot of $|E_{out}|^2$ against an intense $|E_{in}|^2$ as calculated from the corrected form of eqn (18) as given by Ibrahim (1989); the chosen parameters α (=330) and Δ (=860) are explained in caption to Fig. 3.

The answer is that Maxwell's equations are always linear equations. The nonlinearity arises in the atoms driven nonlinearly. Thus the polarizabilities $\alpha(\omega)$ depend on the incident field intensity but that part of the theory which is essentially the calculation of the fields travelling with wave numbers k or k_o is unchanged.

In the case of the optical bistability problem there are a number of subtleties. One is that the dipole field gains a dependence on points **x** inside V which is not simply $\exp[i\mathbf{k} \cdot \mathbf{x}]$; the 'inversion' of atoms (density of atoms put into excited states) becomes important and proves to depend on the position **x** inside V. Nevertheless, the extinction theorem still applies (as it must).

Perhaps the most spectacular feature of the nonlinear extinction theorem result (18) is its possibilities for 'dynamical chaos' (for an introduction to this new realization in dynamics, see, for example, Baker and Gollub (1990)). Ewald was concerned with his 'dynamical theory'; the smooth but nonlinear dielectric which leads to (18) with m dependent on $|E_{in}|^2$ is also a dynamical theory and, as in Ewald's theory, the dynamical system is in a state of steady oscillation. As mentioned above, the Fabry–Perot interferometer action described by (18) is multistable, and Fig. 4 shows an example of this multistability. In the language of modern dynamics, this curve in Fig. 4 is a curve of dynamical 'fixed points'. Fixed points are equilibrium points but may be stable (adjacent points move always towards them) or unstable (adjacent points try to move away). From a related analysis of the 'ring cavity', which included, however, Maxwell boundary conditions at the surface, Ikeda (1979) showed that instabilities arise on the monotonically *increasing* parts of the multistable equilibrium curve and not only on the decreasing parts. These dynamical instabilities lead to a succession of 'bifurcations' of the motion and eventually to the dynamics of a 'strange attractor'. The 'strange attractor' was first discovered by E.N. Lorenz (1963) (a different Lorenz!) in a theory of Bénard hydrodynamical convection, but his *equations* can be identified in the laser. This is an example of the now well-known dynamics called 'dynamical chaos', which has been much studied in laser physics.

The point now is that the analysis based on the extinction theorem which leads to (18) can be extended to show that such dynamical chaos can also be reached in the Fabry–Perot interferometer!

Ewald discovered a 'fact of Nature' in his extinction theorem. This perhaps explains its elegance and its all-embracing character.

References

Agarwal, G.S., Patternayak, D.N., and Wolf, E. (1971). *Optics Commun.*, **4**, 260–3.

Ahmad, Faiz (1973). Some problems in quantum optics. Ph.D. thesis, University of Manchester.

Armstrong, J.A., Bloembergen, N., Ducuing, J., and Pershan, P.S. (1962). *Phys. Rev.*, **127**, 1918–39.

Baker, G.L. and Gollub, J.P. (1990). *Chaotic dynamics*. Cambridge University Press.

Birman, J.L. and Sein, J.J. (1972). *Phys. Rev.*, **B6**, 2482–90.

Bloembergen, N. and Pershan, P.S. (1962). *Phys. Rev.*, **128**, 606–22.

Born, M. and Wolf, E. (1970). *Principles of optics*, 4th edn. Pergamon, Oxford.

Born, M. and Wolf, E. (1980). *Principles of optics*, 6th edn. Pergamon, Oxford.

Bullough, R.K. (1962). *Phil. Trans. Roy. Soc. (London)*, **A254**, 357–440.

Bullough, R.K. (1970). *J. Phys.*, **A3**, 726–50.

Bullough, R.K. (1977). Solitons. In *Interaction of radiation in condensed matter*, Vol. 1, pp. 381–469. International Atomic Energy Agency, Vienna.

Bullough, R.K. and Hassan, S.S. (1983). The optical extinction theorem in the nonlinear theory of optical multistability. In *The Max Born centenary conference proceedings* (September 1982) (ed. M.J. Colles and D.W. Swift), Vol. 369, pp. 257–363. SPIE, Billingham, WA.

Bullough, R.K. and Thompson, B.V. (1970). *J. Phys.*, **C3**, 1780–4.

Bullough, R.K., Hassan, S.S., and Tewari, S.P. (1983). In *Quantum electronics and electro optics* (ed. P.L. Knight), pp. 229–32. Wiley, New York.

Chaichian, M., Ellinas, D., and Kulish, P. (1990). *Phys. Rev. Lett.*, **65**, 980–3.

Darwin, C.G. (1924). *Trans. Camb. Philos. Soc.*, **23**, 137–67.

De Goede, J. and Mazur, P. (1972). *Physica*, **58**, 568–84.

Dirac, P.A.M. (1927). *Proc. Roy. Soc. (London)*, **A114**, 243–65.

Einstein, A. (1910). *Ann. Phys. (Leipzig)*, **33**, 1275–98.

Ewald, P.P. (1912). Dispersion und Doppelbrechung von Elektronengittern (Kristallen). Doctoral thesis, Royal Ludwigs-Maximilian University, Munich.

Ewald, P.P. (1916a). *Ann. Phys. (Leipzig)*, **49**, 1–37.

Ewald, P.P. (1916b). *Ann. Phys. (Leipzig)*, **49**, 117–43.

Ewald, P.P. (1917). *Ann. Phys. (Leipzig)*, **54**, 519–97.

Ewald, P.P. (1925). *Fortsch. Chem. Phys. Phys. Chem.*, **18**, 491–518.

Ewald, P.P. (1937). *Z. Kristallogr.*, **A97**, 1–27.

Fixman, M. (1955). *J. Chem. Phys.*, **23**, 2074–9.

Fresnel, A. (1832). *Mém. Acad.*, **11**, 393ff.

Gibbs, J.W. (1881). *Vector analysis*. Privately printed, New Haven, CT.

Green, G. (1828). *An essay on the application of mathematical analysis to the theories of electricity and magnetism*. Published through subscription at Nottingham.

Hoek, H. (1939). Algemeene theorie der optische activiteit van isotrope media. Ph.D. Thesis, University of Leiden.

Hynne, F. and Bullough, R.K. (1984). *Phil. Trans. Roy. Soc. (London)*, **A312**, 251–93.

Hynne, F. and Bullough, R.K. (1987). *Phil. Trans. Roy. Soc. (London)*, **A321**, 305–60.

Hynne, F. and Bullough, R.K. (1990). *Phil. Trans. Roy. Soc. (London)*, **A330**, 253–313.

Ibrahim, M.N.R. (1989). Numerical investigation in the refractive index theory of multistability. M.Sc. Thesis, UMIST, University of Manchester.

Ikeda, K. (1979). *Optics Commun.*, **30**, 257–60.

James, R.W. (1954). *Optical principles of the diffraction of X-rays*. Bell, London.

Jones, D.L. (1974). Some problems in quantum optics and scattering theory. Ph.D. thesis, University of Manchester.

Kramers, H.A. and Heisenberg, W. (1925). *J. Phys.*, **31**, 681ff.

Lorentz, H.A. (1880). *Ann. Phys. Chem.*, **9**, 641ff.

Lorentz, H.A. (1909). *The theory of electrons*. Leipzig. (Reprinted in 1952: Dover, New York).

Lorenz, E.N. (1963). *J. Atmos. Sci.*, **20**, 130–41.

Lorenz, L. (1880). *Ann. Phys. Chem.*, **11**, 70ff.

Mazur, P. (1958). *Advances in chemical physics*, Vol. 1, p. 309–60. Interscience, New York.

Mazur, P. and Mandel, M. (1956). *Physica*, **22**, 289ff.

Oseen, O.W. (1915). *Ann. Phys. (Leipzig)*, **48**, 1–56.

Patternayak, D.N. and Wolf, E. (1972). *Optics Commun.*, **6**, 217–20.

Rosenfeld, L. (1951). *Theory of electrons*. North-Holland, Amsterdam.

Rosenfeld, L. (1965). *Theory of electrons*. Reprinted with new preface. Dover, New York.

Sein, J.J. (1969). Ph.D. thesis, New York University.

Sein, J.J. (1970). *Optics Commun.*, **2**, 170–3.

Yvon, J. (1937). *Actual. Sci. Ind.*, No. 543.

D
Ewald speaks for himself

1913: Contributions to the theory of the interferences of X-rays in crystals, *Phys. Z.* (1913), **14**, 465–72 (Translated by Hellmut J. Juretschke)

Ewald's thesis on crystal optics was presented in February 1912. When he heard of the work of Laue, Friedrich, and Knipping, he realized that the X-ray experiments could be interpreted by his theory, as his mathematical analysis was valid for electromagnetic waves of any wavelength. This paper, with its construction of the reciprocal lattice and the sphere of reflection, then resulted.

Translator's preface

This translation aims at combining a fair rendering in modern idiom of the original with the nuances and ambiguities typical of the beginning of any new field in which the appropriate vocabulary is yet to be agreed on. It is faithful to the structure and expression of the German original, perhaps excessively so, at the expense of a freer literary interpretation, to allow a scholarly evaluation of its content, and preserve the historical flavour. This will also make it easier for an interested reader to carry out a close textual comparison with the original.

I would like to acknowledge the extensive corrections and suggestions of D.W.J. Cruickshank, G. Hildebrandt, A.F. Moodie, and H.K. Wagenfeld on an early draft of this translation. The incorporation of their comments has crucially shaped the later versions; of course, the final responsibility, including errors or misreadings, remains fully my own.

The occasional errors by the printer or slips by the author, ambiguities in the text, or changes in notation, have been corrected or amended, and identified with an appropriate translator's footnote designated by [TN ...]. I have also taken the liberty to include within the same scheme some commentary by D.W.J. Cruickshank based on this translation of Ewald's paper, which puts certain specific aspects in better historical context. Chapter 6 by D.W.J. Cruickshank includes a further discussion of Ewald's 1913 paper and the sphere of reflection.

This 1913 publication is a short preliminary version of Ewald's three monumental papers in *Ann. Phys.*, **49**, 1–38 (1916); **49**, 117–43 (1916); and **54**, 519–597 (1917), which include most of the same material, but worked out in much more detail. If needed in the context of other translator's notes, these will be referred to below as Ewald Part I, Ewald Part II, etc.

The figures in this translation are reproductions of those published in 1913. The lettering is not easily legible, but nothing has been redrawn to preserve the authentic appearance of the original illustrations of Ewald's reciprocal lattice and sphere of reflection. The reader should note that the vertical axis changes from x in Figs. 1 and 2 to $n\pi/c$ (equivalent to z) in Figs. 3, 4, 5, and 7.

The reader will observe that Ewald refers to 'M. Laue' and not to 'M. von Laue'. As Ewald explained many years later in his obituary for Max von Laue (*Acta Cryst.* (1960)., **13**, 513–15), Laue's father received the hereditary nobility in 1913. It was as Max von Laue that Laue received the Nobel Prize for Physics in 1914.

Contributions to the Theory of the Interferences of X rays in Crystals

P.P. Ewald

Currently there are two points of view on the interpretation of the interference effects described by Laue.[1] The explanation proposed by Laue ascribes the maxima of beam intensity emerging from the crystal slab to the cumulative action in the interior of the crystal of the individual oscillators (dipoles), the regular arrangement of which allows the crystal to be viewed as a three-dimensional diffraction grating. On the other hand, following Bragg, the orientation of the interference patterns can be adequately described on the assumption that the incident X-ray undergoes an optical reflection from each crystallographically allowed plane (lattice plane of the space lattice). On detailed examination, there is no difference between these two approaches, as has already been shown by G. Wulff[2] and M. Laue[3] (see also § 5). The connection between the phenomena of X-rays and ordinary light passing through a crystal can be deduced easily from the equations of §§ 3 and 4 [TN1, 2].

Further, in §§ 5 and 6 conclusions will be drawn about the nature of the interference beams and about the shape of the spots. § 7 is devoted to their intensity. However, no attempt is made to analyse a real photograph to deduce the space lattice responsible for it.

§ 1. Let the dipoles of the space lattice be located at the points.

$$X = 2al, \; Y = 2bm, \; Z = 2cn \quad (1)$$
$$(l,m,n \text{ integers})$$

with X, Y, Z denoting rectangular coordinates. The lattice is therefore of the simplest orthorhombic type (or, respectively, tetragonal or cubic).

Suppose that an X-ray propagates in the crystal in the direction s

$$s = x \cos(s,x) + y \cos(s,y) + z \cos(s,z), \quad (2)$$

as indicated schematically in Fig. 1. Its speed inside the material is nearly that of light c, as there is no noticeable refraction of the beam. However, on theoretical grounds, the exact value c cannot be assigned to the X-ray beam, because, just as in optics, the induced oscillation of the dipoles influences the propagation of the wave. A difference relative to optics derives from the fact that the oscillations

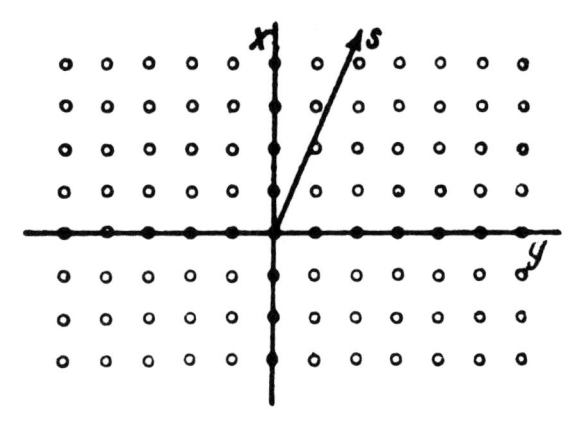

Fig. 1

excited by the X-ray are so weak that the phase velocity is not detectably influenced. Nevertheless, at first we will differentiate below between the propagation velocity in empty space (c) and that in the medium (q) because this allows an easier transition to the optical case.

Close passage of the wave sets each dipole into motion containing oscillations that can be resolved into contributions periodic in time [TN3]. Let one such oscillation have the frequency ω. Then the resulting electromagnetic spherical wave can be described by the Hertz potential

$$\mathbf{a} \frac{\exp[-i\omega(t - \frac{R}{c})]}{R}, \quad (3)$$

where $\mathbf{a}$ is the amplitude of the oscillation, and R is the distance from the dipole to the point of observation. Because of successive excitation of dipoles, either by an X-ray or by an optical wave, the potential of a dipole with co-ordinates XYZ is given by

$$\mathbf{P} = \mathbf{a} \frac{\exp[-i\omega(t - \frac{R}{c} - \frac{S}{q})]}{R}, \quad (4)$$

where S denotes, in analogy with the s of (2),

$$S = X \cos(s,x) + Y \cos(s,y) + Z \cos(s,z),$$

CONTRIBUTIONS TO THE THEORY OF THE INTERFERENCES OF X RAYS

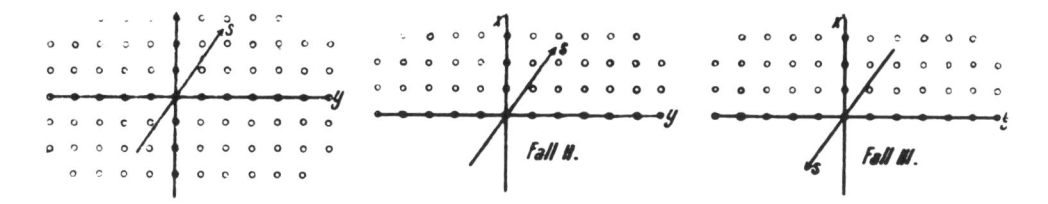

Fig. 2

that is, it measures the distance of the dipole from the plane of zero phase.

With the abbreviations

$$\frac{\omega}{c} = K_o \left(= \frac{2\pi}{\lambda} \right), \frac{\omega}{q} = K,$$

$$K \cos(s,x) = \alpha, \quad K \cos(s,y) = \beta, \qquad (5)$$

$$K \cos(s,z) = \gamma,$$

and omitting the time-dependent factor, the total potential is

$$\mathbf{a}\Pi = \mathbf{a} \sum \frac{\exp(iK_o R + iKS)}{R}, \qquad (6)$$

which is to be summed over all dipoles [TN4]. It should be recalled that the electric field strength **E** follows from the potential by the operation

$$\mathbf{E} = \text{rot rot } \mathbf{a}\Pi = \text{grad div } \mathbf{a}\Pi - \mathbf{a}\Delta\Pi \text{ [TN5]}. \quad (7)$$

§ 2. At this point one has to distinguish several cases, depending on whether one deals with the propagation of the X-ray in the crystal interior, with its incidence onto, or with its emergence from the crystal.

I. In the first case, of course, one can take the lattice of (1) to fill all of space, and correspondingly carry out the sum in (6) over l,m,n ranging from $-\infty$ to $+\infty$. This is exactly the problem solved in my dissertation,[4] so that the result of the summation can be taken over from there.

II. The incidence of an X-ray on the crystal slab is obtained when the lattice occupies only a half-space. Then the sum in (6) ranges for l,m from $-\infty$ to $+\infty$, but for n from 0 to $+\infty$ (the upper half-space is filled by the crystal).

III. The exit from the crystal can be studied if the beam direction in II is reversed, i.e. α, β, and γ in (5) are replaced by $-\alpha$, $-\beta$ and $-\gamma$.

Schematically, we obtain the following three pictures (Fig. 2).

§ 3. Let us start with case I because in principle it is the simplest, and leads directly to the electromagnetic proper modes of the space lattice. For this case the potential (6) becomes (compare also with eqn (7), § 6 of my dissertation) [TN6]

$$\Pi = -\frac{\pi}{2abc} \sum_{l,m,n}^{-\infty \ldots +\infty} \frac{\exp\left(-i\frac{l\pi - a\alpha}{a}x - i\frac{m\pi - b\beta}{b}y - i\frac{n\pi - c\gamma}{c}z\right)}{K_o^2 - \left(\frac{l\pi - a\alpha}{a}\right)^2 - \left(\frac{m\pi - b\beta}{b}\right)^2 - \left(\frac{n\pi - c\gamma}{c}\right)^2}.$$

(8)

This potential is precisely in the form of a sum over successive proper modes, such as, for instance, given by the acoustic potential of a source of sound in a rectangular enclosure.[5] The individual proper mode is given by [see TN6]

$$u_{l,m,n}$$
$$= \exp\left(-i\frac{l\pi - a\alpha}{a}x - i\frac{m\pi - b\beta}{b}y - i\frac{n\pi - c\gamma}{c}z\right)$$
$$= e^{+iKs} \cdot \exp\left[-i\left(l\pi\frac{x}{a} + m\pi\frac{y}{b} + n\pi\frac{z}{c}\right)\right]. \quad (9)$$

It consists of a factor of the same form as the exciting X-ray beam (compare with the Ansatz (4) or (6)) [TN7], and a second factor which repeats periodically in the elementary parallelepipeds of the lattice (i.e. the component of a Fourier series). The mode can be interpreted as a plane wave whose direction cosines are proportional to

$$\frac{l\pi - a\alpha}{a} : \frac{m\pi - b\beta}{b} : \frac{n\pi - c\gamma}{c}. \qquad (9a)$$

Depending on whether a visible light beam or an X-ray beam is propagated, that is, depending on the frequency ω associated with K, one or the other factor in $u_{l,m,n}$ is of primary interest. Light waves have wavelengths long compared with the lattice spacing $2a$, so that

$$2\pi a/\lambda = aK_o \ll 1 \ (1/200) \tag{10}$$

and this applies equally to the products $a(K,\alpha,\beta,\gamma)$. For X-rays, on the other hand, these quantities are certainly significantly larger than one (20–200 times). We therefore distinguish:

1. $K \ll \pi/a$, an optical wave

The part that is periodic in the elementary parallelepipeds contributes no wavelengths comparable with optical wavelengths. On the other hand, the factor e^{iKs} produces waves of long wavelength which, relative to the terms in the Fourier series, are best described as an extended modulation [TN8]. Optically, the only significant contribution from the entire Fourier series derives from the mean value in a unit cell, represented by the *000* term of the series. Hence the optical potential is [TN9]

$$\Pi = \frac{\pi}{2abc} \cdot e^{-iKs} \frac{1}{K^2 - K_o^2}. \tag{11}$$

The higher terms in the sum contribute amplitudes negligible in comparison with the *000* term, as $K \ll \pi/a$.

2. $K \gg \pi/a$, an X-ray beam

Here the terms in the sum (8) that have the largest amplitude are those closest to resonance, that is, those for which the integral indices l,m,n most nearly satisfy the equation

$$\left(\frac{l\pi - a\alpha}{a}\right)^2 + \left(\frac{m\pi - b\beta}{b}\right)^2 + \left(\frac{n\pi - c\gamma}{c}\right)^2 = K_o^2. \tag{12}$$

This condition selects from the infinite number of proper harmonics of the space lattice those which, for a given frequency of the X-ray, have the largest intensity; in cases II and III of the half-crystal, these are the beams that pass through the boundary plane to appear as interference beams on the outside.

In connection with the geometrical interpretation of this condition, it must be stressed that in the case of X-rays K_o and K are practically the same. This is deduced from the experimental absence of even the least refraction of X-rays; extrapolation of the index of refraction based on a dispersion formula (e.g. that of Planck) yields magnitudes such as

$$v^2 = \left(\frac{K}{K_o}\right)^2 = 1 - (4 \times 10^{-8}), \tag{13}$$

representing an undetectable deviation from 1. We can therefore assume

$$\alpha^2 + \beta^2 + \gamma^2 = K^2 = K_o^2. \tag{14}$$

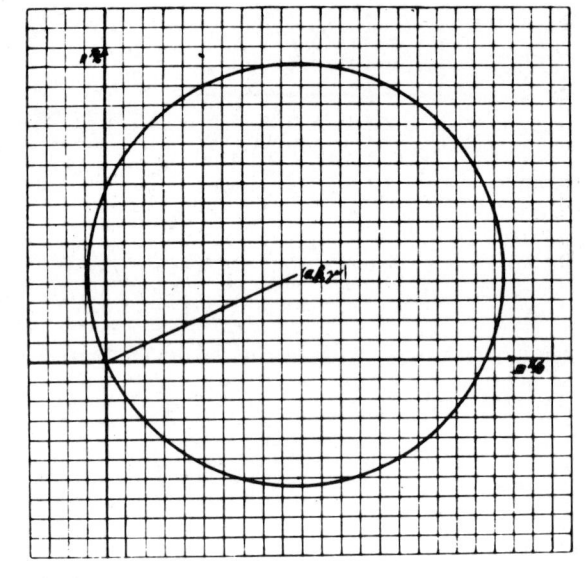

Fig. 3

Now condition (12) implies the following *construction*: Consider a lattice with spacings π/a, π/b, π/c ('reciprocal lattice') and around the point (α,β,γ) construct a sphere which passes through the origin of the lattice. If other lattice points $[l_0,m_0,n_0]$ also lie on the sphere's surface, there will exist in the crystal waves with very large intensity having the same direction as the lines connecting $[l_0,m_0,n_0]$ with the origin of the sphere.

As can be seen from (8), the direction cosines of the line connecting $[l_0,m_0,n_0]$ with α, β, γ in the reciprocal lattice are the same as the direction cosines of the wave normal of the corresponding wave propagating in the crystal. It follows from (12) and (14) that the origin of the lattice always lies on the surface of the sphere.

It is worth noting that the waves of largest intensity are at the same time the only waves of the potential (8) which propagate inside the crystal with the velocity of light c. For all other terms in (8) the sum of squares of the factors of x,y,z in the exponential differs from K_o^2.

It is possible, using Fig. 3, to read off immediately the well-known predictions about the symmetry at normal incidence, etc. Nevertheless, at this point it seems most useful to proceed to cases II and III, to investigate the fields outside the crystal and to determine the influence of the boundary plane $z = 0$.

§ 4. Case II. Incidence of an X-ray beam upon the crystal.

If the normal oscillations of the 'half-crystal' are summed according to exactly the same Ansatz (6) as for the dipole oscillations, one has to distinguish between an external and an internal region. The boundary between these is not, as one might at first assume, the last plane $z = 0$, but because of the periodicity of the field the crystal appears to be extended by a spacing $2c$. It is therefore necessary to count in addition to the upper half-space a layer of thickness $2c$ of the lower half-space as part of the crystal. Accordingly, the following results for the external region are valid and exact for all z:

$$z + 2c < 0, \quad (15)$$

and this holds at any distance from the boundary plane [TN10].

The potential is[6]

$$\Pi = \frac{\pi}{4ab} e^{ic\gamma} \sum_{l,m}^{-\infty\ldots+\infty} \frac{\exp[-i\frac{l\pi - a\alpha}{a}x - i\frac{m\pi - b\beta}{b}y - iv_{l,m}(z+c)]}{v_{l,m} \cdot \sin c\,(v_{l,m} + \gamma)}, \quad (16)$$

with

$$v_{l,m} = \sqrt{\left[K_o^2 - \left(\frac{l\pi - a\alpha}{a}\right)^2 - \left(\frac{m\pi - b\beta}{b}\right)^2\right]}. \quad (16a)$$

Each individual term in the sum again represents a plane wave, propagating with velocity c. The amplitude can attain a maximum value depending on which of the two factors in the denominator is close to zero.

1. $v_{l,m} = 0$, in other words

$$\left(\frac{l\pi - a\alpha}{a}\right)^2 - \left(\frac{m\pi - b\beta}{b}\right)^2 = K^2. \quad (17)$$

These terms denote plane waves running parallel to the boundary plane. One can designate them as surface waves because they do not seem to be able to exist apart from the surface. They have not been detected experimentally, and it is theoretically plausible that at the boundary of the crystal surface their energy goes over into diffracted energy while the surface waves are being destroyed. To establish the impossibility of detecting these waves it is necessary that only a bounded area of the crystal surface is illuminated by an aperture-limited X-ray beam [TN11].

2. $\sin c(v_{l,m} + \gamma) = 0$, which means

$$\sqrt{\left[K_o^2 - \left(\frac{l\pi - a\alpha}{a}\right)^2 - \left(\frac{m\pi - b\beta}{b}\right)^2\right]} = \frac{n\pi - c\gamma}{c}. \quad (18)$$

This is precisely the condition (12), which now justifies the assertion of § 3 that *those proper modes of the space lattice that are in resonance can transfer to the outside*.

In (16a) the positive root is to be taken for v_{lm}, so that the waves move away from the boundary into the external region. According to (18), this means that only those values of $n\pi/c$ are admissible that are larger than γ. In using the geometrical interpretation of Fig. 3 one should therefore take only those waves l_0, m_0, n_0 that travel downwards ($n_0\pi/c > \gamma$) [TN12]. This condition uniquely specifies the interference pattern associated with a frequency ω, that is, a fixed value K_o, in front of the crystal.

Case III. Emergence of an X-ray beam

The potential for the external region is obtained from (16) by reversal of the beam direction (α,β,γ):

$$\Pi = -\frac{\pi}{4ab} e^{-ic\gamma} \sum_{l,m}^{-\infty\ldots+\infty} \frac{\exp[-i\frac{l\pi + a\alpha}{a}x - i\frac{m\pi + b\beta}{b}y - iv_{l,m}(z+c)]}{v_{l,m} \sin c\,(v_{l,m} - \gamma)}, \quad (19)$$

with

$$v_{l,m} = \sqrt{\left[K_o^2 - \left(\frac{l\pi + a\alpha}{a}\right)^2 - \left(\frac{m\pi + b\beta}{b}\right)^2\right]}. \quad (19a)$$

The observable waves of largest intensity are found by a construction analogous to that of Fig. 3, in which the point $(-\alpha,-\beta,-\gamma)$ is chosen as the origin of the sphere of radius K_o. As the root (19a) is again taken as positive, one now needs $n\pi/c > -\gamma$; these waves are represented on the sphere drawn in Fig. 3 by points lying on its *lower* half. Figure 4 makes it possible to read off the pattern obtained at the front and at the back of the crystal, using an arrangement of directions differing somewhat from that of Figs 2 and 3. Specifically, the arrows are now drawn to include both forward and backward scattering by the incident primary beam from the crystal slab sketched in the figure.

§ 5. Using Fig. 4, what general statements can be made about the interference patterns? Let us distinguish between *two cases*: in one, the frequency, i.e. K_o, is held fixed and the direction of incidence is varied. In this variation the point (α,β,γ) moves on the sphere of radius K_o constructed around the origin of the reciprocal lattice. Alternatively, in the second case the direction of the beam is fixed while the

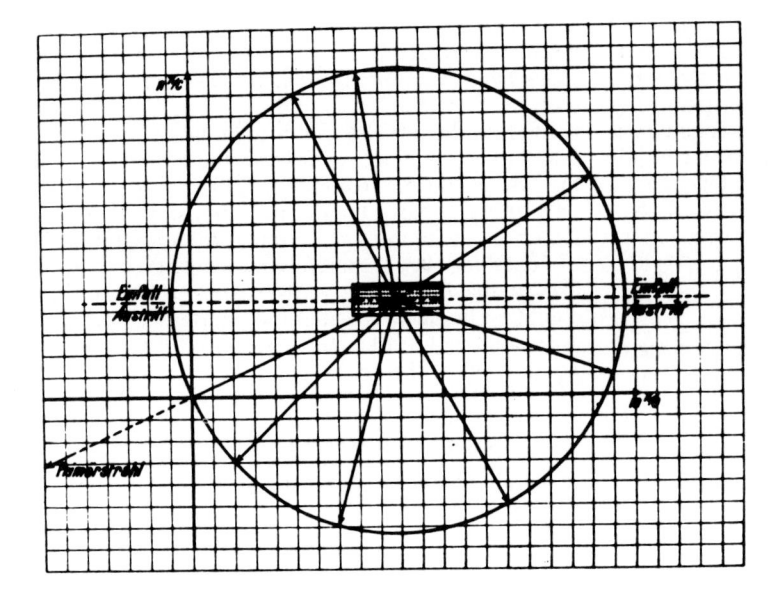

Fig. 4

wavelength of the X-rays (or their Fourier composition) varies within a certain range. Finally, the simultaneous existence of both types of variation will be considered separately in § 6.

In this discussion special attention should be given to explaining the 'reflected' beam, whose origin is not entirely obvious. Although it is immediately apparent from Fig. 4 that there is always an interference maximum in the direction of the incident beam (obscured in photographs by the unscattered beam), the existence and the strong intensity of the reflected beam follows only after a series of considerations.

Case 1. K_o fixed, variable direction of the primary beam

1. The figure permits one immediately to read off the *reciprocity theorem* that connects the directions of the incident and diffracted beams; if the diffracted beams are characterized by their angles relative to the crystal lattice, *a geometrically identical interference pattern results if the incident X-ray beam is interchanged with any one of the diffracted beams.*

That this interchange alters the intensity distribution, and the rule for this alteration, follow from the decomposition into components carried out below in § 7.

2. A reflected beam exists only for particular locations of the point (α,β,γ); namely, with χ as the angle of incidence, only if (α,β,γ) lies in one of the (symmetry) planes $z = $ const. of the reciprocal lattice:

$$\gamma = K_o \cos \chi = h\pi/2c, \quad (20)$$

where h is an integer. Introducing the 'wavelength' $\lambda = 2\pi/K_o$, the angle of incidence must obey the condition

$$\cos \chi = (h/2)(\lambda/2c).$$

The angular difference between two neighbouring directions for which reflection is at all possible must therefore be

$$\Delta\chi = \lambda /(4c\sin\chi) \quad (20a)$$

[TN13].

A similar formula was used by E. Hupka and W. Steinhaus[7] to calculate the wavelength from streaks in the 'reflected' interference spot, with $\Delta\chi$ denoting the angular difference between the streaks, that is, between those directions in which the reflection appears especially strong. On the other hand, other sources explain these streaks as caused accidentally by cleavage planes inside the crystal.

Case 2. Fixed direction, variable K_o

The true time dependence of the electric field strength of an X-ray can scarcely be that of the single harmonic oscillation so far considered. Rather, alone the concept that the X-ray momentum is created by the deceleration of a cathodic particle suggests that the X-ray spectrum must be represented by a Fourier expansion in which many terms have an appreciable amplitude. Let us now assume that *the wavelength of the incident radiation varies continuously over a certain range.* Then a corresponding range ΔK_o exists, for $K_o = 2\pi/\lambda$. Let us further suppose in what follows

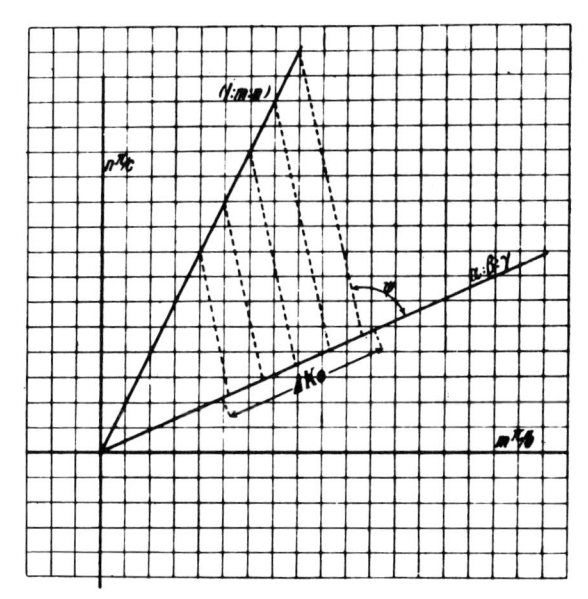

Fig. 5

that the intensity of the primary X-ray spectrum does not vary too rapidly with wavelength. Using Fig. 4, one is easily convinced of the following.

1. As only the ratio of the sphere radius K_o to the lattice spacing influences the location of the interference spots, the interference patterns of similar space lattices are equal as long as

$$(\pi/c)/K_o = \lambda/2c \qquad (21)$$

is constant. Because, in fact, it is observed[8] that substances with the same space lattice and different molecular volume show the same interference patterns, this can be considered a *proof* for the proposition that *the spectrum of the primary X-ray extends continuously over a certain range*. It is equally supported by the invariance of the interference patterns under changes in temperature.[9]

2. *Each interference beam contains waves of differing wavelengths*.[10] Specifically, if one has found an interference spot, $[l_0, m_0, n_0]$ one can then construct the straight line $(l:m:n)$ between it and the origin, which crosses all lattice points that can produce interference beams in the same direction

$$\left(l_0\frac{\pi}{a} - \alpha\right) : \left(m_0\frac{\pi}{b} - \beta\right) : \left(n_0\frac{\pi}{c} - \gamma\right)$$

once the direction $\alpha:\beta:\gamma$ of the primary beam is given (see Fig. 5).

As to the intensity of the interference spot, it depends on how many lattice points are located at positions along the straight line for which the corresponding values of K_o lie within the given interval ΔK_o. The larger the number of such points, the larger will be the intensity of the common spot produced by these beams. The number of points per unit length along the line $l:m:n$ is proportional to $(L^2 + M^2 + N^2)^{-1/2}$, if L, M, N are the smallest integers of the desired ratio; the length of the straight line appropriate to these points is $2\Delta K_o \cos(\Psi/2)$, where Ψ denotes the angle between the incident and the interference beam. As a result, the intensity of the image is proportional to[11]

$$2 \cos \Psi/2 \cdot \frac{\Delta K_o}{\sqrt{(L^2 + M^2 + N^2)}}. \qquad (22)$$

For the reflected beam $\Psi/2$ is equal to the angle χ.

3. Beams in the direction

$$\left(l\frac{\pi}{a} - \alpha\right) : \left(m\frac{\pi}{b} - \beta\right) : \left(n\frac{\pi}{c} - \gamma\right)$$

can be interpreted, in the sense of Bragg's theory, as optical reflections of the incident beam by a plane normal to the straight line $(l:m:n)$. This plane has the indices (L, M, N) and it is evident from (22) that all crystal planes with low indices appear to reflect particularly well.

§ 6. *The shape of the interference spots*

One can deduce from this shape that a single X-ray beam of well-defined direction and wavelength is insufficient to explain the observations. Rather, each spot is the superimposition of a sequence of individual maxima produced by various components of the incident intensity.

According to expressions (16) and (19) for the potential in the outer region, plane waves emerge from the boundary layer. Suppose that the incident beam is limited to a circular aperture, and that it truly possesses only a *single* direction ('plane wave momentum'); in that case, all interference spots should be circular if the photographic plate is parallel to the surface. These circles would then have the same diameter as the aperture, as shown in Fig. 6. This agrees with the latest observations by Laue,[12] who found that the spot shape becomes more circular the further away the X-ray tube is from the crystal. In the original pictures of Laue, Friedrich, and Knipping[13] the spot is elongated, more or less elliptical, with the long axis at right angles to the radial direction from the centre of the whole figure. Such a shape can be explained as a superimposition of many maxima, using the construction of Fig. 4, as follows. Initially, let the figure be drawn for normal incidence (Fig. 7). This leads to the interference point P. Now let the direction of incidence vary within a small cone $d\Omega$; does this lead to a cone of neighbouring interference beams around the original

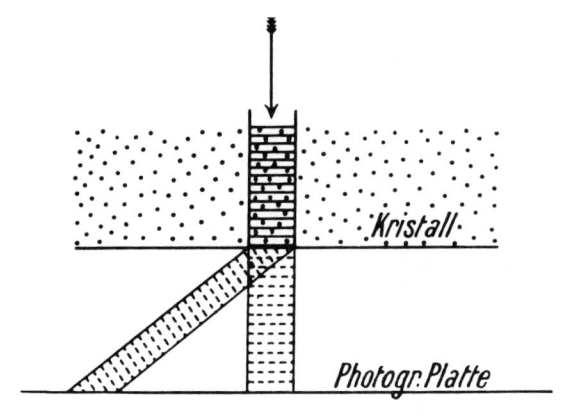

Fig. 6

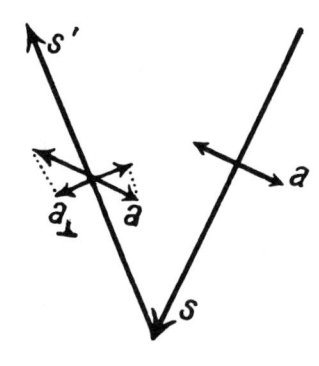

Fig. 8

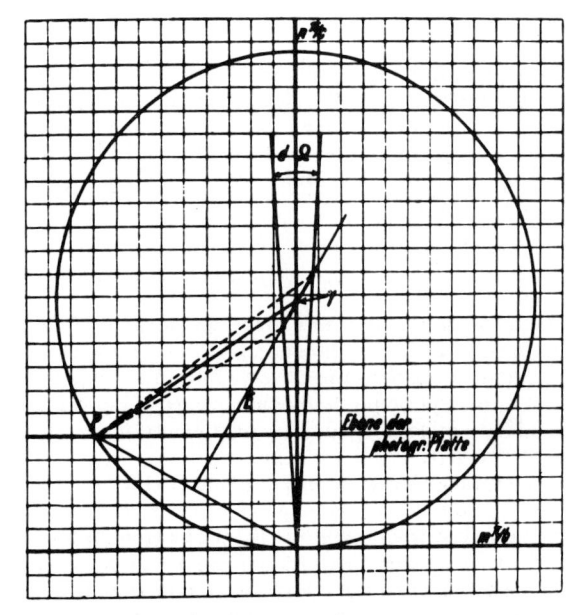

Fig. 7

one? The question is whether there exist in Fig. 7 positions of the origin of the sphere (α,β,γ) which again lead to P as the interference point.

To find such positions of (α,β,γ) one has to look for those points within the cone $d\Omega$ that are equidistant from the origin and from P. This is satisfied on the plane E (see Fig. 7) which cuts the cone $d\Omega$ in an elliptical cross-section. From each point within the ellipse there emerges a beam directed towards P, and the totality of these beams forms a circular cone. The intersection of this cone with the plane of the photographic plate is an ellipse whose small axis lies in the plane of the incident and interference beams. Each point within the ellipse is the centre of a circular interference spot, and therefore the experimental spots are elliptical, and decrease only slowly in intensity towards their edge.[14]

§ 7. So far the discussion has been tied to the Hertz potential of (8), or (16) and (19). Now it remains to determine what consequences arise from the transition to the electric field strength **E**, following the rule (7).

The potentials Π have to be multiplied by the vector amplitude **a**, the amplitude of the dipole oscillations. Therefore the potential of any one of the terms in the sum is essentially of the form

$$\mathbf{a} \cdot \exp(-iKs'), \qquad (23)$$

where s' is the direction of an interference beam. If this potential is used to construct **E**, it turns out that the component $\mathbf{a}_\parallel$ of **a**, along the direction of s, makes no contribution to **E**. Only the other component, $a_\perp$ is effective.

It is plausible to assume that the dipoles oscillate at right angles to the incident X-ray beam (s), as this beam is by far the most intense. To compute the intensity of an interference beam (s') the dipole amplitude **a** has to be decomposed into the component $(\mathbf{a}_1)$ oscillating in the plane (ss') and the component $(\mathbf{a}_2)$ along the normal $(s \times s')$ (see Fig. 8). This latter component contributes fully to s', whereas the former does so only with a factor $\cos(s's)$. Hence the intensity of the beam s' is [TN14]

$$\mathbf{a}_2^2 + \mathbf{a}_1^2 \cos^2(s's). \qquad (24)$$

One can see from the interference patterns that the incident X-rays (or the corresponding dipole oscillations) cannot be polarized, because, if they were, the intensities of two interference beams making the same angle $(s's)$ with the primary beam (corresponding interference spots in different quadrants) should be different, as the decomposition into components $\mathbf{a}_1$ and $\mathbf{a}_2$ is not the same for the two beams.

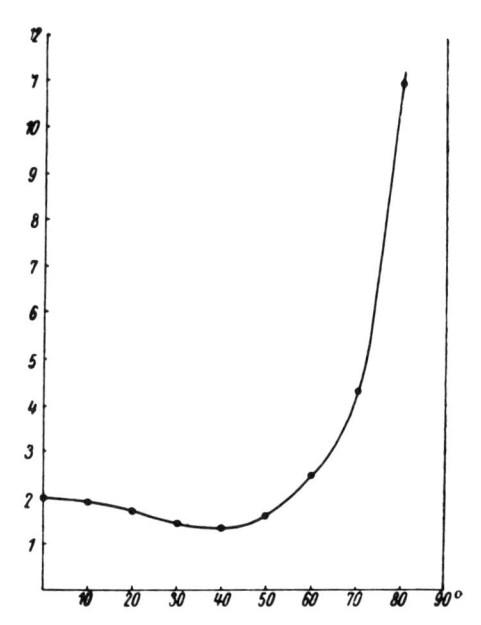

Fig. 9

With $\mathbf{a}_1 = \mathbf{a}_2$, the intensity of a beam is thus proportional to

$$1 + \cos^2(s's). \quad (25)$$

It is worth while to construct the dependence of the reflected beam intensity on the angle of incidence. For the reflected beam we have (compare (20))

$$n\pi/c = 2\gamma,$$

and, according to (18),

$$v_{lm} = \gamma = K_o\cos\chi,$$

Following from (16), therefore, the amplitude is proportional to $1/\cos\chi$. Furthermore, (22) shows that the number of wavelengths (values of K_o) included in the reflected beam is proportional to $\cos\chi$. Taking into account (16), (22), and (25), the intensity is proportional to

$$(1 + \cos^2 2\chi)/\cos\chi. \quad (26)$$

This function is plotted in Fig. 9.

The reflectivity increases sharply with the angle of incidence away from the normal $\chi = 0°$. Although it does not vary markedly between 0° and 60°, at 80° it has reached five times its value at 60°. Actually, at glancing incidence, refraction at the edges of the surface also becomes important (just as with the surface waves in § 4, case II, sub-case 1), a factor which has not been taken into account in this theory, and which would prevent further increase in intensity. In this connection, one should compare the reflection of a light wave from an infinite and a bounded mirror. Experiments currently being undertaken by Mr Friedrich to measure the intensity of the reflected beam have not yet reached the stage of quantitative evaluation, but in the photographs there is an unmistakable rapid decrease of intensity at angles of incidence around 70°.

§ 8. Relation to other theories; characteristic emission from the anticathode

The above considerations contain the same ideas as the theory of Laue and differ from it only by idealizing the problem (laterally unbounded crystal, but, on the other hand, validity of the formulae arbitrarily close to it). Therefore they yield the same results which Laue presented in his first publication. Equation (9a) for the direction cosine of the interference beam is identical with Laue's equations (7), if these are specialized to an orthorhombic lattice. Eliminating the direction cosines α,β,γ in Laue's (7) yields exactly our (12). In addition, all further geometrical consequences of both theories should be identical. In particular, Laue also finds the condition (20) for the appearance of the reflected beam from a monochromatic incident X-ray.

Formulae for the intensity have been obtained by Bragg[15] and by Ornstein[16] by taking into account the density of molecules in the reflecting planes. Laue retains in the intensity expression the unknown function $\Psi(\alpha,\beta)$ which describes the radiating intensity of the individual particle as a function of direction. § 7 of this work employs the natural directional dependence of the simple dipole. Equation (26) for the strength of the reflected beam rests on the assumption that the spectral distribution of the incident radiation is sufficiently constant.

It appears, however, according to Bragg's latest results,[17] that it is precisely this assumption that is often not met, because of the occurrence of a very strong and nearly monochromatic characteristic emission from the anticathode. The interferences caused by it should be treated according to case 1, § 5; at the moment there are very few experimental data. The only comment to be made is that Bragg observed the reflection of the characteristic radiation at a series of incident angles about 4° apart. According to (20a), this observation implies a ratio of wavelength to lattice spacing of 1/10; Laue required for his explanation a series of ratios between 0.04 and 0.14. The characteristic radiation of the anticathode would therefore fall within the already known spectral range. Additional information about the homogeneity of the characteristic radiation should be deducible from the intensity distribution within the interference spots [TN15].

(Received 8 May 1913)

Footnotes

1. Laue, M., Friedrich, W., and Knipping, P. (1912). *Sitz.-Ber. Bayr. Akad. Wiss.*, 303, 363.
2. Wulff, G. (1913). *Phys. Z.*, **14**, 217–20.
3. Laue, M. (1913). *Phys. Z.*, **14**, 421.
4. Dispersion and double refraction of electron lattices (crystals). Dissertation, Munich, 1912.
5. Compare Sommerfeld, *Phys. Z.*
6. The derivation of this formula proceeds using the same methods as for (8). It will be published shortly elsewhere.
7. Hupka, E. and Steinhaus, W. (1913). *Ber. Deutsch. Phys. Ges.*, **15**, 1164.
8. Friedrich, W. to appear shortly in *Ann. Phys.*
9. de Broglie, M. (1913). *Compt. Rend. Sci., Paris*, 1011.
10. See the analogous argument by G. Wulff (1913). *Phys. Z.*, **14**, 220.
11. Compare the formula of L.S. Ornstein (March 1913). *K. Acad. Wetensch. Amsterd.*, 1229, which was reached by entirely different considerations.
12. Laue, M. (1913). *Phys. Z.*, **14**, 421.
13. See [1].
14. Compare with the tables in the first publication of Laue *et al.* The fact that the interference spots are round close to the crystal (see, e.g. W.L. Bragg, *loc. cit.*) is also explicable in terms of the superimposition of many round spots.
15. *Loc. cit.*
16. *Loc. cit.*
17. Roy. Soc. London, Meeting of 17 April. See *Nature*, 24 April 1913.

Translator's notes

[TN1] Although modern English usage applies the term 'diffraction' rather than 'interference' to most of the effects discussed in this paper, I will retain the German 'interference' in most instances. This preserves the historical flavour. It also calls attention to the inherent ambiguity of these terms, in both languages.

[TN2] D.W.J. Cruickshank: Both from the text in the opening paragraph, and from footnotes [14] and [15] in §§ 6 and 8, it appears that the printer dropped a footnote sign after Bragg's name. The missing footnote probably would have been
1b. Bragg, W.L. *Nature*, 1912, **90**, 410; *Proc. Camb. Philos. Soc.*, 1913, **17**, 43–57,
which would be consistent with the material cited in footnote [14] and in the text for footnote [15]. I argue this as follows:
The brief *Nature* note (published 12 December 1912) was quoted by G. Wulff (paper received 3 February) (*Phys. Z.*, 1913, **14**, 217–20) and L. Mandelstam and H. Rohmann, (*Phys. Z.*, 1913, **14**, 220–2) in the 15 March issue of *Phys. Z.*, that is, 2 months before the submission of the Ewald paper. However, W.L.B.'s Letter, although introducing the reflection idea and giving experimental results for cleaved mica, has no formulae and no discussion of spot shapes. The *Proc. Camb. Philos. Soc.* paper was read to the Society on 11 November 1912, and was reported briefly in *Nature* of 5 December (**90**, 402). It was published early in 1913 and does contain a detailed discussion of spot shapes.

[TN3] G. Hildebrandt has kindly pointed out that the German original is ambiguous on whether or not the entire motion can be resolved into such oscillations.

[TN4] Printer's error: the factor Π was omitted on the left side of (6).

[TN5] For historical consistency in maintaining German mathematical notation, (7) retains the original 'rot' for the modern 'curl' operation.

[TN6] Printer's error: the factors x, y, z have been moved into the exponent from their position as factors of the exponents in the original.

[TN7] Because of its acceptance in the English literature, the word 'Ansatz' will not be translated. It includes the meanings of 'assumption', 'postulate', 'proposition', or 'trial formulation'.

[TN8] The original 'Dünung' is literally 'ocean swell'.

[TN9] Ewald uses the exponential e^{iKs} and e^{-iKs} rather indiscriminately with either sign throughout the paper. As this causes no harm, it has not been corrected.

[TN10] In the original, (15) reads $z + c < 0$, for the range of validity of the outer potential, (6). However, the more careful interpretation of the outer potential, (14) in Ewald Part II, § 2, shows that this expression is indeed valid for $z < 0$. The confusion arises because the representation of the *inner* potential for the half-crystal actually extends to $(z + 2c) > 0$ (note the factor 2), so that there is a region $-2c < z < 0$ in which both representations are valid. To call attention to this ambiguity in the original, I have corrected only the factor 2.

[TN11] In other words, there would not be any surface waves travelling beyond the edge of the illuminated spot. It is interesting to note that the entire discussion and interpretation of case II is completely rejected in Ewald Part II, § 7, where it is argued that the mathematical procedure leading to this solution is invalid (poles of integrand in potential are of second order). However, the original point of view is partly revived in Ewald Part III, § 5, in connection with small *imaginary* values of v_{lm} that could lead to surface waves of appreciable amplitude.

[TN12] Printer's error is corrected in the inequality.

[TN13] D.W.J. Cruickshank: As written, the argument in Case 1, sub-section 2, applies when the normal to the crystal 'reflecting planes' is parallel to the z-

axis of the orthorhombic lattice. In general, crystal lattice 'reflecting planes' are not parallel to any planes of the reciprocal lattice (even for an orthorhombic lattice). The general restriction on the possible locations of the sphere centre (α,β,γ) is that the possible (α,β,γ), at distances $2\pi/\lambda$ from the origin, must lie on a set of planes, parallel to the crystal reflecting planes, whose inter-planar spacing is π/d.

The equation below eqn (20),

$$\cos\chi = (h/2)(\lambda/2c),$$

is equivalent to Bragg's law on making the substitutions $\chi = \pi/2 - \theta$, $h = n$, $2c = d$ (inter-planar spacing), giving $\sin\theta = n\lambda/2d$. Equation (20a) follows from the $\cos\chi$ equation on putting $\chi' = \chi - \Delta\chi$, $h' = h + 1$ with the approximations $\cos\Delta\chi = 1$, and $\sin\Delta\chi = \Delta\chi$.

[TN14] Printer's error: In the original the subscripts of the two vectors **a** are interchanged.

[TN15] D.W.J. Cruickshank: Ewald's final paragraph seems to be a last-minute addition before submission to the journal on 8 May 1913. The paragraph makes it seem as though there was only one Bragg. The paper read at the Royal Society meeting of 17 April 1913 was by W.H. Bragg and W.L. Bragg. It was reported (as stated in Ewald's footnote [17]) in *Nature* of 24 April 1913 (**91**, 205) as being written by W.H.B. and W.L.B. A 20-line summary was given—but the summary says nothing about the series of reflections 4° apart. Thus Ewald must have had some other source for information. Unfortunately, the information was either transmitted or understood incorrectly. The Braggs' paper, published on 1 July 1913 in *Proc. Roy. Soc. (London)* (A**88**, 428–38) does not report the reflection of successive orders of a characteristic radiation at 4° intervals. It does report that the three components of a characteristic radiation are spaced by roughly 4° in reflections from some crystal faces. Ewald's application of (20a) to the calculation of λ/d was thus invalid.

Ewald's misinformation possibly derived from incomplete preliminary news of the short paper by W.H. Bragg (*Phys. Z.*, 1913, **14**, 472–3) which was printed immediately following Ewald's paper. The Bragg note, which gives some spacings for three components of a characteristic radiation, was dated 26 April 1913 at Leeds, and was received by the journal on 7 May (manuscript translated into German by Max Iklé).

1968: Personal reminiscences. *Acta Cryst.* (1968), A**24**, 1–3

The January 1968 issues of *Acta Crystallographica*, Section A and Section B, were dedicated to Paul P. Ewald on his eightieth birthday. The opening contribution was by Ewald.

Personal Reminiscences

By P. P. Ewald

The Editor insists on my writing for this issue rather more personal reminiscences than are scattered throughout *Fifty Years of X-ray Diffraction*. Remembering the pleasure I got when reading the personal reminiscences of my friends and colleagues written for that book, I reluctantly comply.

I was born in Berlin in January 1888, and, by the time of my first birthday, had lived under all three emperors that the short-lived German Reich has seen. Berlin was then still a comfortable city, where it could happen that the widowed Empress Frederick on a morning stroll through the Tiergarten Park, seeing my mother taking the toddler to the side of the road, stopped, patted me on the head and said: 'Nice little chap'. This was the last Royal compliment I received.

I was lucky to be born in a family of moderate wealth and ambition, of academic training, and widespread interests. My father was a promising young historian, lecturer at the University of Berlin and collaborator of the *Monumenta Germaniae Historica*, pupil and friend of Waitz, Wattembach, Mommsen and other leading German historians. He died, shortly before I was born, at the age of 37 within a few days of appendicitis, the clinical picture of which was then not yet fully recognized; it was in the same year, 1887, that the first successful appendectomy was performed by an American doctor.

So I grew up in the care of my mother, a remarkable woman of great gifts. When a child she would have loved to study Latin – but that was not done for fear of turning a girl into a bluestocking; later, she longed to study medicine – but it was too early for that, though I believe Clara Tiburtius, the first, and for a long time only female student of medicine in Germany, was her contemporary. With her strong desire for independence she became a portrait painter, the arts being the only professional activity a woman could enter into without social disgrace or an endless fight.

I loved the familiar fragrant atmosphere of oil, turpentine and varnish in the studio, but I never even attempted to paint. (I also failed to find any interest in history, until quite late in my life, though surrounded by a fine historical library; my marks at school in history were without exception 'utterly unsatisfactory'!)

Not having to look after a husband my mother was free to travel, and she took me on extended visits in England, where her parents and she had many friends, and to Paris where she renewed her studies in portrait painting. By the time I was five I spoke a rudimentary English and French, and my mother kept this alive by often speaking to me in these languages. She was also fluent in Italian, but this, unfortunately, I did not learn from her.

My uncles, on my father's side, were, the elder a well-known physician and Head of one of Berlin's big clinics, and the other Professor of Physiology at the University of Strassburg, then belonging to Germany. Naturally I had closest contact with the Berlin family, and in particular with the two cousins who were only a little older than I was. We met for private gym and dancing lessons, made occasional excursions to replenish our aquaria, and, in some years, joined forces on the seaside in the summer vacations.

My real interest in Science began when I was 11 years old. We were staying with friends in Cambridge in whose house we met Dr S. Ruhemann, a pupil of A. W. Hoffmann, who held a Caius College chair of organic chemistry, the first such chair existing in a British University. He took me to his laboratory and I admired the endless rows of small flasks and bottles with colourful chemicals, the Bunsen burners, all the glassware. My enthusiasm came to a climax, however, when Dr Ruhemann blew a perfect four-inch diameter glass sphere over the Bunsen burner, cooled it down and silvered it on the inside. A few days later, on boarding the Hoek boat for the return journey late at night, the sleepy boy, holding the precious object safe in front of him, stumbled on the gangway – with the result that to his great chagrin instead of the perfect sphere only a fragmentary small silver star remained attached to the glass tube. After arriving in Berlin, while my mother was still unpacking, I dismantled the (Auer) gas light over the dining table, attached a long piece of rubber tubing that turned up conveniently, opened the tap and struck a match. There was a loud bang, a two meter flame shot out of the open end of the tubing and a heavy plush curtain that concealed a door was full aflame. Luckily I closed the tap instinctively, and there was a can of water standing on the table, so that the fire was prevented from searing more than the surface of the curtain. Retrospectively I admire the reaction of my mother who, rather than prevent me from further attempts at experimenting, arranged for an old table to be put in an adjoining room and let me continue my glassblowing there in a more cautious way.

In 1900 we moved from Berlin to Potsdam and to a more spacious setting. There I had a small room to myself and soon shared my chemical interests with two friends. For my confirmation the Strassburg uncle gave me a simple chemical balance, and I invested my pocket money in chemicals and glassware. There was no difficulty in obtaining chemicals, except the really poisonous ones, from the corner drugstore, because the 'Drogerie' in those days sold chemicals and not hot-dogs.

The 'Gymnasium' I attended, first in Berlin, then in Potsdam, was a classics school with very little science on the program. I obtained the latter mainly from private reading of journals subscribed to as Xmas gifts, like *Prometheus, Revue Rose*, later *Nature*, and from an excellent synopsis of the technical achievements of mankind in ten volumes, one of which appeared as the standard Xmas and birthday present from an aunt and her husband, the *Buch der Erfindungen, Gewerbe und Industrieen*. Besides, I found in my grandfather's study a series of volumes of *Reports of the Smithsonian Institution*, in which I read eagerly, stretched out on the carpet. Obviously it helped my mental development that I read German, English and French with nearly equal ease (including, of course, Captain Marryat and Jules Verne).

Of the languages taught at school I disliked Latin for its formality and boring literature, and loved the adaptable, warm-hearted Greek. Without my even trying hard, the sound of the epical-lyrical parts of Homer, Sophocles and Euripides (N.B. in the German or Erasmic pronunciation!) stuck to me, and even now I sometimes recite such passages when I cannot readily fall asleep. I believe strongly in knowing poetry by heart so as really to possess it.

I ended the Gymnasium with the final exam which I passed in spite of zero production in history. It so happens that this was the only strict and formal examination I was ever subjected to. The oral exam for my Dr. phil. degree was a soft and friendly affair, and since I received *summa cum laude* for my thesis work, no further examination was required for my habilitation as Privatdozent in Munich.

I started my studies wanting to become a chemist. Through Dr Ruhemann's good offices I was accepted, though a latecomer, as an undergraduate student at Gonville and Caius College in Cambridge. It was there I spent a very happy winter, 1905–1906, trying, in vain, to understand Calculus (Gallop), Inorganic Chemistry (Fenton) and Physics (Bevan); I rowed and later did ju-jitsu, then new at Cambridge, and was an avid reader, listener and sometimes debater at the Cambridge Union Society. The formative value of this two-term experience far exceeds any academic gain. In fact, when I began studying in Göttingen in the spring of 1906 I tried the subjects all over again, with the result of rejecting Chemistry – in particular inorganic chemistry (Wallach) – as being utterly unsystematic; of not being too keen on Physics (Riecke, Voigt); and of not being able to follow Herglotz's course on analytical geometry.

Laboratory work in Chemistry and Physics and exercises in Mathematics were marked, but there were no formal term exams, and a student was left largely to his own evaluation. Evidently I did not do well, but it did not worry me. A friend of the family, the Privatdozent for Philosophy Leonard Nelson, kept an eye on me. He knew that I was waiting for the course on Calculus in order to understand what I had been unable to get from the course in Cambridge. The course, fall 1906 and spring 1907, was to be given by David Hilbert, and Nelson suggested to Hilbert that he should appoint me as the official 'Ausarbeiter', *i.e.* make me write out a weekly account of the lectures as they were being given, to be deposited in the mathematics reading room for the benefit of the other students – an arrangement which the great organizer of the Mathematics Department, Felix Klein, had instituted for most of the courses given in Göttingen. For this work I received my first pay, namely 100 Marks per semester (then £ 5), but the much greater value lay in the weekly help I received from Hilbert's assistant, E. Hellinger, who himself was a devoted teacher. Hilbert's lectures were often garbled, signs wrong or trigonometric functions interchanged, and this had to be straightened out in the written account. These minor blemishes came from the fact that Hilbert prepared his lectures only in a vague way; he created them while delivering them. Usually he spoke at great length of the problem, its difficulties, of various modes of attack one might think of, rejecting one or the other – and finally, when you began to grow impatient with this wasting of the precious time of the period, suddenly there appeared the answer to the problem standing out like a marble statue in a dark park lit up by a ray of sun. What a difference between this and a hackneyed utility course in Calculus! The idea of mathematical invention never left the lecturer and the student became witness to the working of a mathematical mind. In choosing illustrations for the general theorems and methods, Hilbert was not afraid of high-level examples taken from all parts of mathematics and, even in this beginner's course of mentioning recent developments. With the tuition by Hellinger, and through my own intensive occupation with it, I learned nearly all the analysis I know from Hilbert's course.

By now, I had decided to make mathematics my goal.

I had been much intrigued hearing my philosopher friends discuss an approach to mathematics which differed from the one offered in the Göttingen courses. This went back to Weierstrass and they called it epsilontics. Its aim was to eliminate geometrical interpretation in the proofs of analysis by basing them on the firmer ground of algebra. The main proponent of this idea was at the time Alfred Pringsheim in Munich. I enrolled at the University of Munich in the fall of 1907, not heeding the warning of a motherly Frau Professor in Göttingen that ever so many of her young friends had been lured away to Munich by the easy-going life in Southern Germany, never again to return to the citadel of virtue and learning, Göttingen. I must confess that for the next ten semesters I too enjoyed life, friendship and company in Munich without a thought of returning north.

Pringsheim's lectures, delivered with humor and wit and in a very forcible manner, impressed me deeply. Down to the writing on the blackboard they were strictness and neatness themselves. It was an aesthetic pleasure to listen. For two or three semesters I did not miss a single one. But there was something lacking, and to my own dismay I never grew really warm, the way it had been in Hilbert's course.

This happened, however, quite unexpectedly in a course I would never have dreamt of taking in my endeavour to study 'pure' mathematics. An experienced friend, my senior by some years, the Greek physicist D. Hondros, half dragged me to a short two-hour course on Hydrodynamics which A. Sommerfeld had announced. From the first few lectures on, I was captured. The parallelism between the physical motion of a liquid and the vectorial analysis for describing it, the interplay between mathematical formalism and the simple application of Newtonian force and acceleration arguments, opened to me new perspectives in the understanding of Nature and the interpretation of mathematical operations. This I could visualize and remember or recall at will. I felt strongly that *this* was akin to my style of thinking, rather than the algebraic convergence proofs in Pringsheim's lectures. From then on my heart was set on Mathematical Physics, and Sommerfeld's courses and seminars were my main preoccupation; henceforth I considered myself a Theoretical Physicist.

The story of my thesis work with Sommerfeld on the double refraction produced by the anisotropic lattice arrangement of optical resonators in a crystal, and how discussion of this with Max Laue started in him the train of thought leading to his X-ray diffraction experiment – this has been told in *Fifty Years* and need not be repeated. I would like to tell, however, how it came about that I devoted much of my life's work to Crystallography, and in particular to crystallographic publication.

I had always thought of Astronomy as my minor subject and had taken several courses with H. v. Seeliger, the Professor of Astronomy. I never bothered much about my examination, but again my friends were wiser. A year before my thesis work neared its end they told me: 'Look, you know hardly anything of Astronomy, and Seeliger is known to be a stiff examiner. Your thesis has to do with crystals; you still have two semesters time to attend old Groth's course on Crystallography and take his practical class – that will make this friendly old gentleman your examiner'. I followed the advice, with the result that my oral in crystallography consisted of a friendly chat on Bravais, Haüy, Sohncke, Schönflies and Fedorow. This was before Laue's experiment, and before I went back to Göttingen as 'Physics Tutor' to Hilbert. In 1913 I returned to Munich and took over Friedrich's equipment in Sommerfeld's institute for the short period before the outbreak of World War I. I often went for a chat to the crystallographers, Steinmetz and the ever interested and inquisitive Groth.

In 1920 Groth, who was then 76 years old and had edited 55 volumes of his creation, the *Zeitschrift für Kristallographie und Mineralogie*, handed over the editorship to Paul Niggli, with the suggestion of forming an editorial board in the near future. This was put into effect by Niggli in 1924 by appointing M. v. Laue, myself and K. Fajans as co-editors. To celebrate the occasion, Groth invited the four editors and the publisher K. Jacoby to a solemn dinner at his home in Munich. In addressing us, he charged us with keeping alive his endeavour of providing a central, international journal for the publication of crystallographic work. As a sequence to the Munich meeting, reference to mineralogy was omitted from the title of the *Zeitschrift*, and in a leaflet distributed with the next issue and signed by the four editors, the scope of the journal was defined as 'to collect all those papers which attempt to elucidate the Solid State by means of a study of the Crystalline State'. Three years later (1927) the *Zeitschrift für Kristallographie* stressed its international character by accepting manuscripts in English and French. These were often sent to me, and at times more than half the papers in an issue had gone through my hands.

When at the end of World War II there appeared no hope for a speedy revival of the *Zeitschrift*, and on the other hand great pressure developed for publication of held-up research, the old master's plea to safeguard international crystallographic publication could best be fulfilled by the creation of a new journal. It was not without weighing my old obligation to Groth that I undertook the launching of *Acta Crystallographica*. I like to think how Groth, who was always ahead of his times, would be happy to observe the modern development of his favourite subject, and especially its close companionship with chemistry. I am sure he would also be satisfied with the international scope of the progress which takes place.

1969: Introduction to the dynamical theory of X-ray diffraction. *Acta Cryst.* (1969), A**25**, 103–8

The January 1969 issue of *Acta Crystallographica*, Section A, contained the Conference Report for the International Meeting on Accurate Determination of X-ray Intensities and Structure Factors held in Cambridge, England, in June 1968. Ewald contributed this introduction to the dynamical theory.

Introduction to the Dynamical Theory of X-Ray Diffraction

By P. P. Ewald

19 Fordyce Road, New Milford, Conn. 06776, U.S.A.

The general features, terminology, and method of the dynamical theory of X-ray diffraction are discussed, stressing the analogy with the general theory of small oscillations of a mechanical system.

My good friend the organizer of this session prepared me to speak to an audience consisting partly of experts in the applications of this theory and partly to people knowing the theory only by name. I shall dwell, therefore, on the method of the theory rather than on results or applications, of which the other papers of this session will contain examples.

Theories of X-ray diffraction

The geometrical theory, the kinematical theory, the dynamical theory and the developing quantum mechanical theory of X-ray diffraction can be regarded as successive stages of accounting for the same physical phenomena.

(i) In the most primitive stage, the *geometrical theory*, we ask only for the directions under which diffracted rays appear, *i.e.* those directions in which the wavelets scattered at each element of a crystal will reinforce one another without any cancellations. This theory alone leads to the concepts illustrated in Fig. 1, namely the *reciprocal lattice*, the *tiepoint T* determined by making $\overrightarrow{TO} = k_1$ the wave vector of the incident wave, and the *sphere of reflection* of radius k_0 with T as centre. Whenever this sphere passes through a lattice point, such as **h**, the vector $\overrightarrow{Th} = \mathbf{k}_h$ becomes the wave vector of a diffracted wave.

This simplest, geometrical theory leads to the *structure factor* by considering the differences of optical path caused by the dispersion of the atoms throughout the crystal cell. Thereby it opens the way towards crystal structure analysis of a primitive kind, namely by the use of absent reflections.

(ii) In the next stage, the *kinematical theory*, the combined effect of the scattered wavelets in directions other than those of maximum cooperation is taken account of. This requires a relaxation of the strict condition of periodicity of the crystal, a condition which would prevent the crystal from being bounded by a surface or of having a finite size, and which leads to the 'all or nothing' condition that a point of the reciprocal lattice lie exactly on the sphere of reflection. The scattering power – be it electron density $\varrho(x)$ for X-rays, or potential $V(x)$ for electrons – is rendered by a Fourier series in the case of the truly periodic crystal; but for any deviation from periodicity it is rendered by a Fourier integral, and its Fourier transform $F(\eta)$ is a continuous function in Fourier space. For slight departures from periodicity, such as finite size of the crystal, or temperature motion, the transform $F(\eta)$ will still have strong peaks at the positions $\eta = \mathbf{h}$ but each lattice point **h** will be surrounded by a kind of halo of $F(\eta)$ instead of only carrying, in a δ-function like way, a Fourier coefficient $F_\mathbf{h}$. The purely mathematical and physically unreasonable 'all or nothing' condition of intersection of sphere of reflection and lattice point is now replaced by the more generous result of the superposition of the elementary wavelets, namely that the amplitudes of the diffracted waves at great distance are shown, for wave vectors of any direction, by the intersection of the sphere of reflection with the Fourier transform $F(\eta)$ of the scattering density $\varrho(x)$. This extension of the theory leads to the concept of *Lorentz factor* and thereby to a first primitive use of integral intensities for the determination of parameters in crystal structures. It also accounts for the broadening of the diffraction lines of fine powders, and for a host of other properties.

In the kinematical theory all waves are supposed to travel through the crystal with the same phase velocity.

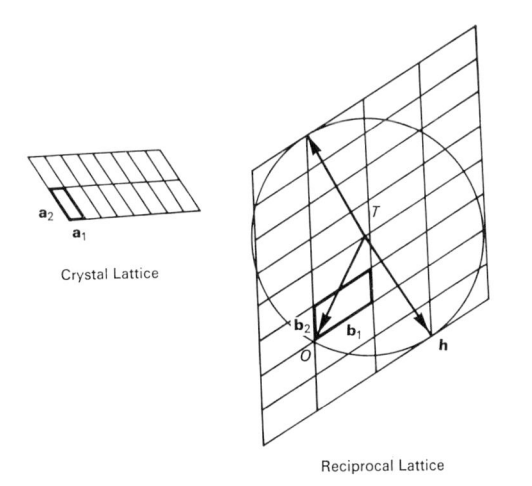

Fig. 1. Crystal lattice, reciprocal lattice, tiepoint T, sphere of reflection and wave-vectors of the primary and two of the secondary rays.

This is generally taken to be c, the light velocity in empty space, and then all wave vectors have the same length $k_0 = v/c = 1/\lambda_0$ where λ_0 is the wave-length in vacuo. In that case the sphere of reflection, which always has radius k_0, passes through the origin O of Fourier space.

The kinematical theory forms the base for all crystal analysis and for the bulk of all work from 1912 on to now. But it has the serious shortcoming of not conserving energy since all energy contained in the diffracted rays is additional to that of the unmodified incident beam. The way to get rid of this defect is apparent from Fig. 1 and the obvious *reciprocity theorem*. This states that if any one of the n strong waves is considered to be the primary wave, the same set of n waves will be generated. Shifting the origin from O to any of the points **h** lying on the sphere makes no difference to the position of T nor of the sphere; only the order vectors will be re-named.

The n rays thus form an inseparable unit which we call the *optical field*; it takes the place of the single plane wave forming the simplest optical field in visible light optics.

(iii) In the *dynamical theory* there are two problems to be solved. The first is to find the conditions under which an n-ray field can exist in, and travel through, the crystal. This is an extension of the optical '*theory of dispersion*' to the short wavelengths of order 10^{-8} cm instead of $5 \cdot 10^{-5}$ cm in the visible range. The second problem is that of connecting the fields inside the crystal to those outside, including the incident wave; the answer is the '*theory of refraction and reflection*'. For the propagation of optical fields inside the crystal, we may assume this to be unbounded, *i.e.* filling all space. There is no outside and no '*incident wave*'. For the latter to appear we have to assume at least one surface to exist – we speak of the '*half-crystal*' filling the space below the plane $z = 0$, or of a crystal slab limited by the planes $z = 0$ and $z = D$.

In the dynamical theory interaction between the scattered or diffracted rays is taken into account; apart from the scattering there is no interaction between the radiation field and the crystal. If temperature motion and absorption are introduced, this is done in an *ad hoc* descriptive manner, which is not very satisfactory even if it is capable of rendering the experimental facts.

(iv) The theory of the future may well be the *quantum theory of diffraction* in which the radiation field and the state of the crystal are considered as parts of a single system. Such a theory was first formulated by M. Born in 1942 and it is being actively developed, in modern terms of Feynman diagrams by Ohtsuki (1964), Kuriyama (1966) and others. In quantum language the kinematical theory represents the simple collision of photon and crystal; the dynamical theory gives the multiple collision of photon and crystal. In both cases the crystal remains in its ground state. In the quantum theory multiple scattering occurs together with a possible change of state of the crystal, by collision of the photon with a phonon or with an electron, leading to photoeffects or to the formation of excited states as a preparation for the emission of characteristic radiation or for Auger effects. If this theory can be carried through in a realistic way, it would make the *ad hoc* assumptions for the incorporation of temperature effects, Compton effect and absorption obsolete.

Analogy between the dynamical theory and the theory of small oscillations in mechanics

Before returning to the 'dynamical' theory let us consider a well-known theory in particle dynamics, the theory of small oscillations of a mechanical system about a position of stable equilibrium. To fix the ideas, consider a system of n equal pendulums coupled by a taut string from which they are hung at equal distances. These pendulums are the counterparts of the n rays forming the optical field. In comparing the mechanical and the optical problems, time dependence in the first corresponds to space dependence in the latter. Thus the problem of exciting the n-wave optical field in a half-crystal by a wave incident on the surface $z = 0$ corresponds to finding the motion of the row of pendulums after an impact on one of them at time $t = 0$. How do we find this motion for times $t > 0$?

The standard method is first to study the *proper modes* of the system, namely those motions which are self-consistent in the sense that each part of the system moves without the application of external forces under its intrinsic forces (gravity in the case of pendulums) and the forces transmitted to it by being coupled to its neighbours. Such proper modes can only be found for certain frequencies of the motion, called the *proper frequencies* $v^{(j)}$; they are the (always real) roots of an equation of order n, the *secular equation* or *equation of dispersion*. The set of frequencies $v^{(1)} \cdots v^{(n)}$ forms the '*spectrum*' of the mechanical system. For each proper frequency $v^{(j)}$ the amplitudes of the n pendulums, $A_1^{(j)}, A_2^{(j)} \cdots A_n^{(j)}$ stand in a fully determined ratio, and the motion can further be 'normalized' by making

$$\sum_{h=1}^{n} |A_h^{(j)}|^2 = 1;$$

the motion is then called a *proper mode*.

The optical analogue to this part of the solution of the mechanical problem is the *theory of dispersion*. In it, we find a self-consistent field of n waves, *i.e.* one in which each wave gives off as much energy to the others, as it receives from them *via* the coupling by means of scattering by the atoms. The amplitudes of the waves therefore remain constant during the progression of the field through the crystal. This corresponds to the constant amplitude simple harmonic motion of each pendulum in a proper mode.

Such a state of self-consistency can not be established for an arbitrary tiepoint, but only for tiepoints lying on a surface of degree $2n$, the *surface of dispersion* (SurfDisp). (The factor 2 arises because a transverse wave has two independent amplitude components; it also holds for the pendulums if they are allowed vibrations in the x and y directions.) The Surf Disp is thus the analogue to the 'spectrum' of the row of pendulums; to indicate that a tiepoint lies on this surface, we speak of a *'proper tiepoint'*. It represents n waves of well-determined amplitude ratios which form the self-consistent optical field or *'proper X-field'*. Having determined the Surf Disp and proper X-field we know all there is to be known about fields inside the crystal.

We have now to tackle the problem of starting the motion of the pendulums from rest by a blow given to the 'primary' pendulum at $t=0$. This blow imparts a certain initial velocity to the number one pendulum, without directly affecting its position or the positions and zero velocities of the others. After the initial impact is over, the subsequent motion of the system can only consist of the superposition of its proper modes, since external forces are no longer acting. Optically, the corresponding situation is that the external *'incident'* wave creates at the surface of the half-crystal a *'primary'* wave of the same energy or amplitude. As this wave travels towards the interior ($z>0$, corresponding to $t>0$ for the pendulums) it is subjected to the coupling by radiation to the other directions and it gives off energy to the other $n-1$ waves which start with zero amplitude at $z=0$. Since the incident wave has been converted into the primary wave at the surface, it is not influencing the optical field inside the crystal any more, so that the field at $z>0$ can only be the superposition of proper X-fields.

We have thus to superimpose (normalized) proper fields with such amplitudes that at $z=0$ the resultant amplitude is E_0, namely that of the incident wave, in the 'primary' direction, and zero in all secondary directions. Such a condition must hold at all points of the crystal surface and thereby allows, in each of the n directions of diffraction, the combination of only such plane waves as keep in step with one another along the surface, that is, for which the scalar product $\mathbf{u} \cdot \mathbf{k}_h^{(j)}$ where $\mathbf{u}$, a position vector in the surface, has the same value for all j. As Fig. 3 shows, this is achieved by combining proper fields represented by tiepoints $T^{(1)}$, $T^{(2)} \ldots T^{(2n)}$ which lie on the same normal to the crystal surface as the point A representing the incident wave. Each tiepoint $T^{(j)}$ is assigned a weight $W^{(j)}$ by the above conditions, and these weighted proper tiepoints represent the solution of the problem of adapting the internal wave-field to the conditions of incidence.

Two-ray fields within the crystal

Let us illustrate the general procedure by a discussion of the most important situation, namely when there exists only *one* secondary ray. If either the primary, or the secondary ray were travelling alone through the crystal, their wave vector would have to have length K_0, where $K_0 = nk_0 = n/\lambda_0$; λ_0 is the wavelength in free space and n the refractive index which for a single ray is given by the Lorentz formula of dispersion according to which the optical density is

$$n^2 - 1 = \frac{Ne^2/m}{\omega_0^2 - \omega^2}, \quad (1)$$

with $N=$ number of electrons per unit volume, e^2/m the ratio for electrons, ω_0 the (circular) proper frequency of the resonator or dipole, and $\omega = 2\pi\nu$ the circular frequency of the optical field. For X-ray frequencies ω_0/ω is usually very small, and the optical density is negative and very small, so that (1) can be replaced by

$$n - 1 = -(\tfrac{1}{8}\pi^2)(Ne^2/m)/\nu^2 = \sim 10^{-5} \text{ to } 10^{-6}. \quad (2)$$

The introduction of K_0 instead of k_0 accounts for the general refraction in the crystal and thereby for the *'deviations from Bragg's Law'* as first observed by Stenström in 1919. The geometrical locus for the foot of the wave vector of the (isolated) primary ray is thus a spherical surface of radius K_0 with the origin as centre, the *'Lorentz Sphere'* of ray 1 and similarly for the (isolated) secondary ray the Lorentz Sphere of radius K_0 about the point $\mathbf{h}$ of the reciprocal lattice. The two spheres intersect in a circle with centre at the midpoint of the vector $\mathbf{h}$, and near this circle lie the tiepoints of the 2-wave fields in which the two rays are coupled by mutual scattering. This coupling leads to a splitting of the two spheres as indicated in Fig. 2:

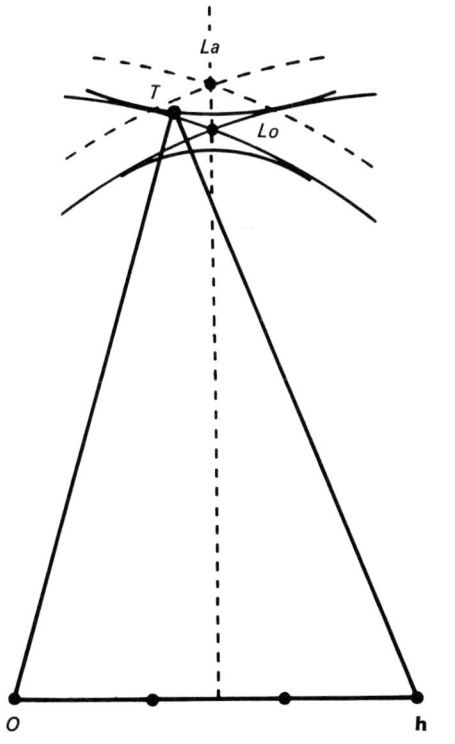

Fig. 2. The Lorentz and Laue spheres about the reciprocal lattice points **O** and **h**; branching of the Lorentz spheres near the Lorentz point Lo.

the tiepoints for the simultaneous 2-ray field are bound to a ring-shaped surface consisting of 4 sheets of hyperbolic cross section. Two of the sheets – the outer ones – belong to the mode where the electric force E is normal to the plane of the two rays, and the other two to those modes for which E lies in that plane. Only the two outer sheets are shown in the figures.

For any plane of the rays 1 and 2 there exist two points of reference: the *Lorentz point Lo* which is at distance K_0 from points O and $\mathbf{h}$ of the reciprocal lattice and is, therefore, the point where the Lorentz spheres intersect in that plane: and the *Laue point La* where the corresponding spheres of radius k_0 intersect.

The distance $|LaLo| = (k_0 - K_0)\cos\theta = k_0(1-n)\cos\theta$, where θ is the Bragg angle; it is therefore about 10^{-5} to 10^{-6} times smaller than the length of the wave vectors. Drawn to the scale adopted for showing the splitting of the Lorentz spheres the wave vectors should be about a mile long. The spheres or circles about O and $\mathbf{h}$ can therefore be replaced by their tangents at the Lorentz point, and these form the asymptotic planes or lines of the SurfDisp. Any point T on the surface is a *proper tiepoint*; the field represented by it is a *proper mode* of the optical field; the wave vectors of its two plane waves are fully determined by T, and so is the amplitude ratio of the two waves since each amplitude is inversely proportional to the excess of the length of the wave vector over the free-space value k_0. Thus all we can know about the proper modes of the 2-wave optical field in the crystal is contained in the SurfDisp.

Half-crystal; two-ray internal field adapted to incident field

In making the transition to the half-crystal, we have to distinguish two cases. In the first, the Laue case, the crystal surface lies on one side of both rays, *i.e.* both rays are directed towards the interior of the crystal. In the second case, the Bragg case, the surface lies between the directions of the primary ray and the secondary ray, so that the latter travels in the crystal towards its surface.

In either case the optical field inside the crystal is a superposition of proper modes, such that the amplitude E_0 incident on the crystal surface is taken over, at $z=0$, by the 'primary wave' and that, in the Laue case, the secondary wave begins with zero amplitude at the surface. In the Bragg case we cannot obtain a second boundary condition without introducing a lower surface of the crystal. In the simplest case this is parallel to the entrance surface, so that the crystal is a slab of thickness D. The condition is then that the secondary ray start with zero amplitude at the lower surface [*cf.* Fig. (4*b*)].

Since these conditions have to hold at *any* point of the surface, the waves in direction 1 have to remain in step with the incident wave along the whole surface, and the waves in direction 2 also have to have a common trace along the surface (the upper one in the Laue case, the lower one in the Bragg case). The wave vector of the trace is the resolved part of the spatial wave vector along the surface. The existence of such surface conditions therefore leads to a unique selection of tiepoints $T^{(1)}T^{(4)}$ which may be combined, namely those lying on the same normal to the surface as the point A representing the incident wave ($A = $ 'Anregungspunkt'). For with this choice, the wave vectors $\overrightarrow{AO}$ $\overrightarrow{T^{(1)}O}\overrightarrow{T^{(2)}O}$ all have the same resolved parts along the surface, and $\overrightarrow{T^{(1)}\mathbf{h}}$, $\overrightarrow{T^{(2)}\mathbf{h}}$ stand in a similar relation; besides the same holds for the other case of polarization for which the hyperbola is not shown in Fig. 3.

The point A in Fig. 3 corresponds to incidence on the reflecting planes under an angle θ smaller than the Bragg angle θ_B since θ_B would be the angle of incidence if A coincided with La. We have

$$|\Delta\theta| = |\theta - \theta_B| = |\overrightarrow{ALa}|/k_0 .$$

As A moves on the Laue circle, the distance $T^{(1)}T^{(2)}$ varies. It has its smallest value, namely the vertex dis-

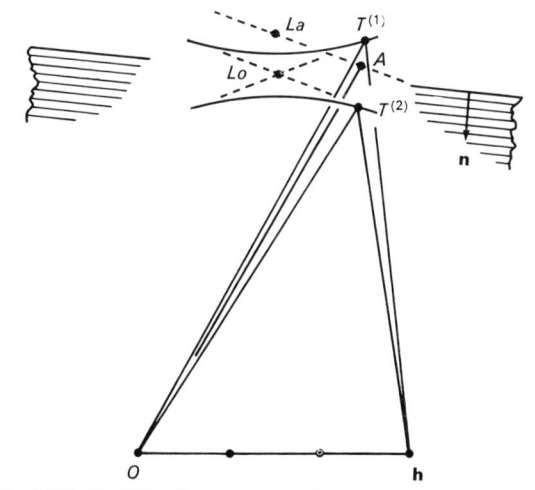

Fig. 3. The SurfDisp for two rays; selection of proper tiepoints for a half-crystal cut off with surface normal **n**.

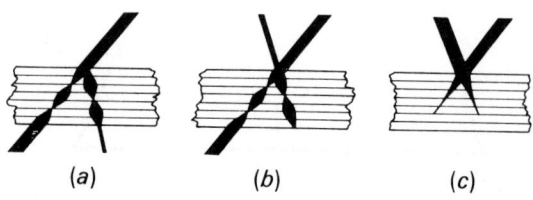

Fig. 4. Wave-fields in the crystal slab, (*a*) Pendellösung in the Laue case, (*b*) Pendellösung in the Bragg case, (*c*) primary extinction and total reflection in the Bragg case.

tance of the hyperbola, when rays 1 and 2 are equally inclined to the reflecting surfaces and A coincides with La. This minimum distance is $K_0|n^2-1| |\alpha_h|/\cos\theta$ where α_h, the '*coupling coefficient*' of the two rays, is the Fourier coefficient of order h in the development of the distribution of the polarizability, or scattering power, or electron density, in the crystal cell divided by the total polarizability. This was loosely called the structure factor in the kinematical theory, and we see that the vertex distance is proportional to its first power. The value given above holds when the electric field vector of the waves is normal to the plane of rays 1 and 2 ('*σ polarization*'); in that case the coupling is stronger than for '*π polarization*' when each ray is affected only by the component of the E field of the other ray normal to it. $|\alpha_h|$ is then to be replaced by $|\alpha_h| \cos 2\theta$ where 2θ is the angle between rays 1 and 2.

The superposition of the proper modes produces two plane waves in direction 1 with vectors $\overrightarrow{T^{(1)}O}$ and $\overrightarrow{T^{(2)}O}$ and two plane waves in direction 2 with wave vectors $\overrightarrow{T^{(1)}h}$ and $\overrightarrow{T^{(2)}h}$. The difference between the vectors of each pair is given by the vector $\overrightarrow{T^{(1)}T^{(2)}}$ and is very small compared to the lengths of the wave vectors themselves. Thus in each direction we obtain beats, and, since $\overrightarrow{T^{(1)}T^{(2)}}$ is normal to the crystal surface, these beats are of constant amplitude in planes parallel to the surface. It is by means of these spatial beats that the flux of energy is transferred from direction 1 to direction 2 and *vice versa*, in the same manner that energy oscillates from one to the other of coupled equal or nearly equal pendulums. This type of solution was therefore called originally '*Pendellösung*', and this name has been accepted internationally. The full analogy however, of the procedure of the dynamical theory to that of the general theory of small oscillations in mechanics was not recognized until recently and is published here for the first time. In the hands of N. Kato and others, the fringes obtained by the Pendellösung type of optical field near the exit surface of a slightly wedge-shaped crystal slab have been used for precise determination of structure factors $|F_h|$ – a method understandable in the light of the above remarks on the smallest $T^{(1)}T^{(2)}$ distance which leads to the longest beats.

The Pendellösung type of solution is illustrated in Figs. 4(a) and (b) for the symmetrical Laue and Bragg cases. The amplitudes of the fields in directions 1 and 2 are indicated by the thickness of the rays. Energy, of course, is conserved; though the direction of the energy flux varies with depth, the integrated flux component parallel to z is the same everywhere – provided of course that absorption is absent.

Whereas in the Laue case the proper tiepoints $T^{(1)}$ and $T^{(2)}$ lie on different branches of the hyperbola, they are on the same branch in the Bragg case. As the point of excitation moves from A in Fig. 5 towards greater glancing angles θ, $T^{(1)}$ and $T^{(2)}$ approach one another and the length of the Pendellösung beats increases, without, however, there occurring any fundamental change in the type of solution until $T^{(1)}$ and $T^{(2)}$ coincide. The first beat then changes, for direction 1, into a linear decrease and for direction 2, into a

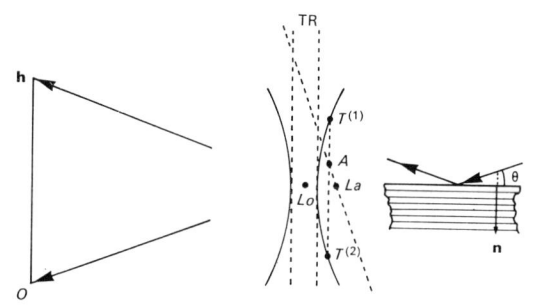

Fig. 5. Bragg case; regions of Pendellösung and of total reflection (TR) for symmetrical Bragg case.

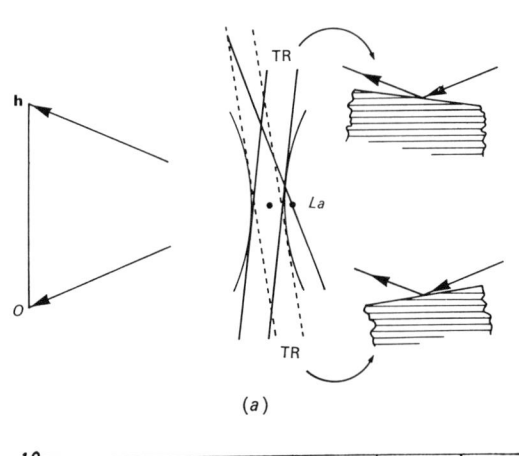

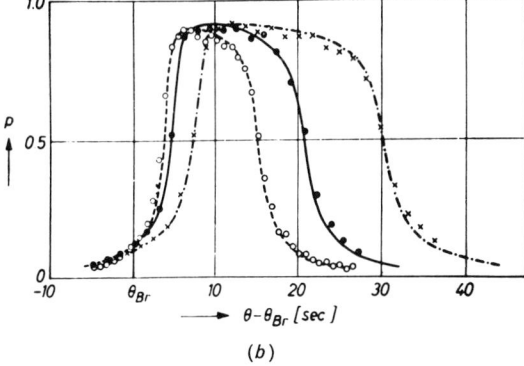

Fig. 6. (a) Bragg case; dependence of width of region of total reflection and of deviation from Bragg's law on the inclination of the crystal surface. (b) Experimental reflection curves showing this dependence (Bubakova, R., *Czech. J. Phys.* 1962 B **12**, 776).

linear increase, with increasing depth – *cf.* the behaviour of a pendulum starting its oscillations at time $t=0$ under the influence of a periodic force having as frequency the proper frequency of the pendulum.

As A moves upwards into the region defined by the tangents to the hyperbola parallel to the surface normal n, the intersections with the SurfDisp become complex, indicating that the wave vectors have a real and an imaginary part. The latter produces an exponential increase or decay of the field, as it proceeds in the crystal, of the form $\exp(\pm\kappa z)$ which leads to Darwin's '*primary extinction*', whereas the real part corresponds to a straight displacement of the tiepoint from one tangent point to the other. For points A between the two tangents all of the energy flux of the incident wave must emerge again as the 'reflected' wave 2 from a sufficiently thick non-absorbing crystal, since the primary extinction weakens the field to zero amplitude at the lower boundary of a crystal slab [Fig. 4(*c*)]. The region between the tangents is therefore the '*region of total reflection*' TR. As Fig. 6(*a*) shows, its centre is displaced from the Bragg angle which is the angle of incidence when A coincides with La.

The displacement as well as the width of TR depend on the direction of the crystal surface. Measured in θ, they are largest when incidence on the surface of the crystal is nearly glancing, and the reflected ray leaves the surface at a steep angle [Fig. 6(*a*)]. This effect is well shown in the reflection curves published by Bubakova (1962) [Fig. 6(*b*)]. It was first observed by Bergen Davis & Terrill (1922) and von Nardroff (1924) in their 'rocking curves' and has been used in reverse for converting a wide incident beam into a narrow monochromatized reflected beam [*e.g.* Renninger (1961)].

The reflection curves in Fig. 6(*b*) also show the rounding-off of the curve produced by the variation of the width of the region of total reflection for the σ and the π polarizations and the asymmetry of the curve owing to absorption.

The dynamical theory was formulated in its essential parts in 1917, but it lay dormant for some thirty years with only a few applications. It came into prominence when the art of growing perfect or near-perfect crystals was developed, and when the discussion of electron diffraction and electron microscope pictures demanded some such theory because of the much stronger interaction of matter with electrons than with X-rays. The discovery of the Borrmann effect in 1941 and of its use as a means for the study of dislocations and other imperfections in crystals heightened the interest in the theory. In 1930 the theory was re-cast in a slightly different form by M.v.Laue and it has been used in that form by most workers. The presentation given above reverts more closely to the form and the ideas of the original publication.

References

General references
 EWALD, P. P. (1916). *Ann. Phys. Lpz.* **49**, 1 and 117.
 EWALD, P. P. (1917). *Ann. Phys. Lpz.* **54**, 519.
 EWALD, P. P. (1926). *Handbuch der Physik*. Berlin: 1st ed. XXIV. See pp. 254–268.
 EWALD, P. P. (1935). *Handbuch der Physik*. Berlin: 2nd ed. XXIII. See pp. 285–301.
 EWALD, P. P. (1937). *Z. Kristallogr.* A**97**, 1.
 EWALD, P. P. (1968). *Acta Cryst.* A**24**, 5.
 JAMES, R. W. (1948). *The Optical Principles of the Diffraction of X-rays*. London: Bell.

Laue's form of dynamical theory
 BATTERMAN, W. B. & COLE, H. (1964). *Rev. Mod. Phys.* **36**, 681.
 LAUE, M.VON (1930). *Ergeb. exakt. Naturwiss.* **10**, 133.

Quantum theory of X-ray diffraction
 BORN, M. (1942). *Proc. Roy. Soc.* A**180**, 397.
 OHTSUKI, Y. H. (1964). *J. Phys. Soc. Japan*, **19**, 2285.
 ASHKIN, M. & KURIYAMA, M. (1966). *J. Phys. Soc. Japan*, **21**, 1549.
 AFANASIEV, A. M. & KAGAN, YU. (1968). *Acta Cryst.* A**24**, 163.
 MOLIÈRE, G. (1939). *Ann. Phys. Lpz.* **35**, 272.

Asymmetric Bragg reflection
 BERGEN DAVIS & TERRILL, H. M. (1922). *Proc. Nat. Acad. Amer.* **8**, 357.
 BUBAKOVA, R. (1962). *Czech. J. Phys.* **8–10**, 710.
 NARDROFF, R, VON (1924). *Proc. Nat Acad Amer.* **10**, 384.
 RENNINGER, M. (1961). *Z. f. Naturf.* **16***a*, 1110.
 RENNINGER, M. (1967). *Adv. X-ray Anal.* **10**, 32.

Pendellösung
 KATO, N. (1961). *Acta Cryst.* **14**, 526, 627.
 KATO, N. & LANG, A. R. (1959). *Acta Cryst.* **12**, 787.

Borrmann effect
 BORRMANN, G. (1941). *Phys. Z.* **42**, 157.
 BORRMANN, G. (1950). *Z. Phys.* **127**, 297.
 EWALD, P. P. (1958). *Acta Cryst.* **11**, 888.
 EWALD, P. P. (1965). *Rev. Mod. Phys.* **37**, 46.
 KATO, N. (1958). *Acta Cryst.* **11**, 885.

1979: A review of my papers on crystal optics 1912 to 1968. *Acta Cryst.* (1979), A**35**, 1–9

The January 1979 issue of *Acta Crystallographica*, Section A, contained papers from the Dynamical Diffraction Symposium held at the University of Oklahoma in March 1978 to honour Paul P. Ewald in his ninetieth year. As a basis for his opening talk Ewald prepared this review.

A Review of my Papers on Crystal Optics 1912 to 1968

By P. P. Ewald

108 *Sheldon Road, Ithaca, New York* 14850, *USA*

(*Received* 19 *April* 1978; *accepted* 13 *September* 1978)

Abstract

The theory of the diffraction of X-rays by crystals was developed by Ewald as part of a unified study of the interaction of light of all wavelengths with crystals, beginning with the work for his 1912 thesis and extending to his papers in 1968. The formulation of the problem in terms of the interaction of electromagnetic radiation with a periodic array of dipoles is placed in its historical perspective and is compared with Laue's version based on the assumption of a continuous electron density distribution. The Borrmann effect, hinted at in 1917, is derived readily from consideration of the dispersion surface.

Introduction

On this happy occasion of a symposium celebrating my ninetieth birthday may I be excused for giving a review of my own papers? The main reason for this is that very few have read my original papers. Yet I see some merits in them as compared to later expositions of the same subjects. Besides, my work has been attempting to establish the unity of classical optics throughout the entire range of wavelengths from infrared to X-rays. This general aspect has received little resonance.

The papers I am going to comment on fall into two groups. There are four main papers under the title *Zur Bergründung der Kristalloptik* (*Foundations of Crystal Optics*; Ewald, 1916*a*,*b*, 1917, 1937; quoted hereafter as *Optics* I–IV). Of these the first is a slight remodelling of my Munich PhD thesis of 1912 *Dispersion und Doppelbrechung in Elektronengittern* (*Kristallen*) [*Dispersion and Double Refraction in Lattices of Electrons* (*Crystals*)] (Ewald, 1912), while the later papers develop the theory further, reaching a satisfactory generality in the fourth paper of the series. Surrounding this backbone of my work are shorter papers on topics of detail, some of which I shall mention when I have to comment upon them.

The optical field from assumed dipole oscillations

In 1910, when I began work on my thesis, there was no quantitative proof for the internal periodicity of crystals. My teacher Sommerfeld had the idea that a proof might be obtainable by investigating whether oscillating electrons or dipoles, when placed in an anisotropic periodic array, a lattice, would produce observable double refraction by the mere fact of that arrangement. It was hoped that if this were so, then the observed double refraction could be used to gain information on the internal structure of the crystals. Sommerfeld handed me the reprint of a paper by Planck (1902) on the theory of dispersion, and with that I went hiking up the Rhine Valley for the summer vacations.

The paper by Planck was quite incomprehensible to me. I could not disentangle the 'incident wave', the 'total field', and the 'field of excitation'. But I came to the conclusion that my first aim would be to find out what kind of field would be generated by the dipoles if they were set in motion by a plane optical wave travelling through the lattice with some unknown velocity q; that is, according to a 'wave of dipole excitation' described by a wave vector $\mathbf{K}$ whose length is v/q, where v is the frequency. Each dipole emits a spherical wavelet which travels away from it with the speed of light *in vacuo*, c, or with a wave constant $k_0 = v/c$. Because fields can be simply superimposed, to find the sum of all wavelets is a purely wave-kinematical exercise.

This summation kept me busy for about a year. Assuming an orthorhombic lattice with translations $\mathbf{a}_1$, $\mathbf{a}_2$, $\mathbf{a}_3$, the positions of the dipoles are given by $\mathbf{X}_l = l_1 \mathbf{a}_1 + l_2 \mathbf{a}_2 + l_3 \mathbf{a}_3$ and each spherical wavelet proceeds according to the distance between the source and the field point $\mathbf{x} = x_1 \mathbf{a}_1 + x_2 \mathbf{a}_2 + x_3 \mathbf{a}_3$, namely $r(l,x) = \{\sum_i [(x_i - l_i)\mathbf{a}_i]^2\}^{1/2}$. The irrationality of this expression prevents a straightforward summation.

A first helpful suggestion came from Sommerfeld. He showed me the development of the function expressing a spherical wave, namely $(\frac{1}{2}\pi r) \exp(2\pi i k r)$, into an integral over the product of a Bessel function of the planar distance ρ and an exponential function whose argument is linear in the third coordinate x_3 or z. He suggested that if it were possible to linearize ρ by using a similar integral technique, then it would be easy to sum the wavelets.

After weeks I found the proper integral technique and when I met Sommerfeld to show him this result, he said 'Wait, I also have to show you something' – and with that he showed me the same integral I had found. It is remarkable that neither of us spoke of the Fourier integral. At that time one spoke of Fourier developments, mainly in cosine and sine series, and the Fourier integral, in spite of its having been known for a long time, was still a bit strange; the term Fourier transform had yet to be invented, or to be commonly used in Germany.

The field generated by the dipoles was now easily calculated; it kept step with the 'wave of dipole excitation'.

The field of excitation and double refraction

My triumph at arriving at this field did not last long. I was now confronted with the task of extracting from this field that part that came from any individual dipole for which one sought to establish the motion. For it seemed obvious that a dipole would not be activated by its own field, but only by the fields radiated to its position by all the other dipoles. This, certainly, was the assumption upon which Planck and H. A. Lorentz based their theories of dispersion. The extraction of one dipole field seemed, however, hopeless because it was thoroughly mixed up with all others. Here a remark of Debye (at the time Sommerfeld's assistant) at a ski meeting in Mittenwald helped. He recalled a method Riemann had used in a similar case. Debye said 'Take the expression for the total field, get the denominator of the sum into the argument of an exponential function by introducing a new integration, and then, by interchanging summation and integration you obtain as the integrand a Θ function. To this you can apply the Θ transformation formula, and that will once more allow you to recognize the contribution of the individual dipole'. The method worked! I could now apply the field of excitation to the individual dipole and relate to it the amplitude of oscillation by the dipole's only physical constant, its polarizability.

There resulted the following situation: the optical field generated by the dipole oscillations had to be such that it just sustained these very oscillations. This condition could be fulfilled by the proper choice of the assumed wave vector **K** of the optical field, and with it of the refractive index $n = |\mathbf{K}|/k_0$. I called the situation obtained in this way one of 'dynamical balance'; since the work of Hartree it is now called a 'self-consistent' regime.

I was now able to conclude my thesis by a numerical calculation of the double refraction of a dipole lattice. On the recommendation of P. von Groth I used the axial ratios of anhydrite. The result was: no correlation between calculated and observed principal refractive indexes; however, indexes and double refraction of the same order of magnitude as observed; furthermore, the correct angular dependence of the optical properties as given by the various crystal-optical surfaces. Thus the characteristic optical properties of crystals were caused, at least partly, by the anisotropy of their internal arrangement.

I often feel embarrassed when the now generally accepted method of summation of potentials in crystals is given my name. True, I extended it later to non-orthogonal axes and I gave an explanation of the way it produces very rapid convergence – but essentially the method seems to go back to Riemann.

General insights gained in the thesis

My thesis brought with it several innovations for the then accepted theory of dispersion. This appelation was, in a way, a misnomer. Dispersion in the strict sense is the spreading of a pencil of light into a spectrum; it results from the dependence of the refractive index on the frequency, itself a consequence of the variation of the polarizability of the resonators with frequency. This dependence is nowadays the main objective of the quantum theory of dispersion which is based on our knowledge of the atom. How the polarizability affects the passage of light through a body is a second, now often neglected, aspect of the old designation 'theory of dispersion'. This part would better be called 'theory of light propagation'. Both the speed of light in a medium and its variation with frequency are *internal* properties of the medium. There was therefore in my thesis no room for the introduction of an 'incident wave' which played a role in the previous theories. In fact, my work established that selecting the proper phase velocity was the condition required for obtaining a self-consistent state of an optical wave in a medium containing resonators. This state was a free vibration of the medium in the same sense that in mechanics one speaks of the free vibrations of an unbounded string, plate or any other system where the internal forces balance and no external force is required.

Both the older theories and mine are essentially linear, which means that fields can be superimposed without their interaction. How then was it explained by the older theory that an incident wave falling from outside on a medium bounded by a plane surface could not be detected inside the medium? It would have phase velocity c, would not be refracted, and would thoroughly destroy the self-consistent state inside the medium. My answer was: this wave does not enter the medium; it is destroyed at the surface by the mere fact of the limitation of the array of dipoles.

As this statement was rather revolutionary, I was keen to get Laue's opinion on it. But his thoughts took another turn and he asked repeatedly 'What would happen in the case of very short waves?' I pointed out that the transformation of the optical potential was valid for all wavelengths and that his question could be easily discussed, but that I was preoccupied with finishing the write-up of my thesis and would have to leave it to him.

This concludes the report on my thesis which was handed in early in 1912 and published later in that year. The same material was slightly shortened and brought up to date when it was republished as *Optics* I in *Annalen der Physik* of 1916.

The optical potential in two forms

Let me make the first achievement of my thesis more precise, although in a more modern form, because it contains the key to all later papers.

Field quantities like the electric-field strength are best obtained by simple processes of differentiation from a vector potential which was introduced by H. Hertz in deriving the field of a single dipole of moment

$$\tilde{\mathbf{p}} = \mathbf{p}_0 \exp(-2\pi i v t), \quad (1)$$

namely

$$\mathbf{z}(\mathbf{x}) = (\tilde{\mathbf{p}}/4\pi r) \exp(2\pi i k_0 r); \quad (2)$$

here r is the distance of the fieldpoint $\mathbf{x}$ from the dipole and $k_0 = v/c$ is the constant of wave propagation in free space.

We have been considering dipole moments phased according to a progressive wave of an unspecified wave vector $\mathbf{K}$; therefore, the amplitude $\mathbf{p}_0$ now becomes $\mathbf{p}_0 \exp[2\pi i(\mathbf{KX})]$, where $\mathbf{X}$ is the position of the dipole and $(\mathbf{KX})$ is the scalar product of the two vectors. The distance from dipole to field point $\mathbf{x}$ is then $r = |\mathbf{x} - \mathbf{X}|$ and the total Hertz potential of the optical field at point $\mathbf{x}$ appears as a sum over all sources of wavelets, that is over the three integers (l_1, l_2, l_3) contained in $\mathbf{X}_l$:

$$\mathbf{z}(\mathbf{x}) = \sum_l \frac{P_0 \exp(j \mathbf{K} \mathbf{X}_l)}{4\pi |\mathbf{x} - \mathbf{X}_l|} \exp(j k_0 |\mathbf{x} - \mathbf{X}_l|) \quad (j \equiv 2\pi i). \quad (3)$$

In its transformed shape the same $\mathbf{z}(\mathbf{x})$ is found to be a sum of plane waves:

$$\mathbf{z}(\mathbf{x}) = v_a^{-1} \sum_h p_0/[4\pi^2(K_h^2 - k_0^2)] \exp[j(K_h x)]; \quad (3')$$

here the summation is extended over all points $\mathbf{h} = (h_1\mathbf{b}_1 + h_2\mathbf{b}_2 + h_3\mathbf{b}_3)$ of the 'reciprocal lattice'; the wave vectors $\mathbf{K}_h$ connect the 'tie point' T (*Ausbreitungspunkt*) to the points $\mathbf{h}$ of the reciprocal lattice, and the tie point is defined as the point $-\mathbf{K}_1$, where $\mathbf{K}_1$ (which above was called simply $\mathbf{K}$) is the assumed wave vector of the dipole phasing. The $\mathbf{b}_i$ are the reciprocal vectors to the $\mathbf{a}_i$.

The transition from the source-conscious form (3) to the plane-wave form (3') can nowadays easily be made by applying the technique of Fourier transforms; in the original papers complex integrals were evaluated by the method of residues.

The reciprocal lattice was not yet explicitly envisaged in the thesis; its importance became manifest only in connection with X-ray diffraction. Also it was sufficient to take the simplest model of a crystal for clarifying the optics, namely a simple orthorhombic lattice. Generalization to non-orthogonal axes occurs only in *Optics* III, while the restriction to a basis of a single dipole could be dropped only in *Optics* IV.

The impact of Laue's discovery

Laue, Friedrich and Knipping found X-ray diffraction in crystals round about Easter 1912. I had become an assistant in Göttingen, and I first heard of the discovery some seven weeks later when Sommerfeld came to deliver a lecture on this discovery to the Göttingen Physical Society. Scientific news travelled slowly in those days.

The evening after Sommerfeld's lecture, I finally sat down to discuss the case of short wavelengths. It was hereby that the reciprocal lattice became an essential tool and that the 'sphere of reflexion' (Bragg) or '*Ausbreitungskugel*' was an obvious construction (Ewald, 1913).

Naturally, interest soon became centered on X-ray diffraction. Methods of indexing had to be worked out. One of the early problems was why many more spots did not appear on the Laue diagrams. In a discussion of the original Laue diagrams of zinc blende which Sommerfeld prepared roughly for the second Solvay Conference in 1913 and which I worked out, this question was answered; here the concept and name of the structure factor appears for the first time. In 1914 W. L. Bragg had succeeded in determining what he then called 'the terribly complicated' structure of pyrites, FeS_2, the first example of a cubic structure with non-intersecting trigonal axes and the second type involving a parameter. For the latter, Bragg found a value from the spectrometer readings. By considering the structure factor as a function of the parameter I derived from Friedrich's Laue photograph of pyrites a different and much more accurate value of the parameter – in fact, this may be considered the first precision determination of a parameter (Ewald & Friedrich, 1914).

War service; the extinction theorem

In August 1914 war broke out. The X-ray equipment was transferred from Sommerfeld's institute to a hospital for medical work, and I accompanied it. In 1915 I was attached to the army and went, together with a mobile X-ray station, to the Lithuanian front. It took several weeks before the unit made its way from the Siemens factory in Berlin to Königsberg in East Prussia, where I was waiting for its arrival. During this period of forced leisure I returned to the nagging problem left over from my thesis, namely that of the cancellation of the incident field.

In order to prove the extinction statement, I had to know the optical field generated by a 'half-crystal', that is by a lattice of phased dipoles filling only the space under the plane $z = 0$. This was again a problem of wavelet summation. I tackled it with the familiar methods of evaluating complex integrals by the method of residues. I stuck to the assumption of an orthorhombic lattice because then the integrals could be evaluated separately for the three orthogonal variables.

The result of the summation was that the field inside the crystal consists of the same waves as in the unbounded crystal ('mesowaves'), but that each such wave is accompanied by a 'boundary wave' ('epiwave') of the same amplitude, opposite phase, wave constant k_0 (instead of $|K_h|$) and wave vector $\bar{\mathbf{K}}_h$ such that $\bar{\mathbf{K}}_h$ and $\mathbf{K}_h$ have the same tangential component along the surface. The last condition is an expression of Snell's law of refraction: it assures that the traces of meso- and epiwave move along the surface of the half-crystal with the same speed, so that if extinction occurs at one point of the surface, it is true for all points.

Besides the epiwave inside the half-crystal, there exists an outer epiwave whose wave vector $\bar{\bar{\mathbf{K}}}_h$ is the mirror image of that of the inner epiwave with respect to the surface.

For light, only the term 000 of the sums yields three ordinary plane waves, namely the mesowave and the two epiwaves, which form, respectively, the refracted, the reflected and the extinguishing waves. All other terms in the sums represent 'inhomogeneous waves' whose complex wave vectors have a real part directed along the surface and an imaginary part normal to the surface. The latter produces a rapidly decreasing amplitude along the wave front with increasing distance from the surface. For visible light this inhomogeneity is so strong that the amplitude diminishes from the top layer of dipoles to the next one at depth d by at least a factor of 1000. The inhomogeneous waves mediate the transition from the field outside the crystal to that inside; in fact, there exists a region just beyond the surface, namely $0 < z < d$, where the description of the field can be given by the formulas for either the inside or the outside space. After cancellation of the incident wave, there remained the well known self-consistent optical field inside the crystal and the reflected wave outside. The Fresnel formulas for the amplitudes of these waves resulted.

The investigation supported in all detail the extinction theorem for visible light. It proved invaluable in its more complicated application in the theory of X-ray diffraction. I was happy to mail the manuscript to Sommerfeld for publication as a sequal to the 1912 thesis before my X-ray unit arrived in Königsberg (Ewald, 1916a,b).

Theory of X-ray diffraction

Hardly any fighting went on across the Dwina River in November 1915 when my car and I were installed in a makeshift hospital some 10 miles behind the front. Exhausted troops were sent there from the French front to recover – mostly old men no longer fit for active fighting and prone to break their brittle bones on the icy morass into which the few roads had been turned by the military traffic. I now had time to consider the 'case of short wavelengths' not in the purely geometrical sense, but as the dynamical problem of how an X-ray optical field could travel in the crystal.

The first step was to recognize the importance of the 'reciprocity theorem' which I had already stated in the paper written after Sommerfeld's lecture in Göttingen. If n points of the reciprocal lattice lie on the sphere of reflection, then the corresponding n rays form an inseparable unit, in the sense that none of them can exist without engendering all others. Thus a bundle of plane waves takes the place of the single plane wave in the light case. In the unbounded medium this composite field has to be subjected to the condition of self-consistency. The adapter for achieving this is the choice of the length of the wave vector of one of the rays; by this, the length of all other wave vectors is determined, or as we now say, the common origin of these vectors, the 'tie point' T, is fixed. The refractive index for X-rays differs from 1 by only about one part in a million; this means that the difference in length between any of the wave vectors $\mathbf{K}_h$ and the vacuum value k_0 is of that order of magnitude. There is no general refractive index; each of the n rays has its own. In spite of the nearly equal lengths of the wave vectors, their differences regulate the amplitudes of the waves according to the 'resonance factors' $(K_h^2 - k_0^2)^{-1}$ in the potential (3'). This all-important influence can be understood by a Fresnel zone argument: the smaller the difference $(K_h^2 - k_0^2)$, the less difference there is between the assumed phase velocity q of the dipole excitation and the velocity c with which the spherical wavelets travel; therefore, the greater is the number of dipole-containing planes which send their contributions approximately in the same phase to a field point $\mathbf{x}$, and the greater is the build-up of their amplitudes. Thus in order to balance the amplitudes of the n co-existing beams, so that none gains or loses amplitude (and energy) from the other beams, the lengths of all wave vectors have to be finely adjusted, and with this adjustment the amplitude ratios of the n beams in a self-consistent mode are fixed.

The simple geometrical expression for this intricate balancing is the 'surface of dispersion' on which the tie point must lie. The general equation of the surface is given in *Optics* III for the case of n co-existing rays. The restriction to an orthorhombic lattice has been dropped, but not that of having only one dipole per cell. The general properties of the surface are discussed. It is shown to consist of $2n$ sheets, corresponding to the number of amplitude components of n transverse waves. There are then $2n$ different modes of the optical field. The unrealistic condition of the kinematical theory, that for diffracted rays to exist the sphere must pass *exactly* through the points of the reciprocal lattice, is now softened to the condition of a close approach. Finally the important case of two rays is discussed in detail.

In order to obtain results that can be checked by experiment the case of the half-crystal has to be treated upon whose surface a plane X-ray wave is incident. There are now n homogeneous waves forming the mesofield, each of them accompanied by an internal epiwave of phase velocity c whose wave vector $\overline{\mathbf{K}}_h$ has the same component along the surface as the mesowave vector $\mathbf{K}_h$. In the outer space there are 'reflected' epiwaves whose vectors $\overline{\overline{\mathbf{K}}}_h$ are the mirror images of $\overline{\mathbf{K}}_h$ across the surface, but their amplitudes are not the same. Besides, there are inhomogeneous waves travelling along the surface whose amplitude is restricted to the first few dipole layers. The condition of self-consistency inside the crystal can be fulfilled by superimposing dynamically balanced proper modes with such amplitudes that in $(n-1)$ directions the internal epiwaves cancel and that in the remaining direction, chosen as (000), the sum of the internal epiwaves cancels the incident wave. This is achieved by superimposing fields represented by $2n$ tie points T, T^*, ..., T^{2n*} situated on the $2n$ sheets of the surface of dispersion along the direction of the surface normal, each field taken with a suitable amplitude. The result is that in each direction of diffracted rays a bundle of plane waves of very slightly differing wave vectors $\mathbf{K}_h^*$, $\mathbf{K}_h^{**}$, ..., $\mathbf{K}_h^{2n*}$ is propagated; these waves form beats depending on the depth below the surface. Energy is exchanged between the rays (*Pendellösung*) and a group velocity for the propagation of energy through the crystal can be defined.

The full solution was easy to establish in the case when all diffracted rays travel from the surface of incidence towards the interior ('Laue case'). It was shown that in this case all $2n$ intersections of the surface normal with the sheets of the surface of dispersion are real, that the solution is fully determined, that the primary ray (000) at the upper surface takes over the energy of the incident ray and that the diffracted rays start with zero amplitude before taking out energy from the primary ray. There is a perfect analogy to a system of n coupled pendulums one of which receives at time $t=0$ an impulse while the others are at rest. The initial energy is then transferred through the coupling to the other parts of the system, and if there is no damping beats will continue forever. The analogy of the spatial behavior of the diffracted rays and the time development of the motion of coupled pendulums has been stressed by me on many oc-casions. The method followed in the dynamical theory is the strict analogue to the theory of small oscillations in classical mechanics.

In the non-Laue case at least one upwards-directed diffracted ray leaves the half-crystal by the surface upon which incidence occurs. This may be the directly reflected or any other diffracted ray. In either case this internal ray produces an equally strong epiwave only in the outer half-space. It does not produce a strong internal epiwave, and it therefore leads to no condition of annihilation in the interior. The amplitudes remain indeterminate. This dilemma is physically justified because of the unrealistic assumption that the half-crystal fills the entire lower half-space. Sufficient conditions are obtained for making the problem, and the answer, precise by assuming a second boundary plane parallel to the upper one, that is at depth D. The drawback of this assumption is that the answer now depends on the thickness D of the crystal plate and that it is complicated by the interference effects between the waves reflected at the upper and the lower boundaries, like in the theory of the Lummer–Gehrcke plate. In order to get rid of this unwanted complication and because in actual crystals the thickness D is rarely constant, I averaged over D. There resulted in the case of two rays the well known 'top hat' reflexion curve. This is, in fact, the curve already found by C. G. Darwin in 1913. It shows the angular region of complex wave vectors and therefore total reflexion, bordered by a falling-off intensity in the Lummer–Gehrcke regions. Prins (1930) refined the curve by including absorption and superimposing for the two cases of polarization and in due course this curve was confirmed with ever-increasing resolution on selected crystals.

Comparison of my theory with Laue's version

Laue's version of the theory dates from 1931 (Laue, 1931), that is five years after wave mechanics had changed our ideas of the constitution of atoms and solids and two years after Bragg and West had shown the usefulness of Fourier methods in the determination of crystal structures. The starting point for Laue is the assumption that each cell of the crystal structure is filled with a continuous electron density and dielectric constant, and that Maxwell's equations can be applied to this medium. For the half-crystal a continuous boundary is assumed, across which the usual boundary conditions for the field can be used.

It seems strange to transfer the concept of dielectric constant, derived as it is from large-scale experiments, to the spaces between atoms. True, H. A. Lorentz had shown in his theory of electrons how in a model medium containing only positive and negative point charges the material constants of the Maxwellian theory can be obtained by considering average values of fields over regions containing many point charges. But this kind of averaging is not applicable to the rapidly varying X-ray fields. A similar objection can be raised against the application of boundary conditions to electron-density and dielectric-constant distributions which protrude into the upper half-space.

In contrast to this, the model I used, even if it may be less realistic, is that of an open structure: there is no closed boundary, the external field is superimposed up to any depth in the medium and the condition of self-consistency holds everywhere. It is logically the simpler model.

In spite of the conceptual differences, Laue obtained the same results as I did. His assumption of the dielectric constant was justified by Kohler (1935) on the basis of wave mechanics. Schrödinger's wave-mechanical perturbation theory also showed that each element of volume of electron density reacted to an optical field like a dipole; and with that the models became fundamentally the same, provided I could carry my approach through to a basis containing an arbitrary number of dipoles. I had achieved this already in 1925 for the case of two rays, but not for the general case of an arbitrary number; Laue's theory and discussion did not go farther than this.

My 'habilitation'; Borrmann effect

When World War I neared its end I used my theory of X-ray diffraction as a thesis for being admitted as a lecturer at the University of Munich. Even though Sommerfeld thought that my speculations would never find an application, he accepted the thesis. I had to give a trial lecture and put up a number of statements which I was willing to defend against interventions by the faculty. I had lost contact with physics during the solitary work on my problem and found it difficult to formulate significant statements. The second of my statements was based on the insight I had gained from my work. It seemed rather strange at a time when absorption of X-rays was considered to be entirely conditioned by the amount of matter transversed. It read (in translation): *In case the absorption of X-rays like that of light can be traced to a consumption of energy in the oscillations of dipoles then under some circumstances diffracted X-rays will not suffer any weakening in an absorbing crystal.*

This statement was based on my knowledge that the field for compensating the incident field would be produced by a dipole amplitude which would be smaller the closer the tie point lies to the 'Laue point', *i.e.* to the only point giving rise to diffracted rays according to the kinematical theory. This statement in December 1917 is a prediction of the effect G. Borrmann discovered experimentally in 1941 (Borrmann, 1941); Laue deduced its laws while he was held 'at His Majesty's pleasure' in Farmhall at the end of the European war (Laue, 1949). In 1917 X-ray absorption, like emission, was known to be a typical quantum effect; therefore the carefully worded beginning of the statement. I forgot the statement even after Laue's explanation of the effect until some fifty years later I happened to come across a copy of the invitation to the inaugural meeting.

Preliminary summary

By 1918 the dynamical theory of X-ray diffraction had been successfully established, first for an orthorhombic lattice, later for a general lattice. The theory was, however, still restricted to a basis of a single point-dipole. No structure factor occurred. The general case of n rays had been treated, but only the case $n = 2$ had been discussed in sufficient detail to check with experiment. This comparison had to wait for many improvements in the resolution of the experimental methods and the availability of near-perfect crystals.

Book and *Handbook*; amplification of the reciprocal lattice

In 1921 I left Munich for a chair in theoretical physics at the Polytechnic School in Stuttgart. Besides assembling a full course of lectures in the first years, I wrote a book: *Kristalle und Röntgenstrahlen* (Ewald, 1923). The way I treated the subject was broad and introductory. Later on, I hated to read in it, but judging by the unexpected compliments from crystallographers of many nations and after many years had passed, it must have served its purpose well. Characteristically neither the dynamical theory, nor even intensity, is mentioned in the book; both were in my view subjects about which too little was known at the time. The book contained a list of all known structures which I meant to continue in a second edition. This became the origin of *Strukturbericht*. In spite of its rapid sale the book never achieved a second edition because I was soon writing a full and concise review of the subject for the *Handbuch der Physik*, which was more to my liking. The 1927 edition of this report (Ewald, 1927) was followed in 1933 by a second edition which was brought up to date (Ewald, 1933).

The last manuscript I finished in Munich (Ewald, 1921) was prompted by a request of the Editor of *Zeitschrift für Kristallographie*, Paul Niggli, as he took over from P. von Groth. The 'polar lattice' had been used by the older morphologists only for determining the orientation of the normals to crystal faces with respect to a set of axes. Because of X-ray diffraction one now became interested both in the orientation and the spacing of atomic net planes. The spacing was the inverse of the spacing of a row of points of the reciprocal lattice passing through the origin; this relation required the 'polar axes' to be renormalized to 'reciprocal axes' in the way Willard Gibbs had defined these, and the reciprocal lattice formed by using these axes as translations.

In crystal structures we were now confronted with motifs consisting of many atoms, repeated by the translations. In the obvious point-by-point description of the structure the coordinates of each atom are listed. Could not the structure also be described by referring only to atomic planes? This is a purely geometrical question, and the answer is: yes – but it was very surprising at the time. No new lattice points can be added to those of the reciprocal lattice for they would indicate irrational positions of net planes. But one can add 'weights' to the existing reciprocal-lattice points. These must be invariant against changes in the description of the structure, such as doubling an axis and the basis. To my astonishment the weights necessary to achieve invariance turned out to be identical with the structure factors! The entire row of reciprocal-lattice points passing through the origin suddenly acquired significance: their weights were the Fourier coefficients representing the sequence and population density of the atomic planes normal to the row.

With this result my geometrically conceived extension of the reciprocal lattice tied up with the suggestion by W. H. Bragg (1915) to use Fourier series and their coefficients in determining crystal structures. This method, first tried out by W. L. Bragg and West in 1929 on diopside became more generally applied when the technique of summing Fourier series in two dimensions had been facilitated by the Beevers and Lipson strips in 1934, the same year in which A. L. Patterson showed what use can be made of intensities instead of the unknown Fourier amplitudes. This last problem had been dealt with already in my 1921 paper although in algebraic terms, anticipating convolution and stressing that only the differences of atomic positions are obtainable from the intensities.

The Paris lectures

In 1932 I returned once more to the dynamical theory on the occasion of a lecture series I gave at the Institut Henri Poincaré in Paris (Ewald, 1938). Here again I stressed the continuity of my approach for all parts of the spectrum, treating with greater care the determination of the field of excitation. The equation of motion of a single isolated dipole requires a damping term because of the loss of energy by radiation. This can only come from the action of its own field on the dipole. By a more thorough discussion than in my previous papers the field radiated to a dipole forming part of a lattice from the other dipoles is shown exactly to cancel the damping term in the dipole's equation of motion. There is no loss of energy by radiation for the individual dipole in a lattice if it contributes to the self-propagating plane-wave optical field. This result was already found by H. A. Lorentz for a dipole in a periodic medium. The paper ends with the derivation of the crystal-optical surfaces, the 'surface of normals' for visible light and the 'surface of dispersion' for X-rays.

Optics IV

Up to this point the dynamical theory had been able to deal in a general way only with crystals having a basis of a single dipole. Already in 1925 I had solved the case of a general basis, but only for one diffracted beam ($n = 2$) (Ewald, 1925). For this most important case the result was that for a perfect crystal the diffracted intensity is proportional to $|F_h|$, where F_h is the structure amplitude, and not as for the mosaic crystal to $|F_h|^2$. The reason for this result was published, but not the detailed derivation.

In 1937 I could finally return to the general problem of an arbitrary basis and an arbitrary number n of diffracted rays. The difficulty of this generalization lies in the following fact: As long as there is only a single point-dipole at the origin of the crystal cell all vectors $\mathbf{K}_h = \mathbf{K}_{000} + \mathbf{h}$, where $\mathbf{h}$ is a lattice vector, describe the same oscillations of the dipoles. The difference of description lies only in interpolating a smaller or greater number of sinusoidal waves in the empty space between the dipoles, and this is physically irrelevant. If, however, a second dipole is placed somewhere inside the cell, then it will pick up a different phase according to each $\mathbf{K}_h$. For the scattering of one beam into another there will thus result a different phase relation between the atoms of the basis according to the value of $\mathbf{h}$. Thus for each scattering process there exists a different moment of the entire basis. This is the vectorial structure amplitude

$$\mathbf{S}_h = \sum_s \mathbf{p}^s \exp[-j(\mathbf{h}\mathbf{x}^s)] \quad (j \equiv 2\pi i); \qquad (4)$$

the summation goes over all dipoles in the cell, each sort marked by an index s and shifted by $\mathbf{x}^s$ from the origin of the cell.

Self-consistency of the optical field now requires that

$$\mathbf{S}_h = \sum_{i=1}^{n} A_{h-i} \mathbf{S}_{i \perp \mathbf{K}_i} \frac{k_0^2}{K_i^2 - k_0^2}, \quad (5)$$

where

$$A_m = \sum_s (\alpha_S/v_a) \exp[-j(\mathbf{h}_m \mathbf{x}^s)] \quad (v_a = \text{volume of cell}). \quad (5')$$

It will be seen that A_m is nothing other than the Fourier coefficient in the development of the distributed polarizability in the cell. $\mathbf{S}_{i \perp \mathbf{K}_i}$ is the part of the moment of the cell that is effective in producing a transverse optical field progressing in the direction of the wave vector $\mathbf{K}_i$.

Equations (5) and (5') are linear homogeneous equations for the n structure amplitudes $\mathbf{S}_h$; they can be solved only on condition that one of the vectors $\mathbf{K}$ be chosen properly. After splitting the $\mathbf{S}_h$ into a component transverse to $\mathbf{K}_h$ and another one parallel to it the condition of solubility takes the form of the vanishing of the product of two determinants. This is done in a general way for n rays, whereby the restriction that these be 'strong' rays becomes immaterial. Thus one should be able to take into account points of the reciprocal lattice lying inside or outside the sphere of propagation at some distance from its surface. This may be of value in discussing the optics of the ultrasoft X-ray region which forms the transition to light optics.

Needless to say, the general determinantal condition leads to the old results if $n = 2$, and if n unlimited in the case of a single dipole basis.

The enhanced Borrmann effect

Borrmann & Hartwig (1965) discovered that when three strong rays coexist in a near-perfect Si or Ge crystal the absorption drops even below the value reached in the case of only two rays. The condition of coexistence limits this 'enhanced Borrmann effect' to a small angular region; thus when the crystal is rotated under preservation of the reflecting condition for one set of net planes then within the dark line on the film which is produced by the simple effect there will be found a small darker spot due to the enhancement. The simple interpretation of this spot is that for three beams the surface of dispersion approaches the Laue point even closer than that for two beams. This prompted the papers by Y. Héno and myself (Ewald & Héno, 1968; Héno & Ewald, 1968) [(I) and (II)]. In them the surface of dispersion was analytically and geometrically discussed for the special three-beam case observed. It is shown in (I) that for $n = 3$ one sheet of the surface of dispersion passes at a smaller distance from the Laue point than does any of the three surfaces for only two out of the three beams. Thus, without even introducing absorption, the origin of the enhancement is manifest from the general resonance denominator $(K_h^2 - k_0^2)^{-1}$. In (II) absorption and atomic factors are introduced and a graphical and numerical rendering of the surface of dispersion is given for variable wavelength and, consequently, angular settings. Finally the effective coefficient of absorption, μ_{eff}, is given for two different three-beam cases in germanium for each of the six sheets of the surface of dispersion. The curves show μ_{eff} as a function of λ/a for the cases $n = 1, 2$ and 3. They show strikingly to what extent a perfect crystal can become transmittant in one of its proper modes, while in others it becomes more opaque. The value of this analytical treatment lies in the general survey it provides, and as such it is satisfactory. On the other hand, it is labor saving to use computer methods in future cases.

Retrospect

I am lucky to have spent so many of my best years in doing research I have loved; in having been well prepared for Laue's discovery by my thesis work; by taking an active part in developing the theory of the subject; by gradually simplifying the mathematics and by arriving at an increasingly deeper understanding of the varied aspects of crystal optics.

On the other hand, my interest always centered on the perfect crystal in which I saw the preferred material for exacting optical investigation. Herein lies a strong limitation, and an abstraction which in important aspects is contrary to nature. I am happy to see how others have been, still are, and will be carrying on beyond the limitations I set for myself.

References

BORRMANN, G. (1941). *Phys. Z.* **42**, 157–162.
BORRMANN, G. & HARTWIG, W. (1965). *Z. Kristallogr.* **121**, 401–409.
BRAGG, W. H. (1915). *Philos. Trans. R. Soc. London Ser. A*, **215**, 253.
EWALD, P. P. (1912). Dissertation, Univ. of Munich: Göttingen, Dietrichsche Universitäts Buchdruckerei, pp. 1–46.
EWALD, P. P. (1913). *Phys. Z.* **14**, 465–472.
EWALD, P. P. (1916a). *Ann. Phys. (Leipzig)*, **49**, 1–38.
EWALD, P. P. (1916b). *Ann. Phys. (Leipzig)*, **49**, 117–143.
EWALD, P. P. (1917). *Ann. Phys. (Leipzig)*, **54**, 519–597.
EWALD, P. P. (1921). *Z. Kristallogr.* **56**, 129–156.
EWALD, P. P. (1923). *Kristalle und Röntgenstrahlen*. Berlin: Springer.
EWALD, P. P. (1925). *Phys. Z.* **26**, 29–32. See correction (1926): *Phys. Z.* **27**, 182.
EWALD, P. P. (1927). *Handbuch der Physik*. Vol. 24, pp. 191–361. Berlin: Springer.
EWALD, P. P. (1933) *Handbuch der Physik*, Vol. 23, pp. 207–476. Berlin: Springer.
EWALD, P. P. (1937). *Z. Kristallogr.* A**97**, 1–27.
EWALD, P. P. (1938). *Ann. Inst. Henri Poincaré*, **8**, 79–110.

Ewald, P. P. & Friedrich, W. (1914). *Ann. Phys. (Leipzig)*, **44**, 1183–1196.
Ewald, P. P. & Héno, Y. (1968). *Acta Cryst.* A**24**, 5–15.
Héno, Y. & Ewald, P. P. (1968). *Acta Cryst.* A**24**, 16–42.
Kohler, M. (1935). *Sitz. Berichte Preuss Akad. d. Wiss.* pp. 334–338.
Laue, M. von (1931). *Ergebn. d. Exakten Naturwiss.* **10**, 133–158.
Laue, M. von (1949). *Acta Cryst.* **2**, 106–113.
Planck, M. (1902). *Sitzungsber. K. Preuss. Akad. Wiss.* pp. 470–494.
Prins, J. H. (1930). *Z. Phys.* **63**, 477–493.

1986: The so-called correction of Bragg's law. *Acta Cryst.* (1986), A**42**, 411–13

The November 1986 issue of *Acta Crystallographica*, Section A, was published as a memorial to Paul Peter Ewald, who died 22 August 1985. The opening Editorial concluded with the following paragraph sketching the genesis of this paper by Ewald:

The posthumous appearance of the first paper in this issue calls for an explanation. Paul Ewald sent the manuscript to Professor Ben Post about 1980 or 1981 with the purpose of stimulating an experiment that would determine the possible effect of n-beam interactions on the precise wavelength of X-rays, in view of the general use of perfect crystals such as Si in wavelength measurement. The experiment has not been completed, but the manuscript was considered to have sufficient interest for inclusion here. The eminent referees whose comments clarified the terminology of this paper are gratefully acknowledged...

The So-Called Correction of Bragg's Law

By P. P. Ewald

(Received 12 February 1986; accepted 2 May 1986)

Abstract

The traditional 'correction of Bragg's law' is discussed for any value of the refractive index of the diffracting medium. The finite angular range of reflection, as given by the dynamical theory of interference, is symmetrical about the corrected Bragg glancing angle only if absorption can be neglected. In this case the centroid of the reflected intensity is given by the corrected Bragg law. When absorption has to be taken into account or when, besides the surface-reflected ray, other strong diffracted rays occur inside the crystal, the simple corrected law is not sufficient. For spectroscopic work of precision equal to that achieved for visible light a careful analysis of the incident X-ray appears necessary concerning its collimation (angular width) and polarization. Furthermore, the setting of the diffracting crystal has to be investigated in order to avoid unwanted simultaneous reflections. The ideal would be to combine a recording of the actual reflection curve with the high-precision absolute measurement of the glancing angle that serves as zero for the reflection curve. This requires novel instrumentation.

1. A solitary ray: the refractive index

We write the periodic spatial part of the wave function as

$$\exp(2\pi i(\mathbf{k}\mathbf{x})), \qquad (1.1)$$

where $\mathbf{x}$ stands for the field point, $\mathbf{k}$ is the *wave vector*, $\mathbf{k}\mathbf{x}$ signifies the scalar product of the two vectors, and the length of $\mathbf{k}$ is the *wave constant*

$$k = |\mathbf{k}| = 1/\lambda, \qquad (1.2)$$

where λ is the wavelength.

A wave in free space has wavelength $\lambda_o = c/\nu$ (ν = frequency, c = velocity of light) and wave vector $\mathbf{k}_o$. Inside the medium a solitary X-ray has a phase velocity q, a wavelength $\lambda = q/\nu$, and wave constant

$$K = \nu/q = (c/q)k = nk, \qquad (1.3)$$

where n is the *refractive index* of the medium, calculated by using the additive property of the 'optical density' according to the Lorentz–Lorenz relation

$$D = \frac{(n^2-1)}{(n^2+2)} = \frac{N(e^2/m)}{4\pi^2(\nu_0^2 - \nu^2)}; \qquad (1.4)$$

here, the sum is to be extended over all types of resonators; ν_o is the natural frequency and N is the number per unit volume of each sort of resonator.

For X-rays, n is very little different from 1 and usually smaller than 1, so that it is often convenient to write

$$n = 1 - \delta. \qquad (1.5)$$

Wave vectors have dimension [1/length] and are therefore best visualized in reciprocal space. For solitary waves we construct two spheres about the origin of this space, one of radius k for waves propagating in empty space, the other of radius K for waves in the medium. Any allowed wave in either medium is represented by a wave vector beginning on the corresponding sphere and drawn into the origin of reciprocal space (point 1, Fig. 1). The sphere of radius K is the simplest example of the surface which, in more complicated cases, is called the *surface of dispersion*.

2. More than one diffracted ray inside the crystal

In order to find out whether an X-ray of wave vector $\mathbf{k}_1$ is a solitary beam, or is inseparable from rays of other directions, we use the primitive Laue theory in which all waves are assumed to have phase velocity c. We construct a sphere of radius k_o so that it touches the origin. The vector from the center to the origin is $\mathbf{k}_o$, the wave vector of the primary wave. If the sphere passes through other points of the reciprocal lattice, say points $h_1, h_2, h_3, \ldots$, then the radii to these points are wave vectors which are inseparably connected to the primary vector. In this case a bundle of waves is the optical field, the propagation of which is studied in the dynamical theory. If there are secondary rays, the refractive index discussed in the previous paragraph can no longer be considered a property of the

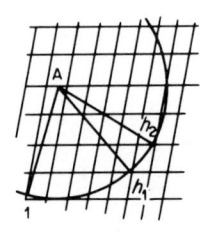

Fig. 1. The reciprocal lattice and the simultaneous wave vectors, with the sphere of reflection.

BOOK REVIEWS

TURNING THE WHEEL OF HISTORY

Michael Hunter, *Science and the Shape of Orthodoxy: Intellectual Change in Late Seventeenth-century Britain*. Woodbridge, The Boydell Press, 1995. Pp. xii+345, £55.00. ISBN 0-85115-594-5

reviewed by Stephen F. Mason, F.R.S.

12 Hills Avenue, Cambridge CB1 4XA

Michael Hunter assembles in his new volume 15 chapters, under four heads, from articles and book reviews published over the period 1971–1992. Hitherto unpublished material appears in an introduction and two of the chapters. The introduction explains the provenance of the chapters and their historical significance in relation to 'A new theory of intellectual change'. The new theory suggests that the intellectual history of Britain during the 17th century may be better understood from the individual case studies Hunter offers, rather than broad generalizations supported by accumulated instances, as provided, for example, by Keith Thomas in his *Religion and the Decline of Magic* (1971).

The outcome of the new approach has a form shaped by the individuals and the aspects of their work chosen for case study. Those selected in the first section of the book as representative of the 1650s, the formative period under the Commonwealth of what was to become the Royal Society, are Elias Ashmole (1617–92), Christopher Wren (1632–1723) and John Evelyn (1620–1706) – individuals who seem to have been chosen primarily for their impeccable royalist and orthodox background, rather than for any outstanding intellectual originality. The chapter on Evelyn is new, but it overlooks John Bowle's comprehensive biography of John Evelyn, published in 1981.

The following sections, each of four chapters, span the decades from the 1660s to the 1700s. The second section, headed 'The Royal Society and the new science', reproduces articles on this theme that appeared between 1979 and 1990, one being a summary of Hunter's book *Science and Society in Restoration England* (1981). The third section, a miscellany of writings from 1971 to 1991 on 'Science and learning', includes reviews of the first volume of *A History of the Oxford University Press* and of a book on Meric Casaubon (1599–1671), who was antipathetic alike to the new science and virtually all of the novelties of his time. His standing derived from the renown of his father, Isaac Casaubon (1559–1614), who had shown that much of the *Hermetic Corpus* was relatively recent, to the *chagrin* of Ashmole and other believers in the high antiquity and authenticity of Hermes Trismegistus and the magical arts.

The final section, headed 'The definition of orthodoxy', contains articles from 1987 to 1992 on the relation of science to heterodoxy; the polemic against astrology written around 1673 by John Flamsteed; the case history of Thomas Aikenhead, a 20-year-old student executed in 1697 at Edinburgh for blasphemy; and a previously unpublished conference paper of 1985: 'The question of witchcraft debated' (1669) by John Wagstaffe. Wagstaffe (1633–77) attributed the invention of the notion of witchcraft and the power of demons to the rulers of the ancient world in attempts to consolidate their political control thereby, and the tradition was taken over and elaborated by the Popes, who devised the concepts of the coven and the contract with the Devil.

The book should prove useful to readers coming to the social history of late seventeenth-century British science for the first time, and the chapter on Wagstaffe will interest veterans familiar with this field.

JOHN FLAMSTEED, OUR ASTRONOMICAL OBSERVATOR

The Correspondence of John Flamsteed, First Astronomer Royal. Volume I (1666–1682). Compiled and edited by Eric G. Forbes and (for Maria Forbes) by Lesley Murdin and Frances Willmoth. Bristol, Institute of Physics Publishing, 1995. Pp. xlix+955. ISBN-0-7503-01473.

reviewed by A.J. Meadows

Information and Library Studies, Loughborough University, Loughborough, Leicestershire LE11 3TU

Recent decades have seen publication of the correspondence of some key figures in the history of British science. The obvious value of such publications is the great saving of effort for historians that results from having the raw material for their work collected in one place. However, this is not their only virtue. The opportunity to see the scientist's thoughts and activities unfolding in chronological sequence is invaluable, as is careful editing and annotation of the material. The latter becomes increasingly important the earlier the period considered, since a proper understanding of the context and nomenclature of the science is then more difficult to achieve.

It is fascinating how much correspondence was being exchanged about science in the seventeenth century, and how much of it survives today. The appearance, in several volumes, of Isaac Newton's and Henry Oldenburg's correspondence has made available major resources of information for historians of seventeenth-century science. To these is now added the first volume of the collected correspondence of John Flamsteed.

Flamsteed's is hardly a household

John Flamsteed, by Thomas Gibson, 1712

name on a par with Newton's, though his appointment as the first Astronomer Royal represents an important landmark in the history of astronomy. Nevertheless, publication of his correspondence is significant for obtaining a balanced viewpoint of late seventeenth-century astronomy. Flamsteed was, par excellence, an observational astronomer. Indeed, the post he held was originally designated Astronomical Observator, not Astronomer Royal. Flamsteed (1646–1719) and Newton (1642–1726) spanned a very similar period of time, but, whereas the latter's prime emphasis was on theory, the former's was on the problems of instrumentation and their use for observation. In this sense, Flamsteed's correspondence, as it appears in this and subsequent volumes, will hold up a mirror to Newton's – not least in respect of the famous quarrel over observational data that developed between the two.

This first volume covers the period 1666–1682, during which time Flamsteed and Newton remained on reasonable terms. The most important event comes near the middle of the volume, with completion of work on the Royal Observatory at Greenwich in 1676. The first full letter included here dates from 1669, when Flamsteed had already begun to establish himself as an observational astronomer. His early letters were all from Derby, where he had been born, and was still living. Flamsteed's father was involved in the brewing industry, and Flamsteed was evidently expected to follow in his footsteps. It was not until the 1670s that it was agreed Flamsteed could go and study in Cambridge. Flamsteed seems to have liked his home well enough, and was prone to argue that Derby had its advantages as an astronomical centre:

> I hope you will not account me culpable for having adapted the calculations to the meridian of a place no more famous than Derby ... the meridian passing over Derby is nearer the middle of England than that of London, and its latitude bisects it nearer than any yet stated.

Flamsteed first met Jonas Moore, Surveyor-General of the Royal Ordnance, on a visit to London in 1670. As the correspondence reflects, this led to continuing contact over questions of instrumentation and observation. Consequently, when the question of determining the longitude came up at court in 1674, Moore recommended that Flamsteed should be asked for his advice. Flamsteed's own memo, reproduced here, notes that

> A Frenchman calling himselfe Le Sieur de St. Pierre about Christmas Last signified that he could find the

Longitude by Astronomicall observations. And procured an order from his Majestie whereby the Right honourable the Lord Brouncker, the Learned Bishop of Salisbury Dr Pell and others were required to Consider of his method and to fit him with such observations as hee demanded ... It pleased these worthy persons according to the power given them to elect mee into theire number, and to refer the observations to be given to my choice.

Sir Jonas Moore (he had been knighted a few years before) appears again in Flamsteed's first reference to the new Royal Observatory: 'The observatory is like to goe well on here. Sir Jonas has accompanyed the King to Portsmouth after his returne I hope it will be begun.' From then on, Flamsteed is in the thick of astronomical observation and of scientific politics: his letters are full of both.

Within a short time of arriving at Greenwich, Flamsteed was being pressed by his mentor to publish the observations he was making there. Flamsteed had no lack of explanations for his failure to publish – more especially, ill health and other duties:

> the last years [observations] I have halfe transcribed. dureing the time I had my Ague but finding how prejudiciall that was to my health I was forced to intermitte it then. afterwards I was hindred by the comeing of the Christchurch boyes before I was recovered ... and now I am visited againe with paines in my feet and legges.

This was in answer to a threat from Sir Jonas Moore to stop his salary unless he started publishing his results. It is apparent from the correspondence that Flamsteed was ahead of many of his contemporaries in recognizing the importance of a long sequence of properly reduced observations. Nonetheless, he was beginning to engender doubts about his willingness to communicate data that were to lead to serious problems for him in later years.

Arguments were common amongst Fellows of the Royal Society, meetings of which were attended fairly regularly by Flamsteed once he had settled in Greenwich. Flamsteed's main animus at this time was reserved for Robert Hooke, a dislike that was not unique among contemporaries: 'but for Mr Hookes vast promises of Inventions I looke upon them onely as boasts or a peece of Contrivance to Magnifie him selfe'.

In other matters besides argument, Flamsteed had much in common with his scientific peers. A case in point is astrology. Many leading members of the Society were dubious of the claims made by astrology, but it remained popular amongst the public at large. Flamsteed noted how much more difficult it was to publish astronomical tables, than astrological.

> I am vext to see our Ephemeredists spend the pages of their almanacks in Astrologicall whimseys tendeing onely to abuse the people and disturbe the publique with anxious and jealous praedictions, whilest the praemonition of Caelestiall appearances which ought to be theire onely charge is wholly contemned or neglected ... I once attempted [to publish an almanac] but because my pages had no praedictions ... they would not passe the presse.

Besides casting light on matters of major scientific importance, there are many reminders throughout the volume both of the close-knit nature of the seventeenth-century scientific community, and of the breadth of interest of its members. For example, one letter from Flamsteed passes on to a friend a method that Sir Christopher Wren had devised to prevent a chimney from smoking. All such entries, great or small, are carefully annotated. In the case of the smoking chimney, we are told that Wren had actually described the idea to the Royal Society ten years before. The editing, as reflected both in the annotations and in the transcription of the letters, is exemplary. Various helpful aids have been included. Letters originally couched in Latin are translated into English; a glossary of astronomical terms current in the seventeenth century is provided; and there are biographical notes for all the people mentioned in the letters.

The letters presented here were collected over a number of years by Professor Eric Forbes. After his tragically early death in 1984, his wife decided to press ahead with production of the letters, gaining financial support for this from various bodies, not least the Royal Society. This first volume indicates how worthwhile all the effort has been. The high standard of editing by Frances Willmoth and Lesley Murdin is matched by the excellent design and production standards achieved by Institute of Physics Publishing. The next volume of Flamsteed letters will be eagerly awaited by historians of science.

THE WONDERFUL GEOMETRY OF DYNAMICS

S. Chandrasekhar, *Newton's Principia for the Common Reader*. Oxford University Press, 1995. ISBN 0-19-851744-0

reviewed by D. Lynden-Bell

Institute of Astronomy, University of Cambridge, The Observatories, Madingley Road, Cambridge CB3 0HA

Copies of Cajori's translation of Newton's *Principia* and of Whiston's *Sir Isaac Newton's Mathematick Philosophy More Easily Demonstrated* (Senex & Taylor, 1716) have rested on my shelves for some years. Nevertheless, I had only dipped into them to compare Mach's criticisms with Newton's discussions of the fundamentals, and to read his proofs of some of the crucial theorems, such as: 'there is no gravitational field anywhere inside a gravitating shell of uniform matter bounded by two similar concentric ellipsoids'.

Reading Chandrasekhar's book has introduced me to much more of Newton's masterpiece. Chandrasekhar's novel approach to understanding *Principia* is to set himself to prove the major propositions by modern methods, and then to follow Newton's proofs. He succeeds in putting Newton's propositions into a far more readable form, and in comparing the relative dullness of the more modern,

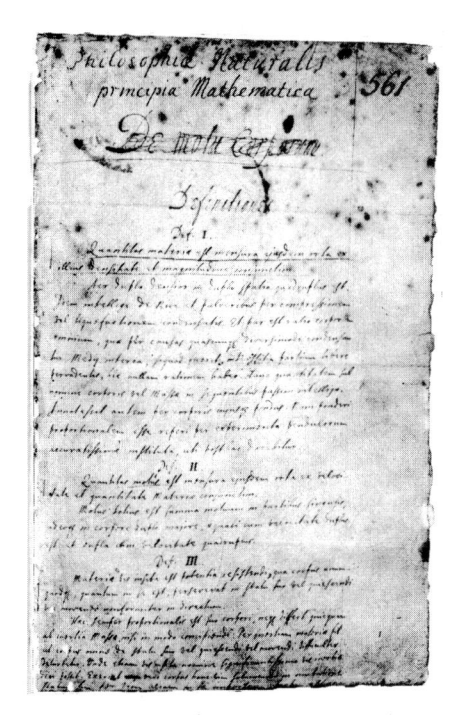

Newton's Principia, first page of manuscript edition

manipulative methods with the greater insight generated by geometry. Some years ago, after much debate, geometry as known to Euclid was dropped from the British school curriculum. At the time I supported the change, although some pure mathematicians bewailed the loss because Euclidean geometry was used to instill the concept of a precise proof. However, much more was lost. Geometry gave the ability to think in pictures, and to picture results derived algebraically. This is crucial to the application of mathematics to the physical world. I now meet some who can manipulate without the ability to understand what the result means, or even why it was asked for. Chandrasekhar's last book can be interpreted as a cry from a superb manipulator of algebra for a return to the more insightful geometrical methods. It will succeed if it reintroduces Newton's geometrical approach to a significant number of practising scientists and schoolteachers who can pass them on to later generations. I think this idea lies behind the subtitle *for the Common Reader*, but Chandrasekhar's common reader is expected to have at least a B.A. in a mathematical science, whereas 279 years earlier Whiston's book required interest and intelligence but less mathematics.

In recommending Chandra's beautifully produced but not inexpensive book to others, it is perhaps worthwhile to point out what I have learned from reading it that 40 years as a mathematical astronomer had not previously taught me.

Firstly, I was delighted to discover that Newton's scientific method did not start from a minute, careful scrutiny of the observations. Rather, he defines his laws of motion, and abstracts from Kepler's law of equal areas in equal times the proof that this implies a centrally directed force. He then makes a general study of many different force laws. The result is such a body of theoretical knowledge about dynamics under their actions that the final comparison with the data in *Principia* book 3 makes an overwhelming case that gravity by an inverse square law between all masses is not only *an* explanation, but, conversely, that no other explanation is possible.

Secondly, I had not previously met Newton's wonderful reciprocal corollary on how the same orbit can be swept out with the same angular momentum when the force is directed either to one centre or to another. This enables him to show that the centred elliptical orbit swept out by a simple harmonic oscillator is directly related to the Keplerian ellipse swept out when an inverse square force is directed toward a focus. Chandra gives a fine supplement on the implications of this lovely theorem to other force laws.

Thirdly, while it is clear that the addition of an inverse cube force is equivalent to a change of angular momentum, Newton proves the lovely theorem that the path traced out relative to rotating axes is precisely the same as that traced in fixed axes without the additional force. This brings me back to Newton's scholium to propositions I and II, the equal areas theorem and its converse the existence of a central force. From Cajori's translation:

> A body may be urged by a centripetal force compounded of several forces in which case the meaning of the Proposition is that the force which results out of all tends to the point S. But if any force acts continually in the direction of lines perpendicular to the *described surface*, this force will make the body to deviate from the plane of its motion; but will neither augment nor diminish the area of the *described surface* and is therefore to be neglected in the composition of forces.

What does Newton mean by '*described surface*'? I suspect he means the surface described by the actual motion, i.e., that perpendicular to the now changing angular momentum h. It is a very remarkable fact that the addition of *any* force along h does not change the motion described in the moving plane perpendicular to h. For Newton's law reads

$$\dot{v} = -V'(r)\hat{r} + K(t,r,v)h.$$

Cross multiplying by r gives

$$\dot{h} = Kr \times h$$

so dotting with h gives h^2 constant and so $r^2\dot{\phi} = h$ in the moving plane. However, dotting our first equation with v and integrating, we still have energy conservation in its original form

$$\tfrac{1}{2}v^2 = \tfrac{1}{2}(\dot{r}^2 + h^2r^{-2}) = -V + E.$$

So, as Newton implies, the motion within this moving plane is that described in the fixed one when the force K is absent. However, the new problem is not finished because we still have to determine the gyrations of h itself. (A pretty case is given by taking $K = \mu r^{-3}$ when h and r sweep around cones and the new force is of the form $v \times B_g$ with $B_g = \mu\hat{r}/r^2$. It would be interesting to see from Newton's working whether he ever considered this gravomagnetic monopole, known to relativists as NUT space, but it is certainly not in *Principia*). Chandra dismisses this scholium in two lines, so perhaps I have read more into it than is really there.

Chandra's book was not written for historians of science, but it would be surprising if they did not find it interesting. Some will be troubled by his approach. Occasionally he has misunderstood Newton, and there is a case where, having done so, he reconstructs a proof that he misrepresents as Newton's. Scientifically, the book succeeds. It makes *Principia* much more approachable, though many will bewail that there was not time to make an index. I am sure it gave Chandrasekhar much joy to see his fine book so beautifully produced, and it serves as a fine reminder of a beautiful spring at Yerkes Observatory, where I went to learn from him the wonders of gravitating ellipsoids. I remember above all his enthusiasm for the beauty that mathematics can reveal in science.

MAN OF MANY MYSTERIES

Joy Hancox, *The Queen's Chameleon: the Life of John Byrom.* London, Jonathan Cape, 1994. Pp. 276, £18.99. ISBN 0224-030477

reviewed by Desmond King-Hele, F.R.S.

7 Hilltops Court, 65 North Lane, Buriton, Hants GU31 5RS

John Byrom as an undergraduate

John Byrom, F.R.S. (1691–1763) was a walking contradiction: a man of great intelligence who often seems extremely stupid; an honourable man who acted duplicitously for much of his life as an ineffectual double agent; a charming man who would ingratiate himself with the powerful and then let them down; a tall man (6 feet 4 inches) who stooped to low methods. His only straightforward activity was as a teacher of the system of shorthand he devised about 1719. Joy Hancox suggests that he saw the system primarily as a cypher for secret messages. When pressed to publish it fully, in 1741, he over-reacted with an extraordinary gambit: he asked the House of Commons to let him bring in a bill that would give him the sole right to teach his system of shorthand for 21 years. The bill was passed within three months, and George II gave the Act his Royal Assent.

John Byrom had a good schooling and entered Trinity College, Cambridge, where he became a Fellow in 1714. A year later he had to take the oath abjuring the descendants of James II. He refused, and resigned his Fellowship. This was his first self-inflicted wound. He went to France and visited the Old Pretender (James's son) at Avignon. Thereafter, one of his guises was as a Jacobite agent, and Hancox traces his links with Thomas Siddal and other Jacobites at Manchester.

Byrom then returned to Cambridge, and in 1719 fell in love with a woman he saw at Sturbridge Fair. According to Hancox, the woman was Princess Caroline, who became Queen in 1727, and Byrom was her lover for eleven years, until she rejected him in 1730. It sounds bizarre, but so was much in the story of the early Georges. King George I, who came from Hanover to take the English throne in 1714, had for 20 years kept his wife shut up in Ahlden Castle because he suspected her of having a lover, Königsmarck, who disappeared without trace. George brought two mistresses to Britain and kept his wife locked up for a further 12 years until she died. He hated his son, later George II, whose children he ordered to be seized and removed to Hanover. Caroline was deprived not only of her children but also, quite often, of her husband, who had a succession of mistresses and returned to her in between.

By all reports Princess Caroline was beautiful, intelligent and very well able to get her own way. Her favourite politician, Sir Robert Walpole, enjoyed a long reign as Chief Minister. Although Joy Hancox offers no hard evidence that Byrom was Caroline's lover, she argues the thesis plausibly and identifies a house in New Round Court near the Strand, rented in the name 'Thomas Siddall'. Byrom was addicted to false names, 'Freeman' being his favourite, and as Caroline lived nearby at Leicester House, the two might have met here, any written messages being in shorthand, of course.

Byrom was well known as a poet in his day, and in several of his poems Hancox finds evidence that he was claiming to be the father of Prince William, Duke of Cumberland, born in 1721, who grew up to be 'the butcher of Culloden' and scourge of the Jacobites. Even if the interpretation is correct, it may be that the poems are fantasy. However, it is true that Caroline detested her elder son Frederick and doted on William. Also, George II was short and dapper, while William was tall and burly.

To descend from the regal to the real, Byrom married a cousin in 1721, leaving her in Manchester while he got on with

William, Duke of Cumberland

his life. He kept her quiet by writing frequent letters with semi-fictional reports of his activities.

Byrom's relations with the Royal Society were also quite bizarre. In the early 1720s he ardently cultivated the friendship of Sir Hans Sloane and Martin Folkes, and he was elected a Fellow in 1724, presumably on the strength of his ingenious shorthand system. As long as Sir Isaac Newton remained President, there was no move to ask George I to be patron of the Society (not because he had put away his wife, but because he would have supported Leibniz). For Sir Hans Sloane, Newton's successor in 1727, the restoration of the Society's links with royalty was a delicate and important matter: he proposed a carefully worded address to the King and Prince of Wales. To his horror, one of the Fellows had the temerity to question 'the form and manner' of the address. Yes, it was Byrom. The dispute escalated, with Byrom accusing Sloane of trying to suppress the right of the Fellows to free speech, and other Fellows accusing Byrom of being a Jacobite. 'They say there have not been so many speeches this hundred years as there have been this month', Byrom wrote, seemingly with some relish. Byrom's motive for starting this furore is obscure: perhaps he just wished to annoy the Fellows. For some years after this, he seems to have withdrawn into his own semi-secret Society, the Cabala Club.

Later in life he lived in Manchester with his wife. But in 1745 he failed to support the Jacobite group there. If he had, he would, like them, have been gruesomely executed. Was he saving his own skin? Was his Jacobitism a pose? Was he really a double agent? The questions multiply.

Joy Hancox's book is the product of many years of research. Byrom's habits of duplicity and secrecy have made her task a daunting one, and she is too often forced into speculation and the weighing of innuendoes. The possible liaison of Byrom and Caroline can never be proved; had there ever been good evidence, Byrom would probably have disappeared like Königsmarck, for it is unlikely that George II gave his royal assent to this act of Byrom's.

Joy Hancox is rather obsessed by machinations, and she accuses Byrom of being involved in the deaths of both George I and Caroline, on the strength of what could be coincidences. Also, she fails to do justice to his poetry, which is sometimes quite powerful. When all is said, however, she must be saluted for trying so hard to unpick a messy mishmash that no professional historian of science would touch with a bargepole.

THE ENLIGHTENING OF SCIENCE

John Gascoigne, *Joseph Banks and the English Enlightenment: Useful Knowledge and Polite Culture*. Cambridge University Press, 1994. Pp. xi+324, Ill., £35.00. ISBN 0521-45077-2

reviewed by Marie Boas Hall, F.B.A.

14 Ball Lane, Tackley, via Kidlington, Oxon. OX5 3AG

Anyone familiar with the Royal Society's portrait of Sir Joseph Banks will readily believe that he dominated a large segment of English 'useful knowledge and polite culture' between his return from the glamorous voyage with Cook to the South Seas and Australia in 1771 and his death in 1820. Very rich, an intimate of George III, and, after one battle in 1783–4, an autocratic President of the Royal Society, he acted as patron, research director and employer to a multitude of botanists, natural historians, proto-anthropologists,

Joseph Banks, by Thomas Phillips, 1815

agriculturalists and 'improvers'. In this work, based upon John Gascoigne's mastery of the epistolary material in the Banks Archive (now in the British Museum of Natural History awaiting complete transcription and publication), together with much printed material, Banks is given his due place in English culture. The author clearly shows that Banks, while not much of a scientist himself, greatly helped to develop much interesting, useful and progressive knowledge about the world and man.

After a brief biographical chapter, Gascoigne presents an exposition of Banks's view of the spirit of his own post-Newtonian age: enlightened because it accepted new ideas (always in moderation); anticlerical, although not areligious; believing in education, except for the masses, and in the always mild improvement of society; careful after 1789 to avoid revolutionary excess. Successive chapters cover natural history (especially botany, Banks's favourite subject), anthropology, 'improvement' as applied to agriculture, industry, and society, all illustrated by accounts of the ideas of Banks and the activities of his associates – assistants, employees, correspondents and friends. The final chapter discusses the impact of the French Revolution upon these men, as well as what became of the ideals of the Royal Society under Banks's long dominance.

Inevitably, since this book is centred on Banks, it is restricted to his views, interests and activities. Hence there is nothing here about post-Newtonian physical science: nothing on astronomy, the problems of longitude determination or the transits of Venus, almost no mention of steam engine technology, nothing on 'Newtonian' experimental physiology, and little on theoretical chemistry. (And p. 115, Chevenix's work on palladium was restricted to trying to prove that it was a compound of platinum). Banks was certainly too young to appreciate Harrison's chronometers or the introduction into England of the Continental calendar. And, certainly, the biological sciences did advance greatly in his lifetime. But one may fairly feel that the author has become a little too Banks-centred in his description of English intellectual life in eighteenth-century England, and a little unfair to the biological scene before Banks.

But this is an admirably informative and revealing account of Banks's wide influence as a scientific patron. His own published contributions were negligible, but he inspired, directed, paid and oversaw the work of countless men to the great benefit of the scientific world. Similarly, although the Royal Society he developed, as a club of gentlemen devoted to generalized learning, was woefully out of date by 1820, and his views had been a limitation on its utility even 40 years earlier, nevertheless the Society over which he ruled benefited greatly from his patronage and prestige, at home and abroad. This is a thorough, informative and detailed scholarly study, a definitive work on its subject, well written and handsomely produced.

DEFINING WHEWELL

Richard Yeo, *Defining Science: William Whewell, Natural Knowledge, and Public Debate in Early Victorian Britain.* Cambridge University Press, 1993. Pp. xiv+280, £35.00. ISBN 0-521-43182-4

reviewed by William J. Ashworth

William Whewell Calendar Project, Trinity College Library, Cambridge

William Whewell

On a wet and dark morning in October 1811 a young man began a long, tedious journey from Lancaster to Cambridge. It was the beginning of a trip that would take William Whewell (1794–1866) from being the son of a carpenter and joiner to being Master of Trinity College. When he first arrived at Cambridge the war with revolutionary France was still raging, and the English physical and mental landscape was gradually changing to meet the needs of a developing industrial economy. For many it was heralding in a new morality expunged of religion and a very real threat to the English Constitution.

It was in this social context that Whewell assimilated into the traditional Anglican culture of the eighteenth century. Within the walls of Trinity College he laboured to protect it from the illusionary and destructive effects of French abstract reason, as well as the growing interests stemming from the new industrial cities such as Manchester. Indeed, he devoted his life to preserving and ensuring that political and intellectual changes did not adversely affect the constitutional marriage between church and state, and the intrinsic role Oxbridge played in this holy alliance.

Whewell has been the subject of a number of recent works and is rightly recognized as an important participant in nineteenth-century debates defining the contours of science, religion and morality. For Whewell the three could not be separated. Richard Yeo sets out to explore Whewell's role in trying to define science within the rich *milieu* of early Victorian culture. For example, debates were fought in theology, science, philology, history, literature and political economy. Whewell was active in varying degrees in all these areas. A discussion of this 'metascientific discourse' occupies much of the first part of Yeo's book.

In this way we begin to get a sense of Whewell's seemingly diverse interests and of a culture still far removed from the specialist disciplinary boundaries of the twentieth century. Yeo is not offering a biography of Whewell but, rather, uses him as someone who drew upon a whole range of media to define science and its cultural meaning. It was this task, argues Yeo, that Whewell began to see as his vocation. He drew upon a huge and diverse range of media (books, reviews, pamphlets, addresses, lectures and sermons) to diffuse his views.

Yeo describes how Whewell developed a philosophy of science to counter a materialist and utilitarian epistemology with one that placed fundamental ideas at the base of the scientific/human mind. We are shown how Whewell's philosophy was then made the foundation of his later work in the moral sciences. His mission was to sever the growing link perceived between the material and moral sphere, with a philosophy which stressed the role of both empiricism and idealism.

For Whewell scientific progress operated through a philosophy of induction in which advances still retained old truths revealed in a new point of view. This was also his outlook on social and political reform. Yeo carefully captures this Burkean conservatism that pregnated all of Whewell's activities, but perhaps could have delved deeper into his Peelite politics. Intellectually, Whewell saw purely deductive habits of thinking as destructive to this view.

His use of history and his onslaught on Ricardian political economy and pure French algebraic analysis were particularly targeted. As Yeo writes in his discussion of Whewell's immensely popular *Bridgewater Treatise* (1833): 'In a book intended to affirm the religious value of science he had divided the scientific community into two kinds of thinkers, with the deductive type, the majority, possessing mental habits that impoverished their religious feeling and their ability to appreciate moral evidence and poetic beauty' (p. 123).

In chapter six Yeo describes how Whewell moved the emphasis from science and the divine design argument to the general moral character of men of science. Whewell turned to history to seek an appropriate scientific morality. He expelled a utilitarian legitimation of science and underlined a moral and intellectual one. Further, great men of science were not bound by strict rules but also used their imagination. Hence, in his metascientific project, Whewell did not seek to ground rules of science, but, rather, 'to generalise the pattern by which these discourses were translated into scientific disciplines' (p. 165).

Yeo's work on Whewell's moral philosophy is particularly welcome – as one of the more neglected areas of Whewellian scholarship. He rightly stresses its central role in Whewell's activities, and reminds us that before 1840 'the key philosophical contest was mainly fought over ethics, not science. This was the conflict between utilitarianism and some version of intuitionism' (p. 180). Whewell's targets were the Cambridge textbooks of John Locke and William Paley.

Yeo shows how moral concerns were apparent from Whewell's earliest reflections, and cannot simply be seen as a later development of his work. He traces the debates and resources Whewell drew upon in moving from within a Baconian framework to an anti-Lockean stance that defined his philosophy of science within a moral constituency.

In part three of Yeo's book we move to the volatile arena of debates in education. During the late eighteenth and early nineteenth century Oxbridge was under siege from attacks stemming primarily from Edinburgh and London. There is a sense in which Whewell's work is incredibly straightforward: it can be seen as a quest to preserve the institution of Cambridge University (and, especially, Trinity College) from these increasingly vociferous onslaughts and from the fangs of political interference.

Whewell's work is here shown to be a reaction and alternative to utilitarian and political radicalism. Science had to be assimilated into the traditional structure of a liberal education. Discovery was not part of a university education, but, rather, 'a balance between new and old knowledge' (p. 222). Whewell's moral philosophy, in particular, was designed to restore this equilibrium through a correct training of the Cambridge mind. Yeo concludes his book by situating Whewell's philosophy as an alternative to Comtean positivism, and an attempt to secure natural theology.

One problem with this book is an overall tension created by concentrating on one man and one place (Cambridge) within the vast geography of the purported boundaries of the book: 'Natural Knowledge, and Public Debate in Early Victorian Britain'. Yeo could have concentrated purely on Whewell, which would have required a narrowing of these boundaries and closer scrutiny of Whewell's social and intellectual circle at Cambridge University, where he spent all his life. Indeed, I would have liked to have seen more on Whewell's science textbooks, numerous sermons and work in philology and literature.

Conversely, Yeo could have focused more on the diverse and opposing groups from Manchester to London, all vigorously competing to define science. Nevertheless, this is a fine book and will no doubt make its mark in discussions concerned with debates on nineteenth-century science.

THE GENTLE PIONEER

P.P. Ewald and his Dynamical Theory of X-ray Diffraction. Edited by D.W.J. Cruickshank, H.J. Juretschke & N. Kato. International Union of Crystallography, Oxford University Press, 1992. Pp. x+161, £40.00. ISBN 0-19-855379-X

reviewed by M.M. Woolfson, F.R.S.

The University of York, Heslington, York, YO1 5DD

Paul Ewald, who died in 1985 at the age of 97, was one of the greats of modern science. Born in Berlin into a comfortable middle-class academic family, he developed a passion for mathematical physics. When Sommerfeld presented him with a list of possible topics for his Doctoral Thesis he chose one related to crystal optics. This work, presented in 1912, could be applied to the behaviour of X-rays in crystals and it is suggested that it was a conversation between von Laue and Ewald early in 1912 that gave von Laue the idea for his famous experimental demonstration of X-ray diffraction.

Ewald served in the First World War and was in charge of a mobile X-ray unit but, fortunately, he found enough free time to further his ideas on how an X-ray optical field could travel in a crystal. After the war he spent three years in Munich and then moved to the Technische Hochschule Stuttgart where he soon became Professor of Theoretical Physics. By 1932 he was Rector of the T.H., but soon afterwards the new Nazi regime began to issue its racist and restrictive edicts. In 1937 Ewald decided to leave Germany and take up a post in Cambridge. From there, in 1941, he went to Queen's College, Belfast, where eventually he obtained a chair. In 1949 he was offered a post at Brooklyn Polytechnic in the U.S.A., in which country he remained for the rest of his life. However, he had previously acquired British citizenship which he never relinquished and of which he was extremely proud.

Ewald's early work was the basis of developments in X-ray crystallographic studies for decades to come in areas completely unknown in 1912. Ewald himself was active for more than sixty years after his seminal doctorate thesis and during this time, and since, many other workers have developed completely new areas of crystallographic science based on his original theory.

Ewald was a champion of the internationalisation of science and of communication through high-quality publication. He realised that crystallography had become an important scientific discipline, spanning many other fields, and he was a prime mover in establishing the International Union of Crystallography in 1947. He had been co-Editor of Zeitschrift für Kristallographie from 1924 until he left Germany but, for a time, Zeitschrift had ceased publication. Mainly though his influence the I.U.Cr. journal *Acta Crystallographica* was set up in 1948, with himself as the first Editor.

This is the background of the narratives to be found in this book by those who knew Ewald and were intimately concerned with his work. Part A is an Introduction by N. Kato on the significance of Ewald's dynamical diffraction theory, which shows clearly how the seedling grew into an oak. Part B is mainly to do with Ewald the man – although disentangling the man and the scientist is sometimes not easy. The contributions are by G. Hildebrandt, H.A. Bethe, Max Perutz, Dorothy Hodgkin, Harmeke Kamminga and Helmut Juretschke. These deal mainly with his period in Europe and his efforts in building the post-1945 international crystallographic community and to a lesser extent with his time in the U.S.A. Throughout these contributions there shines through the image of a man of immense humanity, coupled with humour and kindness – a power for good alone.

Part C gives seven contributions by eminent crystallographers on various aspects of Ewald's work, or derivatives from it. D.W.J. Cruickshank's article is mainly concerned with the development of the reciprocal lattice and the sphere of reflection and David Templeton's on lattice sums and the Madelung constant. The next three articles are on X-ray Topography by A. Authier & B. Capelle, Multiple Diffraction of X-rays and the Phase Problem by R. Collela and the Dynamical Theory of X-ray Scattering by J.M. Cowley & A.F. Moodie, the first recipients of the prestigious Ewald Prize established by the I.U.Cr. in 1987. These three papers relate to important work based on Ewald's original ideas. The final two papers in this section by Helmut Juretschke and R.K. Bullough & F. Hynne are again comments on aspects of Ewald's own work. The final Part D is called Ewald Speaks for Himself and is a set of five papers by Ewald in which he discusses his own work. The book is also provided with a complete bibliography covering the period from 1910 to a posthumous publication in 1986.

The book is well produced and makes an excellent read for those interested either in the science or in the personality of this important scientist. It is in many ways an inspirational book and younger scientists, at the beginning of their careers, might find a model well-worth trying to emulate.

A MAGNIFICENT OBSESSION

G. Venkataraman, *Bhabha and his Magnificent Obsession*
Vignettes in Physics, Sangam Books, London, 1994. £6.95 (Paperback)
ISBN 0 86311 555 1

reviewed by Sir Arnold Wolfendale, F.R.S.,

Department of Physics, The University of Durham, Science Laboratories, South Road, Durham, DH1 3LE

Homi Bhabba

On 22nd January, 1966, Homi Bhabha addressed a vast gathering at the Atomic Energy Establishment in Trombay – held to offer condolences on the death of President Lal Bahadur Shastri. The following day Bhabha's Air India plane crashed into Mont Blanc as it commenced its descent into Geneva airport. Thus perished one of India's most gifted scientists and a man who was a scholar in other fields too. It is ironic that the following day the present reviewer addressed a small group of heart-broken Indian physicists at the site of their joint cosmic ray neutrino experiment in the Kolar Gold Fields.

Venkataram has written a fine book about his countryman, chronicling his early work in Cambridge and elsewhere on cosmic rays, not least the theory underlying their behaviour, his great legacy – the Tata Institute of Fundamental Research (TIFR) in Bombay – and the development of Nuclear Energy in India.

As the TIFR celebrates its 50th anniversary one can look back, by way of Venkataraman's account, at some of its achievements but more particularly at the stage on which this work was done, with Bhabha pushing at every door (most of which opened, such was his 'power') in his pursuit of science and technology of the highest quality. Bhabha's attitude was both that of a patriot, sensitive to India's considerable needs, and a scholar, mindful of India's place in the world.

His patriotism was an object lesson, alas, often forgotten nowadays. Bhabha's letter to Chandrasekhar written on 20th April, 1944, urging him to return from the US to join the TIFR, puts his views with admirable clarity.

> I myself have up to recently felt that after the war I would accept some good job in some University in Europe or America, because a place like Cambridge has an atmosphere which no place in India has at the moment. But, I have recently come to the view that provided proper appreciation and financial support are forthcoming, it is the duty of people like us to stay in our own country and build up outstanding schools of research such as some other countries are fortunate enough to possess.

But it was not to be; Chandrasekhar stayed in the US, where his work blossomed, and India's reputation had to be enhanced *in absentia*.

Venkataram writes well and with a wide age-range in view. In fact, we have a special Preface 'To the young reader' and a review chapter entitled 'A thumbnail sketch'. Although it is true that there is much physics that can only be understood by adult physicists there is also much that will be of interest to all.

Bhabha and his Magnificent Obsession is warmly recommended.

medium; instead, each ray has its own refractive index, and this varies rapidly with its direction.

3. Two beams: geometrical theory corrected for refraction in Bragg reflection

Assume that besides the origin, 1, only one lattice point, h, lies on the sphere of reflection. As long as this condition holds it entails rotational symmetry about the direction of **h** as axis. It is therefore sufficient to consider the meridional plane which contains $\mathbf{K}_1$ and $\mathbf{K}_h$. In Fig. 2, circles of radius k and K are drawn about the points 1 and h, such that $K/k = n$, where n is the refractive index of a solitary wave. Any wave vector of a wave in the interior of the medium starts on the K circle (the smaller one in the case of X-rays). The intersection of the k circles is called the *Laue point*, La; that of the K circles, the *Lorentz point*, Lo. The condition for obtaining two strong rays exists only in the interior of the crystal and the waves are represented by the vectors $Lo \to 1$ and $Lo \to h$. The 'reflecting' planes are normal to **h** and the angle between either of the vectors and the symmetry line of the figure is the *internal glancing angle*, θ. Since $|\mathbf{h}| = p/d$, where d is the spacing of the reflecting planes and p is the order of reflection, we see from the figure that

$$\sin \theta = \tfrac{1}{2}h/K \quad \text{or} \quad \lambda = 2(d/p)\sin\theta, \quad (3.1)$$

which is Bragg's law.

We do not observe the waves inside the crystal and must therefore express this condition in terms of the observable quantities outside, which will be denoted by the suffix 'o'. The condition of refraction (*Snell's law*) states that internal and external waves keep in step along the surface, that is, their wave vectors have the same component along the surface. We therefore draw a line parallel to the vector **h** through Lo and find its intersection with the circle of radius k_o. Connecting this point to the origin shows the vector $\mathbf{k}_o$ of the external wave, where $\mathbf{k}_o$ and the normal to **h** form the *observable external glancing angle* θ_o. If we naively apply Bragg's law to this glancing angle, given the order vector **h**, we find a fictitious wave vector $\mathbf{k}_f$ which has the same direction as $\mathbf{k}_o$ but ends on the symmetry line of the figure.

We wish to find **k** in terms of θ_o, **h**, and **n**. From the figure we obtain the following relations:

$$K = nk \quad (3.2)$$

$$K \cos\theta = k \cos\theta_o \quad (3.3)$$

$$K \sin\theta = \tfrac{1}{2}h = k_f \sin\theta_o \quad (3.4)$$

$$(K)^2 = (\tfrac{1}{2}h)^2 + k^2 \cos^2\theta_o = k_f^2 \sin^2\theta_o + k^2 \cos^2\theta_o. \quad (3.5)$$

Therefore

$$k^2(n^2 - \cos^2\theta_o) = k_f^2 \sin^2\theta_o \quad (3.6)$$

or, in wavelengths,

$$\lambda_o^2 = \lambda_f^2(n^2 - \cos^2\theta_o)/\sin^2\theta_o. \quad (3.7)$$

This equation expresses the wavelength in vacuum, λ_o, in terms of the fictitious wavelength, λ_f, obtained by applying the uncorrected Bragg law to the glancing angle observed on the spectrograph. Relations (3.6) and (3.7) hold for all values of the refractive index n. It is true for $n < 1$ (X-rays) as well as for electrons with $n > 1$, or for visible light in iridescent media. Knowing the order of diffraction h, the wavelength *in vacuo* and the external angle, θ_o, we determine from (3.6)

$$n^2 = \cos^2\theta_o + (\tfrac{1}{2}h/k)^2 = \cos^2\theta_o + (\tfrac{1}{2}h\lambda_o)^2. \quad (3.8)$$

If the refractive index is very close to 1, writing $n = 1 - \delta$ and keeping only the lowest power in δ, we have

$$\lambda_o = \lambda_f(1 - \delta/\sin^2\theta_o), \quad (3.9)$$

a relation Stenström derived in 1919.

4. Two beams in transmission: symmetrical Laue case

This case is represented in Fig. 3. The internal and external wave vectors have automatically the same component along the crystal surface and there is no fictitious wave vector. Bragg's law holds for the observable quantities without correction.

5. Two beams: dynamical theory

According to the dynamical theory, strict compliance with the Bragg law is not required. Instead, a strong secondary beam is generated from a primary one over a small but finite range of angles of incidence. If the primary beam is incident on, and reflected by, the same face of a crystal as in the usual X-ray spectrometer, theory predicts for a non-absorbing crystal a range of several seconds of arc throughout which

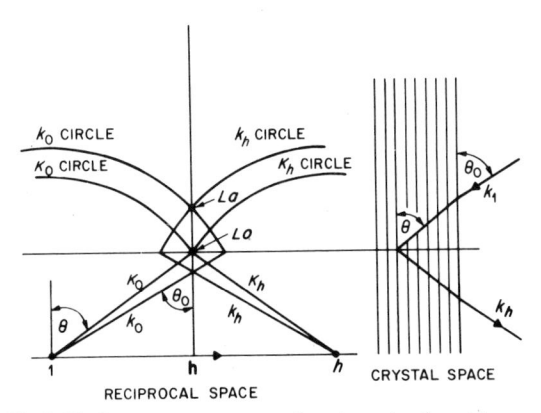

Fig. 2. The Bragg case; wave vectors in reciprocal and crystal space.

an incident ray will be totally reflected. This range is centered on the corrected glancing angle of the geometrical theory. It is flanked, for slightly smaller and greater angles of incidence, by regions of '*Pendellösung*' within which the output drops off rapidly while oscillating between maxima and minima. The entire intensity of the secondary beam, plotted as a function of the glancing angle of the incident beam, *i.e.* the *reflection curve*, is symmetrical about the corrected Bragg angle provided absorption is neglected. The geometrical theory thus leads to the correct wavelength if we assume the glancing angle θ_o to be determined by the centroid of intensity of the (unresolved) line.

In an absorbing crystal, on the other hand, the reflection curve is not symmetrical. Fig. 4 (adapted from a paper by Renninger, 1975) shows the reflection curve for the 422 reflection from a silicon crystal in the Bragg case. The theoretical curve for a strictly plane incident wave has been convoluted with the angular distribution within the ray delivered by the monochromator which shows a spread of $1 \cdot 1''$. This procedure smooths out the *Pendellösung* variations of the reflected intensity. On account of the absorption coefficient (146 cm^{-1}) the curve is no longer symmetrical about the corrected Bragg angle; its centroid is shifted towards smaller angles, that is, the correction discussed so far is too large. The region of total reflection, between $\pm 2''$, is indicated in the figure.

It is true that in X-ray spectroscopy the angular width of the incident beam is much larger than $1 \cdot 1''$ and that therefore the process of convolution would wipe out much of the asymmetry in the recorded lines. But if comparisons of X-ray wavelengths are to achieve the accuracy of those of visible light, then the shape of the reflection curve will have to be taken into account.

6. Influence of simultaneous reflections

It can easily happen that apart from the main reflection of order h further secondary rays of orders h_1, $h_2, \ldots$ are produced, unnoticed by the spectroscopist. This is the more likely the shorter the wavelength. In rotating the diffracting crystal about the normal to the reflecting planes, there may be a few azimuths for which no third beam is generated. In publications of spectroscopic research, little attention has been given to the complete setting of the crystal.

As long as only two rays exist, the cross section of the surface of dispersion is a simple hyperbola for each mode of polarization (say, for the electric vector normal to the plane of the rays). The distance between the vertices of the branches, or sheets of the hyperbolae, determines the width of the region of total reflection, and the central line between the branches gives the deviation from the uncorrected Bragg law.

With the appearance of a third ray, the two-sheet surface is deformed by a third sheet, the details depending on the geometry. Besides, if the beams are not coplanar, the polarization cannot be split in a simple manner; a surface of dispersion consisting of six sheets has to be considered. In a simple case when the third ray lies in the symmetry plane between the first two, this surface has been studied by Ewald & Héno (1968) and Héno & Ewald (1968) with the object of discussing the absorption of the wave fields and, in particular, of accounting for the enhanced Borrman effect. The influence of a third ray on the reflection curve of the first two cannot be determined in general but is not a difficult job for a computer, given the exact data of an experimental case.

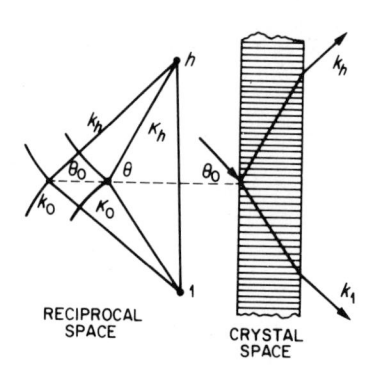

Fig. 3. The Laue case.

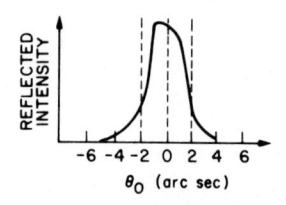

Fig. 4. Reflection curve for absorbing crystal. For the nonabsorbing crystal, total reflection occurs between the dotted lines.

References

EWALD, P. P. & HÉNO, Y. (1968). *Acta Cryst.* A24, 5-15.
HÉNO, Y. & EWALD, P. P. (1968). *Acta Cryst.* A24, 16-42.
RENNINGER, M. (1975). *Acta Cryst.* A31, 42-49.
STENSTRÖM, W. (1919). Dissertation, Lund Univ., Sweden.

P.P. Ewald: Bibliography

Compiled by Hellmut J. Juretschke

The publications list assembled below has been compiled from a number of sources. The main contribution came from Ewald's own typewritten manuscript 'List of Publications by P.P. Ewald', which is fairly complete up to 1965. Additional material came from the 1988 Bethe–Hildebrandt article in the Memoirs of Fellows of the Royal Society, and from the files of Arnold Ewald, Hellmut Juretschke, N. Kato, and Margot Bergmann. H. Wagenfeld supplied an invaluable fleshing out of the references for known and additional book reviews in *Z. Kristallogr*. Finally, the 1988 inventory by A.C. Hengst of the Ewald papers in the Ithaca home provided leads to further references. Altogether, this list certainly includes all his major writings, although the omission of additional minor publications cannot be ruled out.

All listings are ordered by year of publication, otherwise using the *Acta Cryst.* system of reference, wherever possible. Titles have been retained in the language of the original.

The writings have been assembled under three separate headings:

(1) The *scientific publications* include all research papers, major review papers, and major papers on the history of X-rays.

(2) The *occasional writings* encompass more popular articles, anniversaries, dedications, and obituaries, and some personal reminiscences.

Obviously, the division into these two categories involves some arbitrariness, depending on the interest of the reader.

(3) The *book reviews*. This third section is not entirely complete, partly because no systematic file of book reviews has been found, and partly because Ewald's own list was apparently never properly edited. Because in our search, as far as it went, additional book reviews turned up in an astonishing number, the list assembled below cannot be considered definitive. It is certainly representative.

A final biographical comment concerns the initials P.P. During the search, it was discovered that Ewald's thesis apeared under the name Peter Paul Ewald, that all of his scientific papers were signed merely P.P., but that occasional writings after 1948 were increasingly signed Paul P. Ewald, and sometimes even Paul Peter Ewald. After an enquiry to his family, Mrs. Rose Bethe pointed out that Peter does not appear among his baptismal names, that he was always called Paul, but that he added Peter to his name in his earliest writings apparently to be differentiated from his father, also named Paul. In later biographical material Ewald himself listed his name as Paul P.(eter).

Scientific publications

1910   Die Vertikalkomponente des Windes. *Z.f. Flugtechnik und Motorschiffahrt.* **1**, 195–197.
Anemometer zur Messung der Vertikalkomponente der Windgeschwindigkeit. *Phys. Z.* **11**, 1214–1216.

1912   Dispersion und Doppelbrechung von Elektronengittern (Kristallen). Inaugural Dissertation an der hohen phil. Fakultät (II. Sektion) der Kgl. Ludwigs-Maximilians-Universität zu München. Submitted on 16 February 1912. Göttingen: Dietrichsche Universitäts Buchdruckerei, 1–46.
Dispersion and Double Refraction of Electrons in Rectangular Groupings (Crystals). Int. Cong. of Mathematicians, Cambridge, August 1912 (4 pages).

1913   Zur kinetischen Theorie der Materie [Introduction to the Congress in Göttingen]. *Naturwiss.* **1**, 297–299.
Zur Theorie der Interferenzen der Röntgenstrahlen in Kristallen. *Phys. Z.* **14**, 465–472.
Bemerkungen zu der Arbeit von M. Laue: Die dreizählig symmetrischen Röntgenstrahlen-

aufnahmen an regulären Kristallen. *Phys. Z.* **14**, 1038–1040.

1913 Bericht über die Tagung der British Association in Birmingham (10–17 September). *Phys. Z.* **14**, 1297–1307.

1914 Über die Vorzüge der Vektorrechnung. *Naturwiss.* **2**, 217–222.
Die Intensität der Interferenzflecke bei Zinkblende und das Gitter der Zinkblende. *Ann. Phys. (Leipzig)* **44**, 257–282.
Röntgenaufnahmen von kubischen Kristallen, insbesondere Pyrit (with W. Friedrich). *Ann. Phys. (Leipzig)* **44**, 1183–1196.
Interferenzaufnahme eines Graphit-Kristalles und Ermittlung des Achsenverhältnisses von Graphit. *Sitzungsber. K. Bayer. Akad, d. Wiss. Math-phys. Kl.*, 325–327.
Die Berechnung der Kristallstruktur aus Interferenzaufnahmen mit X-Strahlen. *Phys. Z.* **15**, 399–401.

1916 Zur Begründung der Kristalloptik:
Teil I: Theorie der Dispersion. *Ann. Phys. (Leipzig)* **49**, 1–38.
Teil II: Theorie der Reflexion und Brechung. *Ibid.* **49**, 117–143.

1917 Teil III: Die Kristalloptik der Röntgenstrahlen. *Ibid.* **54**, 519–597.

1920 Über die Kontrolle von Kristallstrukturen durch Laueaufnahmen (Rutil, Anatas, Kalkspat) (with A. Kratzer & L. Citron). [Report on the meeting of the Gauverein of the German Phys. Soc. Munich, 3 March]. *Ber. d. D. Phys. Ges.* **1**, 33–36.
Abweichungen vom Braggschen Reflexionsgesetz der Röntgenstrahlen. *Phys. Z.* **21**, 617–619.
Zum Reflexionsgesetz der Röntgenstrahlen. *Z. Phys.* **2**, 332–342.

1921 Das 'reziproke Gitter' in der Strukturtheorie. *Z. Kristallogr.* **56**, 129–156.
Die Berechnung optischer und elektrostatischer Gitterpotentiale. *Ann. Phys. (Leipzig)* **64**, 253–287.
Mitteilungen aus dem Gebiet der Röntgenstrahlen (Bragg's Normalabfall, Mosaikkristall, Seemann, etc.). *Naturwiss.* **9**, 948–950.

1922 Zu Rubens und Hertz Note 'Über den Einfluss der Temperatur auf die Absorption langwelliger Wärmestrahlen in einigen festen Isolatoren'. *Naturwiss.* **10**, 1057–1058.
Die Bedeutung der Röntgenstrahlen für die moderne Naturwissenschaft [Inaugural lecture at T.H. Stuttgart, 14 June 1922]. *Klinische Wochenschrift* **43**, 2147–2150.

1923 *Kristalle und Röntgenstrahlen*. Berlin J. Springer, 327 pages.
Krystallstrukturen. Ch. 155 in *Landolt-Bornstein, Physikalisch-Chemische Tabellen*, Main Vol. Berlin: J. Springer, 863–873.

1924 Über den Brechungsindex für Röntgenstrahlen und die Abweichungen vom Braggschen Reflexionsgesetz. *Z. Phys.* **30**, 1–13.
Die Röntgenstrahlen und der Kristallbau. *Strahlentherapie* **18**, 1–16.
De terugkaatsing en breking van het licht als probleem der elektronentheorie. *Physica* **4**, 234–251.
Allgemeine Ergebnisse über den Aufbau der festen Körper. *Z. Kristallogr.* **61**, 1–17.

1925 Die Reflexion und Brechung des Lichts als Problem der Elektronentheorie. In *Fortschr. d. Chemie, Physik und physikal. Chemie* (ed. A. Eucken) Series B, Vol. **18**, No. 8. Berlin: Gebr. Borntrãger, 491–518.
Über die Symmetrie der Röntgen-Interferenzen. *Physica* **5**, 363–369.
Die Intensitäten der Röntgenreflexe und der Strukturfaktor. *Phys. Z.* **26**, 29–32. [Correction in (1926) *Phys. Z.* **27**, 182].

1927 Strukturbericht 1913–1926 (with C. Hermann). *Z. Kristallogr.* **65**, 4–33.
Gilt der Friedelsche Satz über die Symmetrie der Röntgeninterferenzen? (with C. Hermann). *Z. Kristallogr.* **65**, 251–260.
Der Aufbau der festen Materie und seine Erforschung durch Röntgenstrahlen. In *Handbuch der Physik*, Vol. 24. Berlin: J. Springer, 191–369.

1928 Der Übergang von der Röntgenoptik zur Lichtoptik. In *Probleme der Modernen Physik* (ed. P. Debye). [Festschrift for A. Sommerfeld on his 60th birthday.] Leipzig: S. Hirzel, 134–142.
Untersuchungen Zur Kristalloptik der Röntgenstrahlen (with W. Ehrenberg & H. Mark.) *Z. Kristallogr.* **66**. 547–584.
Die Kristallstruktur der Nichteisen-Metalle. *Metallwirtschaft*, Heft **15**, (April), 2 pages.

1929 Atommodelle; Ergebnisse und Methoden der Atomforschung. *Festschrift of the T.H. Stuttgart on completing its first century (1829–1929)*. Berlin: J. Springer, 92–101.
Some Modern Developments of Wavemechanics, and their Bearing on the Understanding of Crystal Structure. *Trans. Faraday Soc.* **25**, 402–409.
Krystallstrukturen (with C. Hermann). Ch. 155 in *Landolt-Bornstein, Physikalisch-Chemische Tabellen*, Suppl. Vol. Berlin: J. Springer, 595–637.
Der mechanische Aufbau des festen Körpers in atomischer Betrachtung. *In Muller Pouillet's*

Lehrb. d. Phys., 11th Edition, Vol I, 2. Braunschweig: Fr. Vieweg, 925–990.

1930 The Mechanical Structure of Solids from the Atomic Standpoint. Ch. 3 in *The Physics of Solids and Fluids* (eds. G. Prange, L. Prandtl & P.P. Ewald). London: Blackie & Son Ltd., 81–151.

Some Remarks on the General Physical Aspects of Natural Optical Activity. *Trans. Faraday Soc.* **26**, 308–310.

Tagung des erweiterten Tabellen Komites in Zürich, 20–31 Juli 1930. *Z. Kristallogr.* **75**, 159–160.

1931 Strukturbericht 1913–1928 (with C. Hermann). *Ergänzungsband zu Z. Kristallogr. Akad. Verlagsges*, 818 pages. Preface, March 1931.

Bemerkungen zu den Begriffen 'Stase', 'Textur' und 'Phase' bei Friedel [*Z. Kristallogr.* (1931) 79, 1–60]. *Z. Kristallogr.* **79**, 299–307.

Bemerkungen zu der Arbeit C. Hermann. *Z. Kristallogr.* **79**, 338–339.

1932 Der Weg der Forschung (insbesondere der Physik). [Inaugural lecture for the Rektorat, Techn. Hochsch. Stuttgart, 30 April.]

Optique cristalline (Lumière et rayons X)—Interaction des atomes par rayonnement. *Ann. Inst. Henri Poincaré* **8**, 79–110.

1933 Die Erforschung des Aufbaues der Materie mit Röntgenstrahlen. In *Handbuch der Physik*, Vol. 23, part 2. Berlin: J. Springer, 207–476.

1934 Structure of Molecules and of the Ideal Lattice (with J.D. Bernal, A. Muller & J.C. Slater). *Intern. Conf. On Physics, London. Vol. II, Solid State of Matter*. Cambridge: Camb. University Press. Discussion, 50–53.

The mosaic texture of rocksalt (with M. Renninger). *Ibid.*, pp. 57–61.

Aufgaben und Leistungen der Röntgenstrukturforschung für Medizin und Biologie. *Intern. Radiologenkongress, Zürich*, Vol. 2.

1935 Bemerkungen zur vorstehenden Arbeit von A. Hettich [Methoden der Strukturbestimmung. *Z. Kristallogr.* **90**, 473–492]. *Z. Kristallogr.* **90**, 493–494.

1936 Die Röntgeninterferenzen an Diamant als wellenmechanisches Problem (with H. Hönl). *Ann. Phys. (Leipzig)* **25**, 281–308.

Untersuchung linearer Atom Ketten (with H. Hönl). *Ann. Phys. (Leipzig)* **26**, 673–696.

Historisches und Systematisches zum Gebrauch des 'Reziproken Gitters' in der Kristallstrukturlehre. *Z. Kristallogr.* **93**, 396–398.

Die optische und die Interferenz–Totalreflexion bei Röntgenstrahlen, Teil I (with E. Schmid). *Z. Kristallogr.* **94**, 150–164. [Teil II by E. Schmid, *Z. Kristallogr.* **94**, 165–196.]

1937 Zur Begründung der Kristalloptik, Teil IV: Aufstellung einer allgemeinen Dispersionsbedingung, insbesondere für Röntgenfelder. *Z. Kristallogr.* **A97**, 1–27.

The Development of Intensity Interpretation in Crystal X-ray Diffraction. *Current Science (India)* 11–13.

1938 The Force of Excitation in the Theory of Dispersion. *Phys. Rev.* **54**, 893–894.

Elektrostatische und Optische Potentiale im Kristallraum und im Fourierraum. *Nachr. Ges. d. Wiss. zu Göttingen, Mathem. -phys. Kl. Neue Folge. Fachgruppe II*, **3**, 55–64.

1940 X-Ray Diffraction by Finite and Imperfect Crystal Lattices. *Proc. Phys. Soc. Lond.* **52**, 167–174.

1941 A recording automatic balance. *Industr. Eng. Chem. (Analytic Edn.)* **14**, 66–67.

1944 International Status of Crystallography, Past and Future. [Lecture to X-ray Group of Inst. of Phys. in Oxford, 31 March 1944.] *Nature* **154**, 628–631.

1948 Editorial Preface. *Acta Cryst.* **1**, 1–2.

1953 Opening Remarks. In *Structure and Properties of Solid Surfaces* (eds. R. Gomer & C.S. Smith). Chicago: University of Chicago Press, 1–4.

Atomic Theory of Surface Energy (with H. Juretschke). *Ibid.*, 82–117. Discussion, 117–119.

Some Personal Experiences in the International Coordination of Crystal Diffractometry. *Physics Today* **6**, (Dec.) 12–17.

1955 Properties of Solids and Solid Solutions. In *High Speed Aerodynamics and Jet Propulsion* (ed. F.D. Rossini). Princeton: Princeton University Press. *Vol. 1 Thermodynamics and Physics of Matter*, 554–645.

1957 Historical Development of Modern Crystallography. *Norelco Reporter* **4**, 46.

The Solid State. In *Proc. of the Symposium on the Role of Solid State Phenomena in Electric Circuits*. New York: Interscience, 33–40.

1958 Group Velocity and Phase Velocity in X-ray Crystal Optics. *Acta Cryst.* **11**, 888–891.

1959 Vor Fünfzig Jahren. In *Beiträge zur Physik und Chemie des 20. Jahrhunderts. (Dedicated to the 80th birthday of L. Meitner, O. Hahn, and M.v. Laue)* (eds. O.R. Frisch, F.A. Paneth, F. Laves & P. Rosbaud). Braunschweig: Fr. Vieweg & Sohn, 145–146.

Ibid. Simultaneous publication as *Trends in Atomic Physics*. New York: Interscience.

1962 The Origin of the Dynamical Theory of X-ray Diffraction. *J. Phys. Soc. Jap.* **17**, Suppl. B. II, 48–52. (Also in *Proc. Int. Conf. on Magnetism and Crystallography (1961)*, Vol. II.)

1962   *Fifty Years of X-ray Diffraction* (ed. P.P. Ewald). International Union of Crystallography. Utrecht: N.V.A. Oosthoek's Uitgeversmaatschappij, 733 pages.
In addition to being editor, P.P. Ewald contributed the following sections:
Section 1 Introduction, 1–5
Section 2 The Beginnings, 6–80
Section 3 The Tools, 82–118
Section 4, Ch. 15, Dynamical X-ray Optics; Electron and Neutron Diffraction, 248–264
Section 5 In Memoriam
Artur Schoenflies, 351–353
William Thomas Astbury, 354–355
Carl H. Hermann, 357–360
Paul Knipping, 367
Memorial Tablets [for 13 individuals], 368–372
Section 6, Ch. 20, Germany (with E.E. Hellner), 456–468
Section 6, Ch. 25, The World-wide Spread of X-Ray Diffraction Methods, 498–507
Section 8, Ch. 26, The Consolidation of the New Crystallography, 696–706
Structure theories in physical and Fourier space (With A. Bienenstock). *Soviet Physics —Crystallography* **6**, 665–667. [Originally in Russian **6**, 820–824 (1961).]
Symmetry of Fourier Space (With A. Bienenstock). *Acta Cryst.* **15**, 1253–1261.

1965   Crystal Optics for Visible Light and X-Rays. *Rev. Mod. Phys.* **37**, 46–56.
The Dynamical Theory of Diffraction in the Perfect Crystal. In *Int. Conference on Electron Diffraction and Crystal Defects, Melbourne 1965*. Australian Acad. of Sciences. Paper IA-1, 2 pages.

1968   Personal Reminiscences. *Acta Cryst.* **A24**, 1–3.
X-ray diffraction in the case of three strong rays. I. Crystal composed of non-absorbing point atoms (With Y. Héno). *Acta Cryst.* **A24**, 5–15.
Diffraction des rayons x dans le cas de trois rayons forts. II. Influence de l'absorption et du facteur de diffusion atomique (With Y. Héno). *Acta Cryst.* **A24**, 16–42.

1969   The Myth of Myths; Comments on P. Forman's Paper on 'The Discovery of the Diffraction of X-rays in Crystals'. *Archive for History of Exact Sciences* **6**, 72–81.
Introduction to the Dynamical Theory of X-ray Diffraction. *Acta Cryst.* **A25**, 103–108.

1970   Abstract and Appendix A (Developments arising out of Parts I and II on the Foundations of Crystal Optics) in L. Hollingsworth's English translation of *Ann. Phys.* (*Leipzig*) **49**, 1–38, 117–143 (1916), Airforce Cambridge Research Laboratory Bedford, Mass. AFCRL Report 70–8580. Pages iii–iv; A1–A14.

1972   The 'Poststift'—A Model for the Theory of Pole Figures. *J. Less-Common Met.* **28**. 1–5.

1977   The early history of the International Union of Crystallography. *Acta Cryst.* **A33**, 1–3.

1979   A Review of my Papers on Crystal Optics 1912 to 1968. [90th Birthday speech in Oklahoma.] *Acta Cryst.* **A35**, 1–9.

1986   Obituary by J.J. Dropkin and B. Post [Mostly P.P. Ewald's own account]. *Acta Cryst.* **A42**, 1–5.
The So-Called Correction of Bragg's Law. *Acta Cryst.* **A42**, 411–413.

Occasional writings

1904   Radium. (Aus der Welt und Technik). *Potsdamer Intelligenzblatt* 1. Beilage, 1 February.
Tierphotographie. *Potsdamer Intelligenzblatt* 1. Beilage, 4 July.
N-Strahlen. (Rundschau). *Prometheus* **16**, 174–175.

1906   Die Gesetze der Eisbahn. Eine physikalische Plauderei. *Berliner Lokal Anzeiger* 1. Beiblatt, 3 January.
Unser Ausflug nach Cassel. *Göttinger freistudentische Wochenschau* **2**, No. 11.

1921   Erforschung der Kristalle mit Röntgenstrahlen. *Schweizer Chem. Zeitung* 193.
Erforschung der Kristalle mit Röntgenstrahlen. *Natur & Technik, Schweiz. Z. Naturw.* **3**, 9–12.

1928   Die Strukturlehre der Kristalle im Schulunterricht. *Naturw. Monatshefte, biol. chem. geogr. & geolog. Unterricht* **25**, 238.

1930   Die Gastvorträge an der Technischen Hochschule Stuttgart.

1932   Max Born zum 50. Geburtstag. *Metallwirtschaft* **11**, 674.
Zur Entdeckung der Röntgeninterferenzen vor zwanzig Jahren und zu Sir William Bragg's siebzigsten Geburtstag. *Naturwiss.* **20**, 527–530.

1938   Prof. Max Planck: 80th Birthday. *Nature* **141**, 720 [unsigned].
Progress in Physics—Tributes to Max Planck. *Nature* **141**, 981–982 [unsigned].

1942   Pure Science and Politics. In *Science & World Order, Rep. Brit. Ass. for Adv. of Sci.* (Jan. 1942), 24–25.
Dr. Arnold Berliner. Obituary. *Nature* **150**, 284.

1951   Prof. A. Sommerfeld, Obituary. *Nature* **168**, 364–366.

1952 Foreword to *Sommerfeld's Mechanics, Vol. 5 of Lectures on Theoretical Physics*. New York: Academic Press, V–VII.
1953 Paul Niggli, 1888–1953. *Acta Cryst.* **6**, 225–226.
1960 Max von Laue. 1879–1960. *Acta Cryst.* **13**, 413–415. [Correction on page 962].
Max von Laue. 1879–1960. *Biograph. Mem. of Fellows of the Royal Society* **6**, 135–156.
1961 F.J.M. Stratton. *Acta Cryst.* **14**, 204.
1962 Words of Welcome in the Name of the IUCr for the International Conference on Magnetism and Crystallography, Kyoto, 1961. *J. Phys. Soc. Jap.* **17**, Suppl. B II, 21–22.
William Henry Bragg and the New Crystallography. [The First Bragg Lecture: 7 June, Leeds; 13 June, R.I. London.] *Nature* **195**, 320–325.
The New Crystallography. *ICSU Review* **4**, 68–74.
1964 I. Fankuchen. 1904–1964. *Acta Cryst.* **17**, 1091–1093.
1965 R.W. James. Obituary. *Physics Today* **18**, (Jan.) 154.
1967 Peter Debye. *Acta Cryst.* **22**, 947–949.
1968 Erinnerungen an die Anfänge des Münchener Physikalischen Kolloquiums. *Physik. Blätter* **12**, 538–542.
1969 Arnold Sommerfeld als Mensch, Lehrer und Freund. In *Physics of the One- and Two-Electron Atoms* (eds. F. Bopp & H. Kleinpoppen). Amsterdam: North Holland, 8–16.
1970 Sir Lawrence Bragg—Homage. *Acta Cryst.* A**26**, 178.
1971 Lawrence Bragg. *Physics Today* **24** (Oct.) 64–66.
1972 Carl H. Hermann. *Dictionary of Scientific Biography* (ed. C.C. Gillispie). New York: C. Scribner's Sons, 303–304.
1973 Gerhard Borrmann zum 65. Geburtstag. *Z. Naturf.* **28a**, 549.
1974 Physicists I have known. *Physics Today* **27** (Sept.) 42–47.
Bethe, Hans Albrecht. *Encyclopaedia Britannica*, (15th edn), 871–872.
1977 For Achille Papapetrou. *Gen. Relativity & Gravitation* **8**, 539–540.
1979 Max v. Laue: Mensch und Werk. [Lecture on the 100th anniversary of von Laue's birth, Berlin, 2 March 1979.] *Physik. Blätter* **35**, 337–349.
1980 Max v. Laue, 1879–1960. *Acta Hist. Leopoldina* **14**, 23–29.
1983 ACA in 1951–1953 (Accounts of ACA Presidents). In *Crystallography in North America* (eds. D. McLachlan & J.P. Glusker). New York: American Crystallographic Association, 156.
1985 Remembering Peter Debye in Munich. *Physics Today* **38** (Jan.) 9–10.

Book reviews

1914 Researches in Physical Optics, with especial reference to the radiation of electrons. Part I. R.W. WOOD. Columbia University Press, N.Y. *Naturwiss.* **2**, 663–664.
Magnetooptische Untersuchungen mit besonderer Berücksichtigung der magnetischen Zerlegungen der Spektrallinien. P. ZEEMAN. J.A. Barth. *Ibid.*, 664.
Die Lichtbrechung in Gasen als physikalisches und chemisches Problem. ST. LORIA. Fr. Vieweg & Sohn. *Ibid.*, 933.
X Rays, an Introduction to the Study of Röntgen Rays. G.W.C. KAYE. Longmans Green & Co. *Ibid.*, 1004–1005.
1915 Die Interferenz der Röntgenstrahlen. E. HUPKA. Vieweg. *Naturwiss.* **3**, 343.
1920 Die Verwendung der Lauediagramme zur Bestimmung der Struktur des Kalkspats. E. SCHIEBOLD. Abh. sächs. Akad. Wiss., Teubner. *Naturwiss.* **8**, 157.
Die Prinzipe der Dynamik. C. SCHAEFER. Vereinigung wissenschaftl. Verleger. *Ibid.*, 535.
Über die Auffindung der Röntgenstrahlinterferenzen. M. v. LAUE, Nobel Lecture. Müllersche Hofbuchhandl., Karlsruhe. *Ibid.*, 797.
1921 Lectures on the Principle of Symmetry and its Applications in all Natural Sciences. F.M. JAEGER. Elsevier. *Naturwiss.* **9**, 578.
Einführung in die theoretische Physik. A. HAAS. Vereinigung wiss. Verleger. *Ibid.*, 776–777.
Einführung in die theoret. Physik, Bd. 2. C. SCHÄFER. *Ibid.*, 946–947.
1922 Die Entwicklung der Atomtheorie. P. KIRCHBERGER. Teubner. *Deutsche Literaturzeitg.* 310–312.
1924 Atom und Quantentheorie, Vol. II. P. KIRCHBERGER. Teubner. *Deutsche Literaturzeitg.* 1528.
1925 Ein Buch über mathematische Physik. COURANT-HILBERT. Methoden der math. Physik. J. Springer. *Naturwiss.* **13**, 384–387.
Handbuch der Radiologie, Vol. 6. Ed. E. Marx. Akad. Verlagsgs. *Z. Kristallogr.* **61**, 186–187.
1926 Kjemisk Rontgenspektrografic. L. THOM-

ASSEN. Statens Rastoffkomite Publikasion No. 21.

1926 Lehrbuch der Vektorrechnung. J. SPIELREIN. K. Wittmer. *Z. Kristallogr.* **64**, 216.

1927 Die Verwendung der Röntgenstrahlen in Chemie und Technik. H. MARK. J.A. Barth. *Naturwiss.* **15**, 932.
Lehrbuch der Physik: Mechanik und Messmethoden. O.D. CHWOLSON. Vieweg. *D. Lttz.* No. 38, 1872.
Materie, Elektrizität, Energie. W. GERLACH. Th. Steinkopff. *Z. Kristallogr.* **65**, 159.
Handbuch der Experimentalphysik, Vols. 14, 21. Eds. W. Wien & F. Harms. Akad. Verlagsgs. *Z. Kristallogr.* **66**, 318–320.
Vorlesungen über Theoretische Physik an der Universität Leiden. H.A. LORENTZ. Akad. Verlagsgs. *Z. Kristallogr.* **67**, 597.
Materiewellen und Quantenmechanik. A. HAAS. Akad. Verlagsgs. *Ibid.*, 597–598.
Die Differential-und Integral-Gleichungen der Mechanik und Physik. P. FRANK & R.v. MISES. Fr. Vieweg & Sohn. *Ibid.*, 598–599.

1928 Lehrbuch der Kristallphysik. W. VOIGT. Teubner. *D. Lttz.* 2923–2924.

1929 Physica an Techniek der Rontgenstrahlen. A. BOUWERS. Deventer. *Z. Kristallogr.* **70**, 297.
Applied X-Rays. G.L. CLARK. McGraw–Hill. *Ibid.*, 297.
Bibliography of Crystal Structure. J.K. MORSE. University of Chicago Press. *Ibid.*, 297.
Voordrachten over Rontgen-Analyse van Kristallen. N.H. KOLKMEIJER, J.M. BIJVOET. & A. KARSSEN. D.B. Center. *Ibid.*, 298.
An Introduction to Crystal Analysis. Sir W. BRAGG, G. Bell & Sons. *Ibid.*, 298.
Handbuch d. Experimentalphysik, Vol. 7, pt. 2. H. OTT & K.F. HERZFELD. Eds. Wien-Harms. Akad. Verlagsgs. *Ibid.*, 300–302.
Handbuch d. Experimentalphysik, Vol. 7, pt. 1. P. NIGGLI. Eds. Wien-Harms. Akad. Verlagsgs. *Ibid.*, 302–306.
Röntgenographische Werkstoffprüfung. K. BECKER. Vieweg & Sohn. *Z. Kristallogr.* **72**, 107.
Einführung in die Algebra. O. HAUPT. Akad. Verlagsgs. *Ibid.*, 107.
Crystal Structure and Chemical Constitution. Disc. of Faraday Soc. Gurney and Jackson. *Ibid.*, 107.
Handbuch d. Experimentalphysik, Vol. 18, Wellenoptik und Polarisation, Photochemie. Eds. Wien-Harms. Akad. Verlagsgs. *Ibid.*, 108.
Gesammelte Werke. F. NEUMANN. Teubner. *Ibid.*, 108–109.
Intro. à la Physique des Rayons X et Gamma. M. DE BROGLIE & L. DE BROGLIE. Gauthier-Villars et Cie. *Ibid.*, 109.
Magnetic Properties of Matter. K. HONDA. Syokwabo and Co. *Ibid.*, 109–110.

1930 Joseph Fraunhofer's Leben, Leistungen und Wirksamkeit. M.v. BOHR. Akad. Verlagsgs. *Z. Kristallogr.* **74**, 109.
Handbuch der Experimentalphysik, Vol. 20. A. KÖNIG. Eds. Wien-Harms. Akad. Verlagsgs. *Ibid.*, 109–110.
Einführung in die Wellenmechanik. L. DE BROGLIE. Akad. Verlagsgs.; Einführung in die Wellenmechanik. J. FRENKEL. J. Springer. *Ibid.*, 110.
Handbuch der Experimentalphysik, Vol. 8 pt. 2 Akad. Verlagsgs. *Ibid.*, 110–111.
Handbuch der Experimentalphysik, Vol. 22. E. BACK & G. JOOS. Akad. Verlagsgs. *Ibid.*, 111–112.
Stereoscopic Photographs of Crystal Models. W. BRAGG & W.L. BRAGG. Adam Hilger Ltd. *Ibid.*, 433–436.

1931 Handbuch der Experimentalphysik, F. KIRCHNER & A.E. LIND. Eds. Wien-Harms. Akad. Verlagsgs. *Z. Kristallogr.* **76**, 287–288.
The Analytical Expression of the Results of the Theory of Space Groups. R.W.G. WYCKOFF. Carnegie Inst., Washington. *Ibid.*, 288.
Physik des Kautschuks. L. HOCK. S. Hirzel. *Ibid.*, 288.
Les Rayons X. T. THIBAUD. A Colin. *Ibid.*, 477.
Les Applications des Rayons X. J.J. TRILLAT. Presse Universitaire. *Ibid.*, 477–478.
Röntgenographische Untersuchungen. Mikrochemie 8, pt. 2. A. Haim. *Ibid.*, 478.
Die Röntgentechnik in der Materialprüfung. J. EGGERT & E. SCHIEBOLD. Akad. Verlagsgs. *Ibid.*, 478–479.
X-Ray Crystallography. R.W. JAMES. Methuen. *Ibid.*, 574.
Ostwald-Luther Hand- und Hilfsbuch zur Ausführung physiko-chemischer Messungen. C. DRUCKER. Akad. Verlagsgs. *Z. Kristallogr.* **77**, 173.
Kanalstrahlen. P. GOLDSTEIN. Akad. Verlagsgs. *Ibid.*, 174.
Vorlesungen über Wellenmechanik. A. LANDÉ. Akad. Verlagsgs. *Ibid.*, 174.
Was ist Materie? W. BRAGG. Akad. Verlagsgs. *Ibid.*, 174.
Der Aufbau der hochpolymeren organischen Naturstoffe. K.H. MEYER & H. MARK. Akad. Verlagsgs. *Ibid.*, 174–176.

1931 Flüssige Kristalle und Lebewesen. R. BRAUNS. E. Schweitzerbartsche Verlagsbuchh. *Z. Kristallogr.* **80**, 134.
Fortschritte der Röntgenforschung in Methode und Anwendung. In Ergebnisse der technischen Röntgenkunde, Band 2 (eds. F. Körber & E. Schiebold). Akad. Verlagsgs. *Ibid.*, 134–135.
Svante Arrhenius. E.H. RIESENFELD. Akad. Verlagsgs. *Ibid.*, 135.
Les Statistiques Quantiques et leurs Applications. L. BRILLOUIN, Les Presses Universitaires de France. *Ibid.*, 136.
Wissenschaftliche Photographie. E.v. ANGERER. Akad. Verlagsgs. *Ibid.*, 354.
Zwei Dialoge über Raum und Zeit. G. JAFFÉ. Akad. Verlagsgs. *Ibid.*, 354.
Einführung in die mathematische Behandlung der Naturwissenschaften. NERNST-SCHOENFLIES. R. Oldenbourg. *Ibid.*, 354.
Das Experiment. Sein Wesen und seine Geschichte. H. DINGLER. *D. Lttz.* 2006–2012.

1933 Handbuch der Experimentalphysik. Ergänzungswerk 1. W. WEIZEL. Akad. Verlagsgs. *Z. Kristallogr.* **84**, 332.
Statistische Mechanik. R. H. FOWLER. Akad. Verlagsgs. *Ibid.*, 332.
Tables of Cubic Crystal Structure. I.E. KNAGGS, B. KARLIK & C.F. ELAM. A. Hilger, Ltd. *Ibid.*, 332–333.
The Interference of Electrons. P. DEBYE. Blackie & Son Ltd. *Ibid.*, 333.
Spektroskopie der Röntgenstrahlen. M. SIEGBAHN. J. Springer. *Ibid.*, 333.

1934 Kosmologische Probleme der Physik. A. HAAS. Akad. Verlagsgs. *Z. Kristallogr.* **87**, 424.
Abhandlungen zur Strahlenoptik. W.R. HAMILTON. Akad. Verlagsgs. *Ibid.*, 503–504.
Physik für Jedermann. A. HAAS. J. Springer. *Ibid.*, 504.
The Crystalline State I. A General Survey. Sir W.H. BRAGG & W.L. BRAGG. G. Bell & Sons. *Z. Kristallogr.* **88**, 95–96.

1935 Molekül- und Kristallgitterspektren. W. FINKELNBURG et al. Hand- und Jahrb. d. chem. Phys., Vol 9 (eds. A. Eucken & K.L. Wolf) Akad. Verlagsgs. *Z. Kristallogr.* **90**, 96.
Materiewellen und Quantenmechanik. A. HAAS. Akad. Verlagsgs. *Ibid.*, 96.
Teilchenstrahlen. H. MARK. W. de Gruyter & Co. *Ibid.*, 192.
Dokladi Akad. Nauk CCCP, Vols I, II, *Ibid.*, 192.
The Diffraction of X-Rays and Electrons by Amorphous Solids, Liquids, and Gases. J.T. RANDALL. Chapman & Hall, Ltd. *Ibid.*, 287.
Geometrische Elektronenoptik. E. BRÜCKE & O. SCHERZER. J. Springer. *Ibid.*, 287–8.
Ableitung der regelmässigen Systeme nach der Methode von Fedorow. S.A. BOGOMOLOV. Leningrad University. Parts I and II. *Ibid.*, 383.
Ergebnisse der technischen Röntgenkunde, Band 4 (ed. J. Eggert & E. Schiebold). Akad. Verlagsgs. *Naturwiss.* **23**, 786.
Dielektrische Polarisation. O. FUCHS & K.L. WOLF. In *Hdb. d. chem. Physik.* Akad. Verlagsgs. *Z. Kristallogr.* **92**, 157.
Elektronenstrahlen und ihre Wechselwirkung mit Materie. J. HENGSTENBERG & K. WOLF. In *Hdb. d. chem. Physik.* Akad. Verlagsgs. *Ibid.*, 157–158.
Röntgenoskopie und Elektronoskopie von dispersen Systemen, Fäden, Filmen und Grenzschichten. Kolloidgesellschaft (ed. W. Ostwald). Th. Steinkopff. *Ibid.*, 158.
Die Bausteine der Körperwelt. Th. WULF. Springer. *Ibid.*, 158.
The Structure of Crystals. R.W.G. WYCKOFF. Reinhold. *Ibid.*, 158–160.

1936 Kristallplastizität. E. SCHMID & W. BOAS. J. Springer; Distortion of Metal Crystals. F. ELAM. Clarendon Press. *Naturwiss.* **24**, 277–279.
Stereoskopbilder von Kristallgittern. M.v. LAUE, R.v. MISES & E. REHBOCK-VERSTÄNDIG. J. Springer. *Z. Kristallogr.* **94**, 507.
Simplified Structure Factor and Electron Density Formulae for the 230 Space Groups of Mathematical Crystallography. K. LONSDALE. G. Bell & Sons. *Z. Kristallogr.* **96**, 87–88.
Die Atomkerne. C.F.v. WEIZSÄCKER. Akad. Verlagsgs. *Ibid.*, 231.

1937 Materialprüfung mit Röntgenstrahlen. R. GLOCKER. J. Springer. *Z. Kristallogr.* **96**, 395–396.
Röntgenographische Untersuchung von Kristallen. F. HALLA & H. MARK. J.A. Barth. *Ibid.*, 507–508.
Die Denkweise der Physik. G. MIE. F. Enke. *Ibid.*, 508.
Einführung in die Kristalloptik. E. BUCHWALD. Sammlg. Göschen, Vol. 619. *Ibid.*, 508–509.
Technische Oberflächenkunde. G. SCHMALTZ. J. Springer. *Z. Kristallogr.* **97**, 335–336.
Atlas der Analysen-Linien der wichtigsten Elemente. F. LÖWE. Th. Steinkopff. *Ibid.*, 401.

1938 Atomic Structure of Minerals. W.L. BRAGG.

Oxford University Press. *Z. Kristallogr.* **99**, 67.
Einführung in die Quantenchemie. H. HELLMANN. F. Deuticke. *Ibid.*, 68–9.

1938 Lichtzerstreuung. H.A. STUART & H.-G. TRIESCHMANN. *Hdb. d. Chem. Phys.*, Vol. 8. Akad. Verlagsgs. *Ibid.*, 276.
La structure des corps solides dans la physique moderne. L. BRILLOUIN. Hermann et Cie. *Z. Kristallogr.* **100**, 91.
Röntgen-Analyse von Kristallen. J.M. BIJVOET, & N.H. KOLKMEIJER. D.B. Centen's Uitgivers-Maatschappij. *Ibid.*, 91.
Low Temperature Physics. M. & B. RUHEMANN. Cambridge University Press. *Ibid.*, 91–92.

1939 Moleküle und Kristalle. E. FERMI. J.A. Barth. *Z. Kristallogr.* **101**, 272
Mikrophotographie. G. STADE & H. STAUDE. Akad. Verlagsgs. *Ibid.*, 272.
Lehrbuch der Chemischen Physik. A. EUCKEN. Akad. Verlagsgs. *Z. Kristallogr.* **102**, 80–81.
Kurzes Lehrbuch der Physikalischen Chemie. H. UHLICH. Th. Steinkopff. *Ibid.*, 81.
Chemische Physik der Metalle und Legierungen. U. DEHLINGER. Akad. Verlagsgs. *Ibid.*, 81–82.
Kristallchemie und Kristallphysik metallischer Werkstoffe. F. HALLA. J.A. Barth. *Ibid.*, 82–83.
Kontinuierliche Spektren. W. FINKELNBURG. J. Springer. *Ibid.*, 83.

1942 Röntgenstrahlinterferenzen. M.v. LAUE. Akad. Verlagsgs. *Nature* **150**, 450.

1949 The Optical Principles of the Diffraction of X-Rays. R.W. JAMES. G. Bell & Sons. *Science Progress* **37**, 572–573.

1950 Röntgenanalyse van Kristallen. J.M. BIJVOET, N.H. KOLKMEIJER & C.H. MacGILLAVRY. D.B. Centen. *Acta Cryst.* **3**, 322.

1951 On the systems formed by points regularly distributed on a plane or in space. M. A. BRAVAIS, transl. A. J. Shaler. Memoir No. 1 of Crystallogr. Soc. Am. (1950). *Acta Cryst.* **4**, 80.

1952 Fourier Transforms. I.N. SNEDDON. McGraw–Hill. *Acta Cryst.* **5**, 855.
Gmelins Handbuch der anorg. Chemie. Verlag Chemie. *Ibid.*, 856.

Moderne Allgemeine Mineralogie (Kristallographie). W. NOWACKI. Vieweg. *Ibid.*, 860.

1954 Tables for Direct Determination of Crystal Structures. V. VAND. Glasgow University. *Acta Cryst.* **7**, 382.

1955 Der Ultraschall. L. BERGMANN. S. Hirzel. *Acta Cryst.* **8**, 69.
Behavior of Metals under Impulsive Loads. J.S. RINEHART & J. PEARSON. Am. Soc. for Metals. *Ibid.*, 857.

1956 Untersuchungen über die Elektronentheorie der Metalle. S.I. PEKAR. Akad. Verlagsgs. *Acta Cryst.* **9**, 55.
Röntgen-Strukturanalyse von Kristallen. R. KOHLHAAS & H. OTTO. Akad. Verlagsgs. *Ibid.*, 203.

1957 The Precision Determination of the Parameters of the Unit Cell of Crystals by the Asymmetric Method. A.F. IEVINS & J.K. OSOL. Latv. Acad. Sci. *Acta Cryst.* **10**, 388.
Teoría de los Métodos Roentgenograficos del Cristal Giratorio. F. HUERTA. Inst. de Física 'Alonso de Santa Cruz', Madrid. *Ibid.*, 540.
World Dictionary of Crystallographers. W. PARRISH. Philips Laboratories. *Ibid.*, 720.

1958 Molecules and Crystals in Inorganic Chemistry. E.A. van ARKEL. Interscience. *Arch. Biochem. and Biophys.* **73**, 286.
Der Ultraschall (Nachtrag). L. BERGMANN. Hirzel. *Acta Cryst.* **11**, 60.
Einführung in die Vektorrechnung. H. SIRK. Th. Steinkopff. *Ibid.*, 668.
Splitting of Terms in Crystals. H.A. BETHE. Consultants Bureau. *Ibid.*, 755.
Tables et Abaques. J. ROSE. CNRS, *Ibid.*, 900.
Growth of Crystals (1st Sov. Conf. on Crystal Growth) Consultants Bureau. *Ibid.*, 900.

1959 Soviet Research in Crystallography. Consultants Bureau. *Acta Cryst.* **12**, 80.

1961 Elektronenbeugung. E. BAUER. Verl. Mod. Industrie. *Acta Cryst.* **14**, 442.

1962 Crystal Structures. Suppl. V. R.W.G. WYCKOFF. Interscience. *Acta Cryst.* **15**, 173.
Max von Laue. Gesammelte Schriften und Vorträge. Ed. M. Kohler. Vieweg. *Ibid.*, 517.
Max von Laue. Gesammelte Schriften und Vorträge. Ed. M. Kohler. Vieweg. *Endeavour* **21**, 200.

Index of names

This index includes all names appearing in the text on pages 3–148, and in the citations (*page number in parentheses*) at the end of each chapter, or paper. However, it omits all references to P.P. Ewald appearing either in the text, or in connection with citing his publications. The latter have been collected in the Bibliography starting on page 149.

Afanasiev, A.M. (134)
Agarwal, G.S. 106–7, (109)
Ahmad, F. 108, (109)
Albrecht, H. 34
Al-Haddad, M. 17, (22)
Allison, S.K. 15, (22)
Anderson, S.K. 18, (22)
Arif, M. (22)
Armstrong, J.A. 108, (109)
Arrott, A. 6, (23)
Ashkin, M. (134)
Astbury, W.T. 41
Authier, A. 14, 16–17, (22), 66–7, 69, (69), (70)

Baker, G.L. 109, (109)
Balibar, F. 22, 69, (69)
Bando, Y. 89
Barrett, C.S. 64, (69)
Barth, H. 64, (69)
Bartholinus, E. 5
Batterman, B.W. 16, 18, (22), 74, (78), (134)
Becker, P.J. 17, (22)
Beevers, C.A. 142
Bell, J.S. 42
Bellman, R. 80, (89)
Berg, W.F. 64, (69)
Bernal, J.D. 37, 41–2
Bertaut, E.F. 62, (63)
Bethe, H.A. 3–4, 6, 15, 22, (22), 31–2, (34), 35, 40, (43), 79–81, 83–4, (89), 91, (97)
Bethe, M. 21
Bethe, R. 21, 38, 50
Bevan,-. 126
Bienenstock, A. 49, (50), 56, (59)
Billy, H. 76, (78)
Birman, J. 106–7, (109)
Blackman, M. 81, (89)
Bloch, F. 10
Bloembergen, N. 108, (109)
Bollmann, W. 16
Bonse, U. 17–18, (22), 64, (69)
Born, M. 5, 11, 15, (22), 90, (97), 104, 106, (109), 130, (134)

Borrmann, G. 16, 19, 21, (22), 64–6, (69), (70), (134), 141, 143, (143)
Bowen, D.K. 17, (23)
Bragg, W.H. 16, 30, 32, 40–2, 121, 123, 142, (143)
Bragg, W.L. 5, 15–16, (22), 28, 30–1, (34), 35, 39–42, (43), 46, 114, 119, 121, (122), 123, 138, 140, 142
Bravais, A. 39, 55–6, (59), 127
Brill, R. 45
Brillouin, L. 40
Broglie, M. de (122)
Brümmer, O. 16, (22)
Bubakova, R. 133–4, 134
Budde, E. 54, (59)
Buerger, M.J. 37
Bullough, R.K. 6, 11, 92, 99, 103–8, (109), (110)

Campbell, H.N. 16, (22)
Campbell, J.W. 59, (59)
Capelle, B. 17, 69, (70)
Chaichian, M. 107, (110)
Chang, S.L. 20, (22), (23), 74, 77, (78)
Chapman, L.D. 20, (22), 75, (78)
Cimmino, A. 17, (22)
Clarebrough, L.M. (89)
Clothier, R. (22)
Cole, H. 16, (22), (134)
Colella, R. 17, 19–20, (22), 73, 75–7, (78)
Compton, A.H. 15, (22)
Cowan, P.L. 18, (22)
Cowley, J.M. 6, 12, 15, (22), (23)
Crapper, M.D. (78)
Cruickshank, D.W.J. 6, 22, 28, 32, 58, (59), 113, 122–3

Darwin, C.G. 6, 15, (22), (23), 31, 37, 40, 99, 102, 104, 107, (110), 134, 140
Dash, W.C. 16, 20, (23), 67
Davis, B. 30, 134, (134)
Davisson, C.J. 15
Debye, P.J.W. 16, 21, 27, 40, 61, 137
Dederichs, P.H. 16, (23)

De Goede, J. 107, (110)
Dehlinger, U. 31
DesLattes, R.D. 17, (23)
Detaint, J. (70)
Deutsch, M. 17–18, (23)
De Wette, W. 62, (63)
Dickinson, R.G. 59
Dirac, P.A.M. 35, 37, 107, (110)
Dosch, H. 77, (78)
Dropkin, J.J. 22, 27, (34), 50
Ducuing, J. (109)
DuMond, J.W.M. 17, (23)

Ehrenberg, W. 31, (34)
Einstein, A. 33, 104, 106, (110)
Ekstein, H. 72
Ellinas, D. (110)
Epelboin, Y. 17, 65, (70)
Evans, R.C. 43
Ewald, Arnold 35–6
Ewald, Ben 46
Ewald, Clara ('Doter') 27, 35–7, 125
Ewald, Ella 3, 33, 35–8, 44, 48
Ewald, Linde 35–6
Ewald, Lux 35–6
Ewald, Paul (father) 27, 125, 149

Fajans, K. 40–1, 127
Fankuchen, I. 45, 72
Fedorov, E.S. 56, 127
Feibelman, P. 94, (97)
Feldman, R. 19, (23)
Fenton,-. 126
Feynman, R.P. 14, (23)
Fixman, M. 106, (110)
Fokker, A.D. 40
Forman, P. 47, (50)
Forwood, C.T. (89)
Fresnel, A.J. 5, 105, (110)
Friedrich, W. 4, 5, 28, (34), 39, 53, 58, (59), 113, 119, 121, (122), 127, 138, (144)
Frondel, C. 20
Fues, E. 31

INDEX OF NAMES

Gallop,-. 126
Gao, Y. (78)
Germer, L.H. 15
Gibbs, J.W. 32, 54, (60), 101, (110), 142
Gjønnes, J. 83, (89)
Glusker, J.P. 39, 43, (43)
Goeppert-Mayer, M. 90, (97)
Gollub, J.P. 109, (109)
Golovchenko, J.A. (22)
Gong, P.P. 75, (78)
Goodman, P. 15, (23), 83–4, (89)
Graeff, W. 18, (22)
Grassmann, J.G. 55
Green, G. 102, (110)
Groth, P.H.R.v. 39–41, 127, 137, 141
Guigay, J. 17, (23)

Habash, J. (59)
Hahn, O. 33
Hamilton, J. 36
Hamilton, W.C. 6, (23)
Harker, D. 45
Hart, M. 17–18, 22, (22), (23), 72, (78)
Hartree, D.R. 101, 104, 137
Hartwig, W. 19, (22), (70), 143, (143)
Hassan, S.S. 108, (110)
Hauptman, H. 71
Haüy, R. 39, 127
Head, A.K. 84, (89)
Heidenreich, R.D. 12, (23)
Heisenberg, W. 103, (110)
Hellinger, E. 126
Helliwell, J.R. 57, (59), (60)
Henins, A. 17, (23)
Héno, Y. 19, 21, (23), 45, 49, 71, (78), 97, 143, (144), 148, (148)
Henry, N.F.M. (60)
Herglotz,-. 126
Hermann, C. 16, 21, 31, 32, (34), 41
Herring, C. 96, (97)
Herzfeld, K. 40
Hilbert, D. 27, 45, 126, 127
Hildebrandt, G. 3, 5, 21–2, (23), 35, 39–40, (43), 55, 58, 91, (97), 113
Hirsch, P.B. 16, 20, 83–4, 86, (89)
Hjalmar, E. 29
Hodgkin, D.C. 3, 36
Hoek, H. 99, 103, (110)
Hoffmann, A.W. 125
Høier, R. 76, (78)
Hollingsworth, L. 90, (97)
Hondros, D. 127
Hönl, H. 31, (34)
Hosemann, R. 64, (69)
Howie, A. 83–4, (89)
Huang, T.C. (23)
Huang, M.T. (78)
Humble, P. (89)
Hümmer, K. 76–7, (78)
Hupka, E. 118, (122)
Hynne, F. 6, 11, 92, 99, 103–7, (110)

Ibrahim, M.N.R. 108, (110)
Ikeda, K. 109, (110)
Iklé, M. 123
Irmler, M. (22), (70)

Jacoby, K. 127
James, R.W. 15–16, (22), (23), 31, 40, 100, (110)
Jones, D.L. 107, (110)
Jones, R.E. 62, (63)
Jones, R.G. (78)
Juretschke, H.J. 5–6, 13, 20, 22, (23), 34, 53, 61, (63), 76, (78), 90, (97), 97–9, 102, 106, 107
Juretschke, R. 49

Kagan, Y. (134)
Kaiser, H. (22)
Kambe, K. 17, (23)
Kamminga, H. 4, 43, (43), 46
Kappler, E. 64, (69)
Karle, J. 71
Kato, N. 13–14, 16–19, 21, (23), 28, 65, (70), 77, (78), 84, (89), 91, (134)
Kik, G. 33
Kikuchi, S. 15
Kinder, E. 12, (23)
King, H.E, Jr. (78)
Klein, A.G. (22)
Klein, F. 48, 126
Knipping, P. 5, 28, 39, 53, 113, 119, (122), 138
Kochendörfer, A. 31
Kohler, M. 95, (97), 141 (144)
Kohra, K. 17, (23)
König,-. 4
Kossel, E. 15
Kramers, H.A. 103, (110)
Kulish, P. (110)
Kuriyama, M. 6, (23), 130, (134)

Lamla, E. 19, (23), 96
Lang, A.R. 14, 16, 18, 20–2, (23), 64, 66–7, (69), (70), 72, 77, (78), (134)
Laue, M.v. 3, 5–7, 16, 21, (23), 27–8, 30, 32–3, (34), 37–42, 46, 53–6, (60), 65, (70), 73, 90, 92–3, 95, 97, (97), 98, 113–14, 119, 121, (122), 127, 134, (134), 137–8, 140–1, (144)
Lehmpfuhl, G. 83, (89)
Leighton, R.B. (23)
Lipscomb, W.N. 72, (78)
Lipson, H.S. 142
London, F. 31
Lonsdale, K. (60)
  see also Yardley, K.
Lorentz, H.A. 94, 103–4, (110), 137, 140, 142
Lorenz, E.N. 109, (110)
Lorenz, L. 103, (110)
Love, W. 38

McConville, C.F. (78)
McLachlan, D. Jr 43, (43)
Mair, G. (22)
Mandel, M. 106, (110)
Mandelstam, L. 122
Mark, H. 31, (34), 40, 45
Marthinsen, K. 76, (78)
Maxwell, J.C. 105
Mayer, E. 49, (50)
Mazur, P. 99, 104, 106–7, (110)
Meitner, L. 33
Mermin, D. 56, (60)
Meyer, G. 31
Milne, A.D. 17, (23)
Miyake, S. 22, 83, (89)
Moffat, K. (59)
Molière, G. 7, 16, (23), (134)
Mommsen, T. 125
Moodie, A.F. 6, 15, 22, (22), 88, (89), 113
Morton, A.J. (89)
Moyal, J. 36

Nardroff, R.V. 134, (134)
Nelson, L. 126
Newkirk, J.B. 16, (23), 64, (70)
Nicholson, R.B. (89)
Niers, H. 87, (89)
Niggli, P. 41–2, 127, 141
Nijboer, B.R.A. 62, (63)
Nishikawa, S. 15, 20, 59, (60)

Ohtsuki, Y.H. 130, (134)
Opat, G.I. (22)
Ornstein, L.S. 121, (122)
Oseen, C.W. 10, 92, 94, (97), 98, (110)
Ott, H. 31, 40
Overhauser, A.W. (22)

Pashley, D.W. (89)
Pasternak, B. 37
Patel, J. 20
Patternayak, D.N. 106–7, (109), (110)
Patterson, A.L. 56, 142
Pauling, L. 59
Penning, P. 17, (23)
Pershan, P.S. 108, (109)
Perutz, M. 3, 36
Petroff, J.F. 67, (70)
Pinsker, Z. 16, (23)
Planck, M. 27, 116, 136, 137, (144)
Polder, D. 17, (23)
Post, B. 19, 20, 22, (23), 27, (34), 45, 74–5, (78), 145
Prince, N.P. (78)
Pringsheim, A. 27, 126, 127
Prins, J.A. 12, 15, (23), 140, (144)

INDEX OF NAMES

Ramachandran, G.N. 64, (70)
Rayleigh, Lord 45
Rees, A. 12, (23)
Renninger, M. 12, 15, 17, 19, (23), 30–1, (34), 66, (70), 72–3, 95, 134, (134), 148, (148)
Riecke, E. 126
Riemann, G.F.B. 137
Riley, C.E. (78)
Robbins, M.F. (22)
Rohmann, H. 122
Rosenfeld, L. 99, 101–4, (110)
Ruhemann, M. 31
Ruhemann, S. 125–6

Sands, M. (23)
Sauvage, M. 67–8, (70)
Scherrer, P. 16
Schmid, E. 31–2, (34)
Schmidt, M.C. 75, (78)
Schönflies, A. 56, 127
Schrödinger, E. 141
Schwartzel, J. (70)
Schwegle, W. (78)
Seeliger, H.v. 127
Sein, J.J. 106–7, (109), (110)
Seymour, D.L. (78)
Shaler, A.J. 55, (59)
Shen, Q. 76–7, (78)
Shull, C.G. 18, (23)

Snell, W. 10, 104
Sommerfeld, A. 4–5, 27–8, 44–5, 49, 53, 58, (122), 127, 136–9
Sohncke, L. 127
Steinhaus, W. 118, (122)
Steinmetz, H. 127
Stenström, W. 29, 131, 147, (148)
Stephanik, H. 16, (22)

Takagi, S. 6, (23)
Tanner, B.K. 17, (23)
Taupin, D. 6, (23)
Templeton, D.H. 6, 29, 62, (63)
Terrill, H.M. 134, (134)
Tewari, S.P. (110)
Thompson, B.V. 107, (110)
Thomson, G.P. 15
Tiburtius, C. 125
Tischler, J.Z. 74, (78)
Trucano, P. 74, (78)

Uyeda, R. 20

Voigt, W. 126

Wack, D.C. (78)
Wagenfeld, H.K. 22, 32, (34), 61, (63), 87, (89), 95, (97), 113

Wagner, H. 21
Waitz, G. 125
Wallach,-. 126
Waller, I. 40
Wattembach,-. 125
Weckert, E. (78)
Werner, S.A. 6, (22), (23)
West, J. 140, 142
Whelan, M.J. 83–4, (89)
Wilkins, S.W. 17, (23)
Williams, D.E. 62, (63)
Wilson, A.J.C. 3–4, 62
Wilson, E.B. 54, (60)
Wolf, E. 5, 11, (22), 104, 106–7, (109), (110)
Wood, E.A. 21
Woodruff, D.P. 77, (78)
Wulff, G. 114, (122)
Wyckoff, R.W.G. 40–1, 55, 59, (60)

Yardley, K. 41
Yoder, D.R. (22), (78)
Yvon, J. 99, 106, (110)

Zachariasen, W.H. 12, 16, (23)
Zarka, A. 68–9, (70)
Zheng, Y. (70)

Index of subjects

Acta Crystallographica 3, 20, 41, 43, 46–7, 71, 127
Aharanov–Bohm effect 17
Aharanov–Casher effect 17
American Crystallographic Association 46, 48
  Pittsburgh meeting (1975) 20
  Warren prize (1981) 20
Ammersee symposium (1925) 15, 30–1, 37, 40
anhydrite 137
Annalen der Physik 61
austenite 84–5

benzil, structure factor phases 75–6
Bethe's dynamical theory, *see* dynamical theory of electron diffraction
Bloch wave 8
  see also Ewald-Bloch wave
Borrmann effect, *see* dynamical theory of X-ray diffraction
boundary conditions on half-crystal 9–13, 54, 93–7, 116–18, 130, 132, 139–40
Bragg's law 5–6, 28, 40, 123
  deviations from 29–30, 92, 131, 146–8

calcite, 16, 30
Cambridge Crystallographic Laboratory 37
crystal defects
  by electron diffraction 83–5
  dislocations (direct obervation) 16, 20, 64–70
crystal perfection 16–18, 64–70
  Pendellösung method 18
  standing wave method 18
crystal structure analysis
  direct method 71
  proteins 59
crystal structure, London Conference of (1929) 41
crystallographic community 39–43, 45, 47
Current Contents 61

diamond, 30
dielectric constant of crystals, *see* index of refraction

dipole waves in crystals 91–5, 99–107, 136–7
dispersion surface 9–10, 28, 93, 130–2, 139–40, 146
  linearization of 11, 19
  tiepoint 9–10, 93, 131–2, 138–40
dispersion, anomalous 7, 71–2
dispersion, general theory 97
double refraction
  electron waves 12
  visible light 5, 53, 137
  X-rays 11, 14
dynamical theory of electron diffraction 4, 79–89
  Bethe's theory 15–16, 32, 79–81, 83–4
  Borrmann effect 83
  Howie–Whelan equations 84
  pseudopotential approximation 81–4
  two-beam formulation 79–89
  using convergent beams 87–8
dynamical theory of X-ray diffraction 4–21, 30, 32, 40, 44, 53, 73, 77, 90–7, 129–34, 136–43
  Borrmann effect 16, 19, 29, 134, 141, 143
  Borrmann triangle 14, 65
  conceptual structure of Ewald's theory 91
  dispersion relations 8
  dynamical balance 8–10, 137
  eikonal theory 17, 21
  Ewald's construction, *see* sphere of reflection
  Ewald's theory of ray and energy flow 14, 19, 21
  Ewald–Oseen extinction theorem, *see* optical extinction theorem
  excitation error 28, 80
  many-beam case 18, 30, 54, 95–6, 130–1, 139–40, 142–3
  plane wave theory 10, 17
  spherical wave theory 13, 21, 65, 96
  Takagi–Taupin type equation 17
  two-beam case 11, 12, 30, 64–5, 95, 131–4, 141, 147
  see also
    boundary conditions on half-crystal
    dipole waves in crystals
    dispersion surface
    Ewald–Bloch wave
    multiple diffraction
    Pendellösung

rocking curve
self-consistent field
dynamical wave field 64
  see also Ewald–Bloch wave

electron diffraction, *see* dynamical theory of electron diffraction
electron microscopy 16, 83–4
electrostatic energy of crystals 61–3
elementary interference field 92
  see also Ewald–Bloch wave
epiwaves and mesowaves 139
Ewald, P.P.
  academic career 29, 35, 44–5
  ACA presidency 46, 48
  centenary of Einstein, Hahn, Meitner, von Laue 33
  collaborators in Germany 31
  colleagues in Brooklyn 45
  dissertation 4, 27–8, 73, 90, 98, 136–7
  during Nazi regime 32
  as editor 42–3, 46–7, 127
  family life in Cambridge and Belfast 35–8, 42
  Fifty years of X-ray diffraction 5, 33, 47, 57
  Foundations of crystal optics 5, 16, 28, 49, 90–8, 136, 138
  Habilitationsschrift 4, 29, 92, 141
  Handbuch der Physik 21, 30, 45, 141
  as historian 47
  Kristalle und Röntgenstrahlen 29, 40–1, 55, 141
  Max Planck medal 33
  obituaries 3, 22
  scientific activities after 1948, 45
  as teacher 49
  university studies 27, 44, 126–7
  as versifier 48–50
  war service 4, 28–9, 138–9
  and wife Ella 38, 44, 48
  youth 27, 44, 125–6
Ewald–Bloch wave 7, 8, 10, 14, 73, 92, 94
extinction 13, 17
  primary 134

Fabry–Perot interferometer 101, 105, 108–9
Feynman diagrams 83, 130

INDEX OF SUBJECTS

Fourier expansion of spherical waves 53, 136–7
Fourier space 56–7, 91, 129
 symmetry in 56
Fresnel formulae 94, 101, 105, 139
Fresnel's equation of wave normals 5
Fulbright grantees 20

GaAs, phases of structure factors 72–3
gnomonic projection 55–6

Hertz potential 53, 100, 114–17, 120, 138

index of refraction 5, 8, 92, 102–3, 131, 146
International Tables (Internationale Tabellen) 32, 37, 42–3, 45, 55
International Union of Crystallography 3, 37–9, 42–3, 46, 84
 1948 Congress (Cambridge, USA) 3
 1957 Congress (Montreal) 16, 20
 1962 Meeting (Munich) 3, 33
 1987 Congress (Perth) 3, 46
International Union of Pure and Applied Physics 3

kinematical theory of X-ray diffraction 6–7, 93, 129

lattice
 of dipole oscillators 5, 27, 53, 91, 99, 114, 136–7
 polar of Bravais 55–6, 141–2
lattice sums 6, 53, 61–3
 convergence of 62–3
 in real space 62
 in reciprocal space 62
Laue, M.v., discovery of X-ray diffraction 3, 28, 30, 39, 98, 114, 127, 138
 formulation of dynamical theory 7, 16, 32, 73, 95, 134, 140–1
Laue method 57–9
 Ewald analysis of Laue photographs 58–9
 multiple orders problem 57–8
 shape of interference spots 119–20
 unsymmetrical photographs 59
Limoges summer school in X-ray topography (1975) 64, 69
Lorentz factor 129
Lorentz–Lorenz relation 103, 131, 146

Madelung constant 6, 61–3
 and Patterson function 62
magnesium oxide 12, 84, 86
mean field formulation 94
mosaic crystals 15, 16, 31
 standing waves in 77
multiple diffraction 71–8, 79–89, 91
Munich physics colloquium 27

neutron diffraction 17–8

optical extinction theorem 10–11, 27, 92, 98–110, 138–9
 amorphous dielectric 98–106
 conceptual problems 104–6
 incoherent scattering 103–4, 106
 Lorentz polarizing field 101–2
 many-body theory 106–7
 nonlinear optics 107–9
 pair correlation function for smooth dielectric 98–9, 102–4, 106
optical field 130, 136, 140
 see also Ewald–Bloch wave

Pendellösung, and thickness fringes
 with electrons 12, 84–6
 with X-rays, xi, 12, 16, 18, 20–1, 65, 72, 77, 92, 132–3, 140, 148
phase problem 18–20, 71–8, 88
phase velocity 137, 146
point dipoles, Thomson scattering 6, 7
Point Group seminar 45–6
polaritons 107
polarization of interference beams 120
Polytechnic Institute of Brooklyn 44–5
proper wave mode for X-rays 7, 91, 131–2
 see also Ewald–Bloch wave
pyrite 28, 30, 58, 138

quantum theory of diffraction 130
quartz 16, 68–9

ray optics 14
reciprocal lattice 32, 53–8, 129, 138, 142
reciprocity theorem 64, 92, 118, 130, 139
Renninger effect 19, 72–5, 95
resonance factor 139
rocking curve (line profile) 92, 134
 dynamical angular width 12
 dynamical integrated intensity 13
 kinematic angular width 7
 kinematic integrated intensity 7

self-consistent field 8–10, 27, 53, 91, 102–3, 130, 137, 139
shape transforms 6–7, 56–7
Si, dislocations in 67–8
 electron diffraction 82, 87–8
 structure factor phases 74
Solvay Conference (1913) 28, 54, 58, 138
space group theory 41, 56
sphere of reflection 6, 28, 54–5, 73, 92, 116–18, 129, 138, 146
structure amplitude, vectorial 95, 142
structure factor 6, 28–9, 56, 58, 129, 138, 142
 absolute value 18
 experimental determination of phase 74–7
Structure Reports (Strukturberichte) 16, 31, 41–2, 45, 141
synchrotron radiation 16–17, 57–8, 67–8, 71, 74, 77

thickness fringes, *see* Pendellösung
tiepoint, *see* dispersion surface
top-hat profile 12–13, 140
 see also rocking curve
total reflection
 visible rays 11
 X-rays 9, 12, 31–2, 134, 147
two-state physics 14

Umweganregung 19, 73–6, 95–6
UNESCO 46

V_3Si, structure factor phases 75

X-ray optical devices
 channel-cut monochromator 17
 interferometer 17
 monolithic multi-crystals 17
X-ray topography 16–17, 64–70
 and acoustic vibrations 68–9
 double crystal 66
 extinction contrast 64
 formation of images 64
 Pendellösung in 65
 section and projection topographs 66
 types of images 65–6
XRAG Oxford meeting (1944) 37, 42

Zeitschrift für Kristallographie 30, 37, 41–2, 45, 127, 141
zincblende 59